Mechanics and Model-Based Control of Advanced Engineering Systems

AF789529

Alexander K. Belyaev • Hans Irschik
Michael Krommer
Editors

Mechanics and Model-Based Control of Advanced Engineering Systems

Springer

Editors
Alexander K. Belyaev
Inst. Problems Mech. Eng. RAS
State Polytechnic University
St. Petersburg, Russia

Hans Irschik
Michael Krommer
Institute for Technical Mechanics
Johannes Kepler University Linz, Austria

ISBN 978-3-7091-1957-0 ISBN 978-3-7091-1571-8 (eBook)
DOI 10.1007/978-3-7091-1571-8
Springer Heidelberg New York Dordrecht London

© Springer-Verlag Wien 2014
Softcover reprint of the hardcover 1st edition 2014
This work is subject to copyright. All rights are reserved by the Publisher, whether the whole or part of the material is concerned, specifically the rights of translation, reprinting, reuse of illustrations, recitation, broadcasting, reproduction on microfilms or in any other physical way, and transmission or information storage and retrieval, electronic adaptation, computer software, or by similar or dissimilar methodology now known or hereafter developed. Exempted from this legal reservation are brief excerpts in connection with reviews or scholarly analysis or material supplied specifically for the purpose of being entered and executed on a computer system, for exclusive use by the purchaser of the work. Duplication of this publication or parts thereof is permitted only under the provisions of the Copyright Law of the Publisher's location, in its current version, and permission for use must always be obtained from Springer. Permissions for use may be obtained through RightsLink at the Copyright Clearance Center. Violations are liable to prosecution under the respective Copyright Law.
The use of general descriptive names, registered names, trademarks, service marks, etc. in this publication does not imply, even in the absence of a specific statement, that such names are exempt from the relevant protective laws and regulations and therefore free for general use.
While the advice and information in this book are believed to be true and accurate at the date of publication, neither the authors nor the editors nor the publisher can accept any legal responsibility for any errors or omissions that may be made. The publisher makes no warranty, express or implied, with respect to the material contained herein.

Printed on acid-free paper

Springer is part of Springer Science+Business Media (www.springer.com)

Foreword

The present volume is dedicated to Professor Nikita F. Morozov, full member (Academician) of the Russian Academy of Sciences on the occasion of his 80th birthday.

Professor Nikita F. Morozov is an outstanding Russian scientist, who enriched mechanical science with many achievements in the field of solid mechanics, and an eminent teacher, who created a famous school of mechanics.

Professor Morozov was born on July 28, 1932 in Leningrad. His childhood coincided with a harsh time for the country. For the two and a half years of the siege of Leningrad, he was actively involved in the work of fire brigades. Aged 11, in 1943 he was awarded the state medal "For the Defense of Leningrad."

In 1949 young Nikita Morozov joined the Faculty of Mathematics and Mechanics of the Leningrad State University. After graduation and teaching at other universities, he returned to the Leningrad State University, and since 1976 he has headed the Elasticity Department at this university, which is now called St. Petersburg State University.

His fruitful work on a number of scientific problems in the mathematical theory of elasticity nominated him into a series of well-known experts throughout the world. Nowadays his name is associated with the application of rigorous mathematical methods in the theory of elasticity.

Professor Morozov and his disciples led to significant progress in the rigorous mathematical formulation and study of the problems of brittle fracture. A significant contribution was made in the study of the problem of equilibrium and propagating cracks in the theory of damage accumulation. He proposed a new criterion of brittle fracture that is relevant to the problems of high-rate and ultrahigh-rate loading. Some pioneering methods for testing materials that meet the needs of modern industry were suggested on the basis of this criterion.

In recent years, Professor Morozov focused his efforts in the field of the application of solid mechanics to problems of nanomechanics. Much attention is paid by him to the examination of the relationship between the diffusion processes in phase transformations, on one side, and deformation, stability, and destruction, on the other side. He was successful in creating a research team at the Institute of

Problems of Mechanical Engineering of the Russian Academy of Sciences, which includes a lot of talented people. Continuing the best traditions of the schools in mechanics by academicians Novozhilov and Lurie, Professor Morozov organized a weekly seminar at the Institute of Problems of Mechanical Engineering. A broad variety of actual topical problems of mechanics and physics are discussed there, and scientists and researchers consider it a privilege to report their results at his seminar.

Professor Morozov has published more than 250 scientific papers, including 8 monographs and 3 textbooks. He was the scientific advisor of 60 doctoral theses, among which are 8 Doctor-of-Sciences theses. In 2000 he was awarded the State Prize of the Russian Federation in the field of science and technology for his work on nonlinear problems of solid mechanics.

The social-scientific activities by Professor Morozov are extensive and multifaceted. He is vice-chairman of the Russian National Committee for Theoretical and Applied Mechanics, chairman of the Scientific Council of the Russian Academy of Sciences on Solid Mechanics, member of the General Assembly of IUTAM, and a member of the editorial boards of leading national and international journals on mechanics.

While communicating with Professor Morozov, you never think about his age or his numerous regalia. Many young people could envy of his energy, interest in everything new, positive thinking, and focus in the future. His practical wisdom, quick wit, and fine sense of humor are the qualities that leave a strong impression, and some of his phrases have become aphorisms. Despite many responsibilities and concerns related to various activities, he always finds time if somebody approaches him with scientific issues. When he begins discussing science, he instantly turns off all the problems, so it seems that the discussed scientific question is the main interest of his life, if it is not the only one. Talking with Professor Morozov, you feel free and easy. He knows how to tactfully point out the shortcomings of the work, and he is always open to discussion and ready to adjust his mind. Being a talented administrator of science, he knows exactly how to be supportive and not to get in the way, and he is able to captivate with his ideas and to respect the scientific views and interests of his colleagues and subordinates.

His most important personal feature is an exceptional decency. This is the main reason why under his guidance, a number of scientific directions in mechanics are developing, and these areas are equally supported regardless of his personal research interests.

St. Petersburg, Russia
Linz, Austria
August 2013

Alexander K. Belyaev
Hans Irschik

Preface

This book contains the contributions presented during the *International Workshop on Advanced Dynamics and Model-Based Control of Structures and Machines*, which took place in St. Petersburg, Russia, in July 2012. The workshop aimed at bringing together scientists with an outstanding expertise in mechanics and control, with emphasis on the application of advanced structures and machines. The workshop was intended as a scientific event for the area mechanics and model-based control within the Austrian Center of Competence in Mechatronics (ACCM), which served as the steering organization for the workshop.

Mechanics and model-based control are both rapidly expanding scientific fields and fundamental disciplines of engineering. They share demanding mathematical and/or system-theoretic formulations and methods. One of the challenges in mechanics and model-based control is utilizing the ever-increasing computer power with respect to both the simulation of complex physical phenomena in mechanics and the design and real-time implementation of novel control systems. Further challenges follow from the availability of efficient multifunctional materials, the so-called smart materials, allowing the design and implementation of new types of actuator/sensor fields and networks. The key objectives of the workshop were:

- Enabling the interchange of ideas from advanced mechanics of structures and from control theory
- Clarification of expectations of research in the field of mechanics from advanced control theory, and vice versa
- Development of joint research proposals and teams with participation from other countries
- Encouragement of collaborations among industry and universities across the borders of the participating countries

The workshop continued a series of international workshops, which started with a Japan-Austria Joint Workshop on Mechanics and Model-Based Control of Smart Materials and Structures and a Russia-Austria Joint Workshop on Advanced Dynamics and Model-Based Control of Structures and Machines. Both workshops took place in Linz, Austria, in September 2008 and April 2010, respectively. We believe that such workshops will result into the creation of research teams with participation not only from Austria, Japan, and Russia but also from other countries; the latter is enabled by the participation of widely renowned scientists from Europe and Taiwan. Such teams should push the frontiers of mechanics and control of advanced structures and machines to new dimensions.

The undersigned editors of the present book, which is entitled *Mechanics and Model-Based Control of Advanced Engineering Systems*, are happy to present in the following 10 full-length papers of presentations from Russia, 9 from Austria, 8 from Japan, 3 from Italy, one from Germany, and one from Taiwan. It is hoped that these contributions will further stimulate the international research and cooperation in the field.

Linz, Austria — Hans Irschik
St. Petersburg, Russia — Michael Krommer
August 2013 — Alexander K. Belyaev

Acknowledgements

Support of the *International Workshop on Advanced Dynamics and Model-Based Control of Structures and Machines* from the *Austrian Center of Competence in Mechatronics (ACCM)* is gratefully acknowledged. ACCM is a research center of peak performance in mechatronics with high international recognition and reputation. In a dynamic network with numerous national and international partners from science and industry, ACCM builds a bridge between scientific mechatronic research and its realization in products. ACCM is funded as a competence center in the realm of *Competence Centers for Excellent Technologies (COMET)* by the *Austrian Federal Ministry for Transportation, Innovation and Technology*, by the *Austrian Federal Ministry of Economics, Family and Youth*, and by the *Province of Upper Austria*. ACCM is jointly run by the following organizations:

- *Johannes Kepler University of Linz*
- *Linz Center of Mechatronics GmbH*
- *voestalpine*

The help of these organizations is highly appreciated.

The editors also wish to thank the *Institute for Problems in Mechanical Engineering of the Russian Academy of Sciences in St. Petersburg* for serving as host of the workshop and *Mrs. Silvia Schilgerius* from *SpringerWienNewYork* for her support during the preparation of this book.

Contents

Wave Propagation in Functionally Graded Material Bar Due to Collision

Tadaharu Adachi and Masahiro Higuchi

Abstract We mathematically analyzed wave propagations in functionally graded material (FGM) bars collided by a homogeneous bar on the basis of Laplace transformation and calculated by using numerical Laplace transformation and its inversion. Young's modulus in the FGM bar was assumed to be proportional to the square of its density, which was similar to foam materials. Finally in the FG bar with increasing modulus from the impact end to the fixed end, much larger compressive stress and even large tensile stress occurred near the fixed end. In the FG bar with decreasing modulus from the impact end, the stress history varied moder-ately however large tensile stress occurred.

1 Introduction

Functionally graded materials (FGMs) [1, 2] are suggested to be applied materials for impact energy absorption [1–6]. Longitudinal impact response of the FGMs must be clarified to investigate the impact energy absorption characteristics of FGMs.

Longitudinal responses of FGMs have been analyzed by several researchers until now. Chiu and Erdogan [7] analyzed FGM's longitudinal impact problems. The stress history of FGM has been analyzed by Bruck [8], and more recently Abu-Alshaikh and Kokluce [9] showed that elastic wave propagation in FGM could be analyzed by using an approximate model expressed as a laminate of

T. Adachi (✉)
Department of Mechanical Engineering, Toyohashi University of Technology, 1-1 Hibarigaoka, Tempaku, Toyohashi, 441-8580, Japan
e-mail: adachi@me.tut.ac.jp

M. Higuchi
School of Mechanical Engineering, Kanazawa University, Kakuma-machi, Kanazawa, 920–1192, Japan
e-mail: higuchi-m@se.kanazawa-u.ac.jp

Alexander K. Belyaev et al. (eds.), *Mechanics and Model-Based Control of Advanced Engineering Systems*, DOI 10.1007/978-3-7091-1571-8_1,
© Springer-Verlag Wien 2014

thin homogeneous plates. Cui [6] evaluated energy absorption characteristics of FGMs by using a laminated model, as well. Kiernan et al. [10] simulated split Hopkinson bar tests for FGMs by finite element analysis. Han et al. [11] and Santare et al. [12] considered finite elements in a finite element analysis of elastic wave propagation in FGM. Samadhiya et al. [13] used a laminate model to perform a spectral analysis of FGM vibration. Berezovski et al. [14] analyzed dynamic stress in FGMs that had non-uniform dispersion of particles. Liu et al. [15] investigated how to identify the distribution of material properties in FGMs on the basis of elastic wave propagation. The authors recently analyzed the longitudinal impact problem of FGMs mathematically and clarified that the distribution slope of the material properties was significant for understanding stress histories generated in FGMs [16, 17]. The longitudinal impact problems of FGMs have previously been analyzed and considered, as outlined above. In order to clarify FGM impact energy absorption from the viewpoint of impact response, we need to analyze the problem of an FGM colliding with an impactor and consider the relation between the distribution of material properties and the impact response of the FGM. This FGM collision problem has not yet been analyzed.

In the present study, wave propagation in an FGM bar collided with an impactor is analyzed to determine the suitability of FGM as a material for energy absorption. The impact load and stress problems in the FGM bar are solved mathematically by using Laplace transformation. Numerical Laplace transformation and its inversion are formulated and applied to compute the solutions. A material model of the FG bar is assumed to have foam material characteristics.

2 Analysis

2.1 Analytical Collision Model

We analyze wave propagation in an FGM bar subjected to collision with a homogeneous impact bar, as shown in Fig. 1. An FGM bar with a length of L is collided at a velocity of V_0 with an impact bar with a length of L_I. To simplify the problem, the cross sectional areas of both bars are the same and the materials of both bars are linearly elastic in the analysis. The density and elastic modulus in the FGM bar are distributed along the axial coordinate, x. The Young's modulus, $E(x)$ is assumed to be proportional to the square of the density, $\rho(x)$ [18, 19] as

$$\rho(x) = \rho_0\varphi(x)\,,\ E(x) = E_0(\varphi(x))^2, \tag{1}$$

where $\varphi(x)$ is the shape function of the distribution. The shape function is assumed to be

$$\varphi(x) = \Phi_1\left(\frac{x}{L} + \Phi_0\right)^2, \tag{2}$$

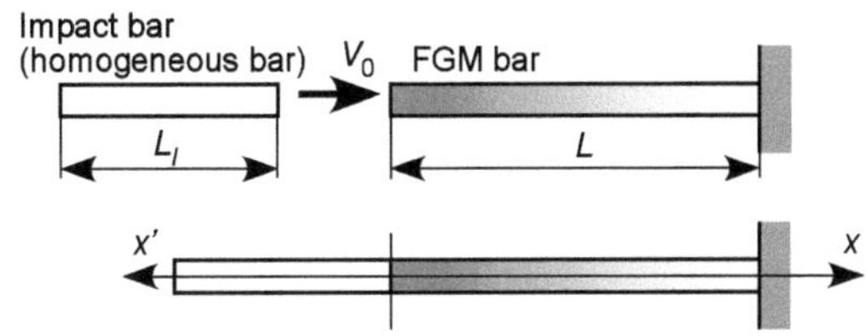

Fig. 1 Collision problem of FGM bar

where Φ_1 and Φ_0 are coefficients determined from the distribution in the FGM bar. The density and Young's modulus in the impact bar are ρ_I and E_I.

The stress-strain relation of the FGM is expressed by Hooke's law.

$$\sigma(x,t) = E(x)\varepsilon(x,t), \tag{3}$$

where σ, ε and t are stress, strain and time, respectively. The strain in the FGM bar is axial displacement, u differentiated by axial coordinate, x. The displacement of the FGM bar at the impacted tip, $U(t)$ must coincide with that of the impact bar at the tip, $U_I(t)$ during contact between the bars if local deformation near the contact area of both bars is neglected.

$$U(t) = V_0 t - U_I(t). \tag{4}$$

The displacement of both bars at the tip can be expressed as the convolution with the step responses at the impacted tips, as

$$U(t) = \int_0^t \frac{F(t-t')}{A} \frac{d}{dt'}\left(\frac{u(0,t')}{\sigma_0}\right) dt',$$
$$U_I(t) = \int_0^t \frac{F(t-t')}{A} \frac{d}{dt'}\left(\frac{u_I(0,t')}{\sigma_0}\right) dt', \tag{5}$$

where $F(t)$ and A are impact force due to the collision and cross-sectional areas of both bars. $u(x,t)$ and $u_I(x',t)$ are the displacements of the FGM bar and the impact bar subjected to step force per unit cross-sectional area, σ_0, which are solved in Sect. 2.3.

2.2 *Step Response of FGM Bar*

The response of the FGM bar is subjected to step force per unit cross-sectional area, $\sigma_0 H(t)$ at the tip $x = 0$. The other end $x = L$ of the bar is fixed.

$$\sigma = -\sigma_0 H(t) \text{ at } x = 0 \text{ and } u = 0 \text{ at } x = L, \tag{6}$$

where $H(t)$ is Heaviside step function. The initial condition is

$$u = \partial u/\partial t = 0 \text{ at } t = 0. \tag{7}$$

The equilibrium of stress for the FGM bar is given as

$$\frac{\partial \sigma(x,t)}{\partial x} = \rho(x) \frac{\partial^2 u(x,t)}{\partial t^2}. \tag{8}$$

By using Laplace transformation, Eq. (8), the solution can be reduced.

$$\frac{E_0}{\sigma_0 L} \bar{u}(\xi, s) = -\frac{1}{\Phi_1^2 \Phi_0^3 s} \frac{\left(\frac{\xi+\Phi_0}{1+\Phi_0}\right)^{\lambda_1} - \left(\frac{\xi+\Phi_0}{1+\Phi_0}\right)^{\lambda_2}}{\lambda_1\left(\frac{\Phi_0}{1+\Phi_0}\right)^{\lambda_1} - \lambda_2\left(\frac{\Phi_0}{1+\Phi_0}\right)^{\lambda_2}}, \tag{9}$$

where

$$\lambda_{1,2} = -\frac{3}{2} \pm \frac{\sqrt{9\Phi_1^2 + 4\Phi_1 s^2}}{2\Phi_1}, \quad \xi = \frac{x}{L}.$$

The Laplace transformation is defined as

$$\bar{u}(s) = \int_0^\infty u(\tau) \exp(-s\tau)\, d\tau, \quad \tau = \frac{C_0}{L} t, \quad C_0^2 = \frac{E_0}{\rho_0}.$$

2.3 Analytical Procedure

By substituting Eq. (9) into Laplace transformed Eqs. (4) and (5), the impact force in Laplace transformed domain can be derived. The impact force is inversed numerically from the Laplace transformed domain to the time domain by using the numerical inversion of the Laplace transformation described in Sect. 2.4. The calculated impact force will increase compressively just after the collision and change to tension owing to the return of the reflected wave to the collision point after complicated fluctuation. However, the impact force must be definitely compressive because the tips of the impactor solely contact at the tip of the FGM bar and are not joined. The bars separate mutually just when the impact force transforms from the compression to the tension. Therefore, the calculated impact force after the force changes to the tension should become zero to express the termination of the

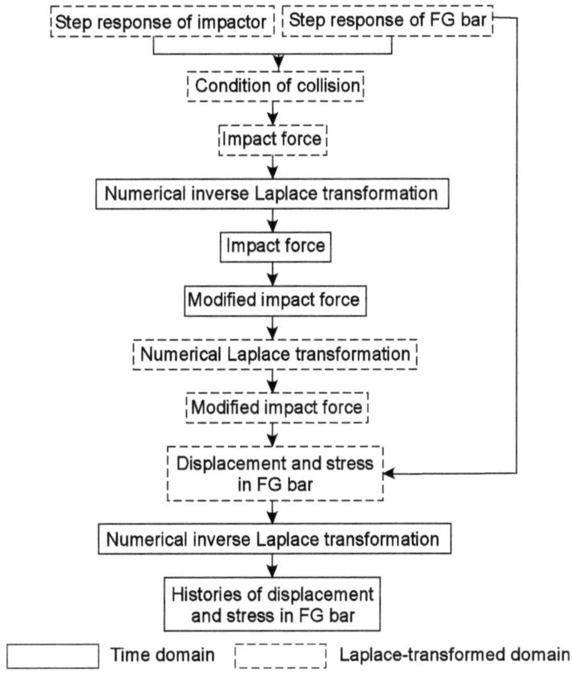

Fig. 2 Computational flow

collision. The impact force $F(t)$ calculated with Eq. (4) and the numerical inversion of the Laplace transformation are redefined as the modified impact force $F_C(t)$ as follows:

$$F_C(t) = \begin{cases} F(t) & t \leqslant t_c \\ 0 & t > t_c \end{cases}, \tag{10}$$

where t_c is the first time that impact force $F(t)$ turns to tension.

After the impact force $F_C(t)$ is Laplace transformed again, the transformed load is substituted into Eqs. (3) and (9) to clarify the stress histories in the FGM bar. The computational flow of the analysis is summarized in Fig. 2.

2.4 *Numerical Laplace Transformation*

The numerical Laplace transformation is formulated on the basis of Krings-Waller's numerical inversion [20, 21] to calculate the analyzed solutions. The pair of numerical Laplace transformation on the basis of Krings-Waller's method can be formulated as follows [22]:

$$\bar{u}\left(\gamma + in \cdot \Delta\omega\right) = \frac{T}{N}\sum_{k=0}^{N-1} u\left(k \cdot \Delta\tau\right)\exp\left(-\gamma k \cdot \Delta\tau\right)\exp\left(-\frac{i2\pi nk}{N}\right)$$

$$u\left(k \cdot \Delta\tau\right) = \frac{\exp\left(\gamma k \cdot \Delta\tau\right)}{T}\sum_{n=0}^{N-1} \bar{u}\left(\gamma + in \cdot \Delta\omega\right)\exp\left(\frac{i2\pi nk}{N}\right). \quad (11)$$

In Eq. (11), γ is greater than the real part of all singularities of Laplace transformed $u(t)$. The parameter s is discretized into N parts along the integration path with consideration for the sampling theorem in Fourier transformation.

$$s = \gamma + i \cdot n\Delta\omega,\ n = 0, 1, 2, \cdots, N - 1, \quad (12)$$

where

$$\Delta\omega = 2\pi \big/ T, \quad i = \sqrt{-1}, \quad \Delta\tau = \frac{T}{N}. \quad (13)$$

T is time range in the calculation.

In actual numerical calculation, N and γ parameters are determined in consideration of the accuracy of results with analyzed time duration T. The actual values of the parameters are determined as $N = 2^{14}$ and $\gamma = 8/T$ for accuracy. Numerical results calculated by Eq. (11) fluctuate dramatically with time and this accuracy decreases in the later analyzed time duration. Therefore in order to ensure necessary accuracy, the results in the latter quarter of the analyzed time are eliminated [21].

3 Numerical Results

We analyzed numerically wave propagations in two FGM bars. One FGM bar (FGM bar A) had Young's modulus distributed increasing from the impact end to the one fixed end (Young's modulus, $E(L)/E(0) = 10$) and the other bar (FGM bar B) had the modulus distributed decreasingly (Young's modulus, $E(L)/E(0) = 0.1$). Therefore, the boundary condition of the FGM bar B inverts the impact end and fixed end of the FGM bar A. The lengths of the FGM bars were the same as the one of the impact bar.

Figure 3 shows stress history at each position of the FGM bar A having the Young's modulus distributed increasingly from the impact end. The history at the impact end ($x/L = 0$) denotes impact load history due to the collision. The histories vary more steeply at more close position of the FGM bar. The history at the fixed end fluctuated intensively. Generally when the impact bar collided to a homogeneous bar with the same boundary condition, stress is compressive at any position of the bar [23]. However in this case large tensile stress generates by stress wave reflected at the fixed end. Figure 4 shows stress history at each position of the FGM bar B having the Young's modulus distributed decreasingly from the impact end. The histories

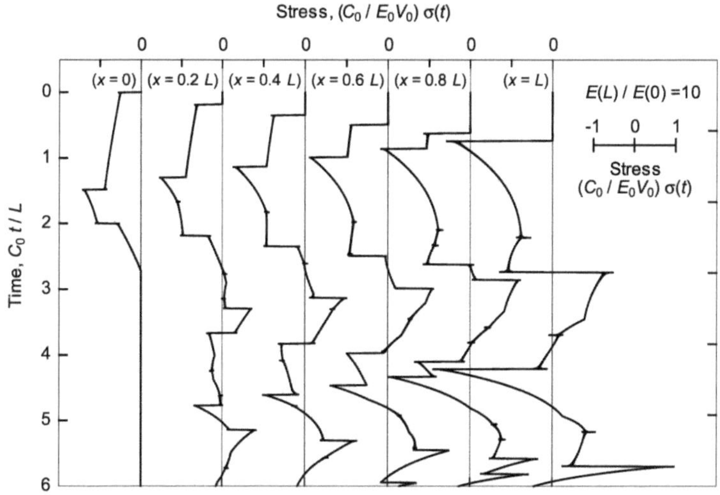

Fig. 3 Stress histories in FGM bar A. Horizontal axis denotes amplitude of stress at each point. $L/L_I = 1.0$

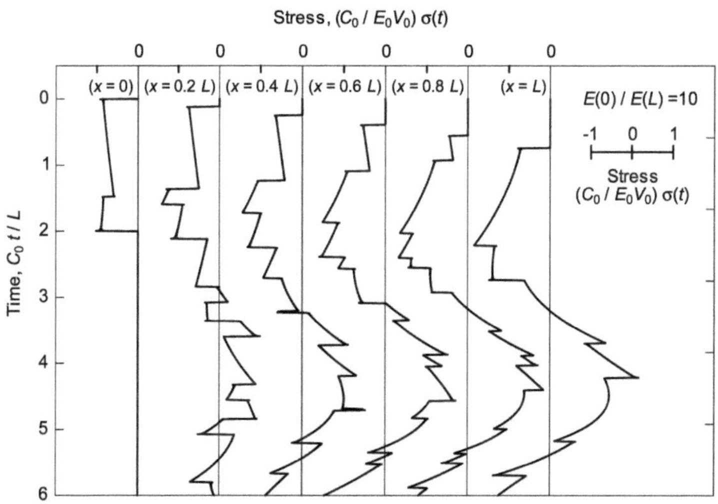

Fig. 4 Stress histories in FGM bar B. Horizontal axis denotes amplitude of stress at each point. $L/LI = 1.0$

varied smoothly with small steps and could be approximated by a sine curve. The histories had much different trends from the ones of the bar A.

Finally the stress histories in the FGM bar collided by the impact bar are much different from the histories for collisions to a homogeneous bars [23] and are found to be strongly dependent on the distribution of mechanical properties: density and Young's modulus. When we apply the FGM to mechanical components for impact energy absorption, slope of the mechanical properties in the component must be considered.

4 Conclusion

Wave propagations in the FGM bars collided by a homogeneous bar were analyzed on the basis of Laplace transformation and were calculated by using numerical transformation and its inversion. Finally in the FGM bar with increasing modulus from the impact end to the fixed end, much larger compressive stress and even large tensile stress occurred near the fixed end. In the FGM bar with decreasing modulus from the impact end, the compressive stress was approximately the same as the one in the homogeneous bar, and the history of the stress varied moderately however large tensile stress occurred.

References

1. Tanigawa, Y.: Some basic thermoelastic problems for nonhomogeneous structural materials. Appl. Mech. Rev. **48**, 287–300 (1995)
2. Noda, N.: Thermal stresses in functionally graded material. J. Therm. Stress. **22**, 377–512 (1999)
3. Adachi, T., Higuchi, M.: Development of integral molding of functionally-graded syntactic foams. In: Irschik, H., Krommer, M., Belyaev, A.K. (eds.) Advanced Dynamic and Model-Based Control of Structures and Machines, pp. 1–9. Springer, Heidelberg (2012)
4. Higuchi, M., Adachi, T., Yokochi, Y., Fujimoto, K.: Controlling of distribution of mechanical properties in functionally-graded syntactic foams for impact energy absorption. Mater. Sci. Forum **706–709**, 729–734 (2012)
5. Adachi, T., Higuchi, M.: Fabrication of bulk functionally-graded syntactic foams for impact energy absorption. Mater. Sci. Forum **706–709**, 711–716 (2012)
6. Cui, L., Kiernan, S., Gilchrist, M.D.: Designing the energy absorption capacity of functionally graded foam materials. Mater. Sci. Eng. A **507**, 215–225 (2009)
7. Chiu, T.C., Erdogan, F.: One-dimensional wave propagation in a functionally graded elastic medium. J. Sound Vib. **222**, 453–487 (1999)
8. Bruck, H.A.: One-dimensional model for designing functionally graded materials to manage stress waves. Int. J. Solids Struct. **37**, 6383–6395 (2000)
9. Abu-Alshaikh, I., Kokluce, B.: One-dimensional transient dynamic response in functionally graded layered media. J. Eng. Math. **54**, 17–30 (2006)
10. Kiernan, S., Cui, L., Gilchrist, M.D.: Propagation of a stress wave through a virtual functionally graded foam. Int. J. Non-Linear Mech. **44**, 456–468 (2009)
11. Han, X., Liu, G.R., Lam, K.Y., Ohyoshi, T.: A quadratic layer element for analyzing stress waves in FGMs and its application in material characterization. J. Sound Vib. **236**, 307–321 (2000)
12. Santare, M.H., Thamburaj, P., Gazonas, G.A.: The use of graded finite elements in the study of elastic wave propagation in continuously nonhomogeneous materials. Int. J. Solids Struct. **40**, 5621–5634 (2003)
13. Samadhiya, R., Mukherjee, A., Schmauder, S.: Characterization of discretely graded materials using acoustic wave propagation. Comput. Mater. Sci. **37**, 20–28 (2006)

14. Berezovski, A., Engelbrecht, J., Maugin, G.A.: Numerical simulation of two-dimensional wave propagation in functionally graded materials. Eur. J. Mech. A-Solids **22**, 257–265 (2003)
15. Liu, G.R., Han, X., Xu, Y.G., Lam, K.Y.: Material characterization of functionally graded material by means of elastic waves and a progressive-learning neural network. Compos. Sci. Technol. **61**, 1401–1411 (2001)
16. Adachi, T., Yoshigaki, N., Higuchi, M.: Analysis of longitudinal impact problem for functionally graded materials. Trans. Jpn. Soc. Mech. Eng. A **79**, 502–510 (2012)
17. Higuchi, M., Yokochi, Y., Adachi, T.: Evaluation on integrated molding of functionally-graded epoxy foams. Trans. Jpn. Soc. Mech. Eng. A **78**, 660–664 (2012)
18. Higuchi, M., Adachi, T., Yoshioka, T., Yokochi, Y.: Evaluation on distributions of mechanical properties in functionally graded syntactic foam. Trans. Jpn. Soc. Mech. Eng. A **78**, 890–901 (2012)
19. Adachi, T., Higuchi, M.: Impulsive responses of functionally graded material bars due to collision. Acta Mech. **224**, 1061–1076 (2013)
20. Krings, W., Waller, H.: Contribution to the numerical treatment of partial differential equations with the Laplace transformation – an application of the algorithm of the fast Fourier transformation. Int. J. Numer. Methods Eng. **14**, 1183–1196 (1979)
21. Adachi, T., Ujihashi, S., Matsumoto, H.: Impulsive responses of a circular cylindrical shell subjected to waterhammer waves. J. Press. Vessel Technol. **113**, 517–523 (1991)
22. Adachi, T., Sakanoue, K., Ujihashi, S., Matsumoto, H.: Damage evaluation of CFRP laminates due to iterative impact. Trans. Jpn. Soc. Mech. Eng. A **57**, 569–575 (1991)
23. Graff, K.: Wave Motion in Elastic Solids. Dover, New York (1991)

Rough Dynamic Response Prediction for Simple Railway Bridges Subjected to High-Speed Trains

Christoph Adam and Patrick Salcher

Abstract This paper describes an efficient approach for prediction of the dynamic peak response of shear deformable railway bridges subjected to high-speed trains, which is based on response spectra. In the proposed response spectra the modal peak response is presented as a function of a non-dimensional modal speed parameter and the bridge span to wagon length ratio. A rough estimate of the dynamic peak bridge response is found by modal combination of the modal peak responses identified from readily available response spectra considering the actual train and bridge parameters.

1 Introduction

If a high-speed train passes a railway bridge with a critical speed, resonance effects may have a severe impact on the train-bridge interaction system. In such a situation the serviceability may be impaired, and critical stresses exceeded, and thus a quasistatic computation of the bridge response is insufficient. Higher modes may contribute significantly in particular to the bridge acceleration, however they cannot be captured with a simplified quasistatic analysis. A large bridge acceleration response leads to instability of ballast, and passengers discomfort. Depending on the fundamental bridge frequency and bridge geometry, Eurocode 1 [1] allows for single-span bridges a quasistatic analysis, if the maximum travel speed is smaller than or equal to 200 km/h. However, comparative analyses of the authors have shown that a quasistatic computation may underestimate the actual dynamic bridge response even in the admitted parameter range of Eurocode 1 [1].

Various mechanical models of different degrees of sophistication have been developed to predict the dynamic response of railway bridges subjected to

C. Adam (✉) · P. Salcher
Unit of Applied Mechanics, University of Innsbruck, 6020 Innsbruck, Austria
e-mail: christoph.adam@uibk.ac.at; patrick.salcher@uibk.ac.at

Alexander K. Belyaev et al. (eds.), *Mechanics and Model-Based Control of Advanced Engineering Systems*, DOI 10.1007/978-3-7091-1571-8_2,
© Springer-Verlag Wien 2014

high-speed trains, see e.g. [2–4]. For example, Cojocaru et al. [5] model the train as an additional elastic beam, which crosses the bridge. Some of the numerical studies have also been validated by experiments [6]. A comprehensive state-of-the-art of the analysis of high-speed train-bridge interaction is provided in the textbook of Yang et al. [7].

Most generally, the assessment of the dynamic bridge response is based on complex numerical models leading to time-consuming time history analysis. In an effort to reduce this effort Hauser and Adam [8] have translated the response spectrum methodology from earthquake engineering into bridge dynamics. This methodology permits for simple bridges a rough, however quick and easy to apply assessment of the peak response induced by high-speed trains, which is particular useful in the initial design phase. Simultaneously, Fink and Mähr [9] have developed independently a similar response prediction concept. Salcher [10] and Adam and Salcher [11] derived for a large number of different characteristic train sets response spectra for both single-span and continuous two-span bridges modeled as Bernoulli-Euler beam. Later, also Spengler [12] has seized this idea studying the effect of high-speed trains on the response of railway bridges. In the present study, a modified response spectrum concept is introduced to include the effect of shear deformations of simply supported bridges for the considered train-bridge interaction problem, compare also with [13].

2 Mechanical Model of the Bridge-Train Interaction System

The bridge and the passing train vehicle represent a rather complex interactive system with time-dependent mechanical properties. Depending on the response quantity to be predicted and on the required accuracy mechanical modeling of this system may be performed with different degree of sophistication [7]. In a detailed model a spring-mass system describes the dynamic behavior of each train car consisting of body, bogies, and viscoelastic connection elements [4]. For example, the contact problem between rails and wheels, and the non-linear behavior of the ballast should be specified appropriately. The numerical solution of the resulting mechanical model is in general computationally expensive, and might come along with numerical stability problems. Thus, detailed system modeling is not efficient in the process of initial bridge design.

Since in the design process of a bridge the properties of all passing trains to be developed during the bridge life cycle cannot be foreseen, analysis should not be performed considering particular characteristics of the vehicle [3]. Consequently, in the simplest approach the passage of a bridge by a high-speed train is considered as a sequence of moving concentrated forces of constant speed v. Each concentrated load represents the static reaction force of a train axle. This model, which is adopted for the present study, disregards the inertia effect of the train, and thus, it leads in general to slightly conservative bridge response predictions [3].

In this paper simply supported single span bridges with a single track are analyzed. In contrast to previous studies [11, 12] the effect of shear deformation is taken into account, which may play a significant role for truss and/or short span bridges. Consequently, it is assumed that lateral bridge vibrations $w(x,t)$ induced by the N axle loads F_i, $i = 1, 2, \ldots, N$ of the considered train, are described sufficiently accurate by means of the partial equations of motion of a shear beam with constant structural parameters (mass per unit length ρA, bending stiffness EI, shear stiffness GA_S) [14]

$$\rho A \ddot{w} + EI w_{,xxxx} - \rho A \frac{EI}{GA_S} \ddot{w}_{,xx}$$
$$= \sum_{i=1}^{N} F_i \left[\delta(x - \xi_i) - \frac{EI}{GA_S} \delta_{,xx}(x - \xi_i) \right] \left[H(t - t_i^0) - H(t - t_i^E) \right] \quad (1)$$

The Dirac delta function $\delta(x - \xi_i)$ describes mathematically the action of the ith axle load with amplitude F_i, which is located at time t at length coordinate $\xi_i = vt - s_i$ (of the bridge). The unit step functions H specify the arrival and departure of F_i at time instants $t_i^0 = s_i/v$ and $t_i^E = (s_i + L)/v$, respectively. v is the constant train speed, s_i denotes the initial location of F_i, and L is the span of the bridge [11].

Modal decomposition of the lateral displacement $w(x,t)$ into the mode shapes ϕ_n, $n = 1, \ldots, \infty$, of the actual boundary value problem, $w(x,t) = \sum_{n=1}^{\infty} q_n(t)\phi_n(x)$, leads to an infinite set of ordinary oscillator equations of motions for the modal coordinates q_n, viscous damping is modally added,

$$\ddot{q}_n + 2\zeta_n \omega_n \dot{q}_n + \omega_n^2 q_n = \frac{1}{m_n} \sum_{i=1}^{N} F_i \phi_n(\xi_i) \left[H(t - t_i^0) - H(t - t_i^E) \right], n = 1, \ldots, \infty \quad (2)$$

The solution of this equation can be found by standard methods of structural dynamics such as Duhamel's integral [15].

The nth mode shape ϕ_n, the corresponding natural circular frequency ω_n, and modal mass m_n of a simply supported shear beam are derived as [14]

$$\omega_n = \frac{n^2\pi^2}{L^2} \left(\frac{EI}{\rho A} \right)^{1/2} \left(1 + \frac{n^2\pi^2}{L^2} \frac{EI}{GA_S} \right)^{-1/2},$$
$$\phi_n = \sin \frac{n\pi x}{L}, \quad m_n = \frac{\rho A L}{2} \quad (3)$$

Structural bridge damping is a fundamental bridge parameter, in particular for excitation at resonance. In Eurocode 1 [1] lower limit values for ζ_n used in this study are defined. They depend on the bridge structure and on span L.

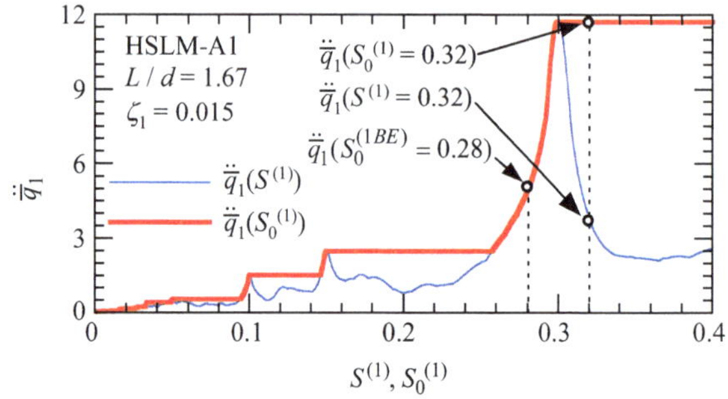

Fig. 1 Comparison of modal peak accelerations $\ddot{\bar{q}}_1(S_0^{(1)})$ and $\ddot{\bar{q}}_1(S^{(1)})$ of the first mode

3 Proposed Response Spectra

In a response spectrum the dynamic peak response of a single-degree-of-freedom oscillator is presented as a function of characteristic excitation and structural parameters. A readily available response spectrum provides the design engineer with a tool to predict the dynamic peak response without performing computational expensive time-history analyses. The characteristic excitation parameters of the considered problem are the ratio of bridge length L to wagon length d, L/d, and the modal speed parameters $S^{(n)}, n = 1, 2, \ldots$. For given length ratio L/d and nth speed parameter $S^{(n)}$ the nth modal peak bridge response is presented in non-dimensional form,

$$\bar{q}_n = \max|q_n(t)| \frac{\rho AL}{F_{\max}}\left(\frac{\omega_n}{2\pi}\right)^2, \quad \ddot{\bar{q}}_n = \max|\ddot{q}_n(t)| \frac{\rho AL}{F_{\max}}, \quad S^{(n)} = \frac{\pi v}{\omega_n L} \tag{4}$$

F_{max} is the maximum single force of the analyzed train model. Since in general the peak response does not occur at the maximum admissible speed $S_0^{(n)}$ but at a lower speed $S^{(n)} < S_0^{(n)}$, coefficients $\bar{q}_n(S_0^{(n)})$ and $\ddot{\bar{q}}_n(S_0^{(n)})$ denote the nth modal peak displacement and acceleration, respectively, in the range $0 \leq S^{(n)} \leq S_0^{(n)}$. As an example, Fig. 1 shows for the specific train load model HSLM-A1 according to [1], a length ratio of $L/d = 1.4$, and viscous damping of $\zeta_1 = 0.015$ the peak acceleration of the first mode, plotted against the corresponding modal speed parameter. The bold line shows the modal peak acceleration as a function of the maximum admissible speed parameter $S_0^{(1)}$ (which enters the response spectrum), while the thin line corresponds to the actual modal peak response at the specific speed parameter $S^{(1)}$.

Based on the mechanical model of the bridge-train interaction system presented before, the authors have derived modal response spectra for simply supported bridges subjected to high-speed trains by series of time history analyses for the HSLM-A load models of Eurocode 1 [1] and for real European high-speed trains. For each train load model and viscous damping specified according to Eurocode 1 [1] these readily available spectra are presented as a function of $S_0^{(n)}$ and L/d.

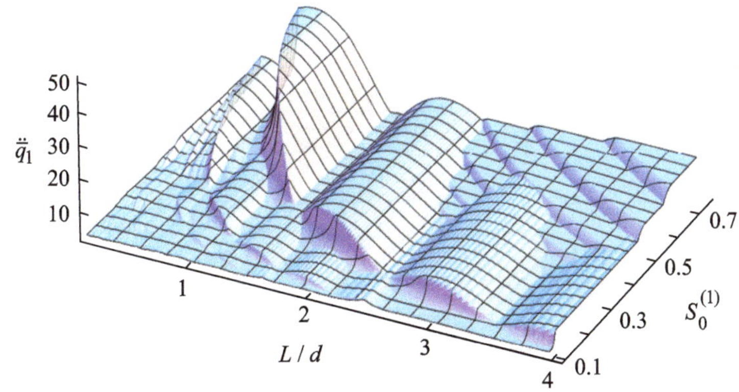

Fig. 2 Acceleration response spectrum of the first mode. HSLM-A1 load model. Damping coefficient $\zeta_1 = 0.015$

Figure 2 shows as a showcase the acceleration response spectrum of the first mode for the HSLM-A1 train set and bridge damping of $\zeta_1 = 0.015$. It can be seen that for certain values of $S_0^{(1)}$ and L/d the gradient of the response surface is very steep, which makes the peak response vulnerable to small parameter variations.

4 Application

Based on the parameters $S_0^{(n)}$ and L/d of the actual considered bridge problem the modal peak responses $\bar{q}_n(S_0^{(n)})$, $\ddot{\bar{q}}_n(S_0^{(n)})$ are identified from the corresponding response spectra. The number of included modes k ($n = 1, \ldots, k$) depends on the considered response quantity. In general, the fundamental mode approximates sufficiently accurate the peak deflection max $|w(x,t)|$, i.e. $k = 1$. When estimating the maximum acceleration max $|\ddot{w}(x,t)|$, three modes should be taken into account ($k = 3$). In all cases, the selection of k should be based on a convergence test taking into account the symmetry and antisymmetry of the mode shapes. Note that the peak response does not occur necessarily at mid-span because of higher mode effects.

The modal peak responses must be superposed to obtain an estimate of the actual maximum bridge response. Since in a response spectrum representation information of the phase shift between the individual modal peak responses is not available, modal combination rules such as the ABSUM method [15]

$$\max |w| (x) \approx \frac{4\pi^2 F_{\max}}{\rho A L} \sum_{n=1}^{k} \left| \frac{\bar{q}_n}{\omega_n^2} \phi_n(x) \right|,$$

$$\max |\ddot{w}| (x) \approx \frac{F_{\max}}{\rho A L} \sum_{n=1}^{k} \left| \ddot{\bar{q}}_n \phi_n(x) \right| \tag{5}$$

and the SRSS method [15]

$$\max|w|(x) \approx \frac{4\pi^2 F_{\max}}{\rho AL}\sqrt{\sum_{n=1}^{k}\left(\frac{\bar{q}_n}{\omega_n^2}\phi_n(x)\right)^2},$$

,

$$\max|\ddot{w}|(x) \approx \frac{F_{\max}}{\rho AL}\sqrt{\sum_{n=1}^{k}\left(\ddot{\bar{q}}_n\phi_n(x)\right)^2} \tag{6}$$

well known from applications in earthquake engineering, are utilized. The SRSS rule provides in general more accurate results than the ABSUM method, but may underestimate the peak response. The ABSUM rule gives always an upper bound of the peak acceleration in the context of the underlying mechanical model.

5 Example

In an example problem the peak displacement and peak acceleration at mid-span of a composite steel-reinforced concrete bridge subjected to the train model HSLM-A1 is assessed. The train and bridge parameters are specified as: $L = 30$ m, $\rho A = 18{,}000$ kg/m, $EI = 1.40 \cdot 10^{11}$ N/m^2, $GA_s = 5.5 \cdot 10^9$ N, $\zeta_n = 0.015$, $v_{max} = 300$ km/h (= 83.3 m/s), $d = 18$ m, $F_{max} = 170$ kN.

Based on these parameters the first three natural circular frequencies of the shear deformable bridge (Eq. 3) are evaluated: $\omega_1 = 27.1$ rad/s, $\omega_2 = 84.5$ rad/s, $\omega_3 = 149$ rad/s. At mid-span the amplitudes of the corresponding mode shapes are: $\phi_1(x = 0.5L) = 1$, $\phi_2(x = 0.5L) = 0$, $\phi_3(x = 0.5L) = -1$. Thus, in a three mode approximation only the first and third mode contribute to the peak response at mid-span. The non-dimensional parameters of this train-bridge interaction problem required for application of response spectra are: $L/d = 1.67$, $S_0^{(1)} \approx 0.32$, $S_0^{(3)} \approx 0.06$. From the two-dimensional representation of the corresponding response spectra shown in Figs. 3 and 4 the following modal peak response quantities for the first and third mode are identified: $\bar{q}_1 = 0.48$, $\bar{q}_3 = 0.23$, $\ddot{\bar{q}}_1 = 12.0$, $\ddot{\bar{q}}_3 = 1.66$. The SRSS combination yields a maximum mid-span peak deflection of $\max w(x = 0.5L) = 0.0081$ m, which is identical to the exact solution (from a complete time history analysis) of the considered beam problem. Note that only the first mode contributes significantly to peak mid-span deflection. Thus, the ABSUM combination rule, $\max w(x = 0.5L) = 0.0083$ m, overestimates slightly the peak deflection, when both the first and the third modal displacements are considered.

Evaluation of the peak acceleration leads to the following outcomes. SRSS: $\max \ddot{w}(x = 0.5L) = 3.80$ m/s^2, ABSUM: $\max \ddot{w}(x = 0.5L) = 4.29$ m/s^2, exact: $\max \ddot{w}(x = 0.5L) = 3.83$ m/s^2, one mode approximation: $\max \ddot{w}(x = 0.5L) = 3.76$ m/s^2. The results show that for this example the ABSUM rule

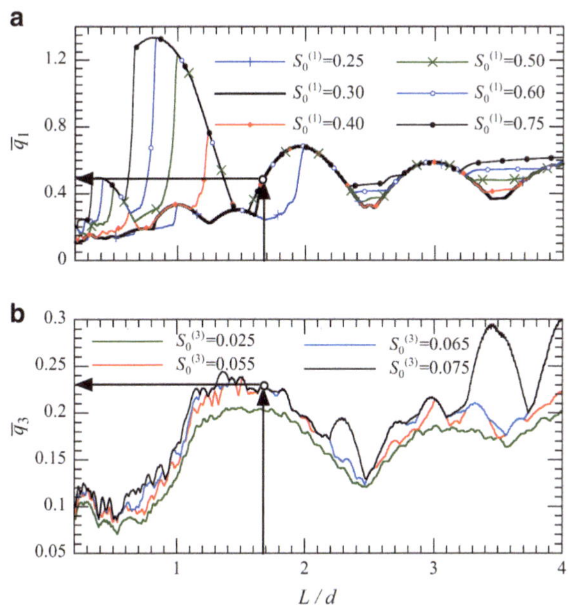

Fig. 3 Response spectra of the modal deflection for (**a**) the first mode, and (**b**) the third mode. HSLM-A1 load model. Damping coefficients $\zeta_1 = \zeta_3 = 0.015$

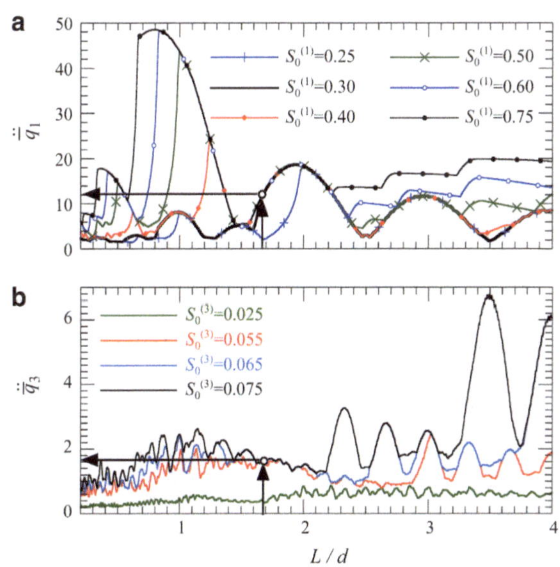

Fig. 4 Response spectra of the modal acceleration for (**a**) the first mode, and (**b**) the third mode. HSLM-A1 load model. Damping coefficients $\zeta_1 = \zeta_3 = 0.015$

overestimates the exact peak acceleration by 12 %, however, the result from the SRSS rule gives a very accurate estimate. The difference between the exact peak acceleration and its one mode approximation is 2 %.

Subsequently, the effect of shear deformation is assessed comparing the derived peak bridge responses with outcomes based on the corresponding Bernoulli-Euler beam (which is rigid in shear). The first natural circular frequencies of the Bernoulli-Euler beam, $\omega_{1(BE)} = 30.6\,\text{rad/s}$, $\omega_{2(BE)} = 122\,\text{rad/s}$, $\omega_{2(BE)} = 275\,\text{Hz}$, show that these quantities are significantly affected by shear. In Fig. 1 the effect of the frequency shift is visualized. Since $\omega_{1(BE)}$ is larger than ω_1, the corresponding maximum admissible speed drops from $S_0^{(1)} \approx 0.32$ to $S_0^{(1BE)} \approx 0.28$. According to Fig. 1 at $S_0^{(1BE)}$ the modal peak acceleration is much smaller than at $S_0^{(1)}$. Thus, the exact peak displacement and peak acceleration based on the Bernoulli-Euler theory, $\max w_{(BE)}(x = 0.5L) = 0.0064\,\text{m}$ and $\max \ddot{w}_{(BE)}(x = 0.5L) = 1.90\,\text{m/s}^2$, respectively, underestimate considerable the response based on the more accurate shear beam theory. It can be concluded that for certain bridges response spectra considering the effect of shear deformation must be utilized for a reliable peak response prediction.

References

1. EN 1991-2.2003: Eurocode 1: Actions on structures. Part 2: Traffic loads on bridges (2003)
2. Fryba, L.: A rough assessment of railway bridges for high speed trains. Eng. Struct. **23**, 548–556 (2001)
3. Museros, P., Alarcon, E.: Influence of the second bending mode on the response of high-speed bridges at resonance. J. Struct. Eng. ASCE **131**, 405–415 (2005)
4. Liu, K., De Roeck, G., Lombaert, G.: The dynamic effect of the train-bridge interaction on the bridge response. In: Papadrakakis, M., Lagaros, N.D., Fragiadakis, M. (eds.) Proceedings of 2nd International Conference on Computational Methods in Structural Dynamics and Earthquake Engineering (COMPDYN 2009), Rhodes, p. 13, 22–24 June 2009
5. Cojocaru, E., Irschik, H., Gattringer, H.: Dynamic response of an elastic bridge due to a moving elastic beam. Comput. Struct. **82**, 931–943 (2004)
6. Lu, Y., Mao, L., Woodward, P.: Frequency characteristics of railway bridge response to moving trains with consideration of train mass. Eng. Struct. **42**, 9–22 (2012)
7. Yang, Y.B., Yau, J.D., Wu, Y.S.: Vehicle-Bridge Interaction Dynamics: With Applications to High-Speed Railways. World Scientific Publishing, Singapore (2004)
8. Hauser, A., Adam, C.: Abschätzung der Schwingungsantwort von Brückentragwerken für Hochgeschwindigkeitszüge (in German). In: Proc. D-A-CH Tagung 2007 der Österreichischen Gesellschaft für Erdbebeningenieurwesen und Baudynamik, 27–28 Sept 2007, Vienna, CD-ROM paper, paper no. 1, p. 10 (2007)
9. Fink, J., Mähr, T.: Simplified method to calculate the dynamic response of railway-bridges on the basis of response spectra. In: Ivanyi, M., Bancila, R. (eds.) Proceedings of 6th International Conference on Bridges across the Danube, Budapest, pp. 245–256 (2007)
10. Salcher, P.: Dynamische Wirkung von Hochgeschwindigkeitszügen auf einfache Brückentragwerke (in German). Diploma thesis, University of Innsbruck (2010)
11. Adam, C., Salcher, P.: Assessment of high-speed train-induced bridge vibrations. In: De Roeck, G., Degrande, G., Lombaert, G., Müller, G. (eds.) Proceedings of 8th European Conference on Structural Dynamics (EURODYN2011), KU Leuven, Leuven, CD-ROM paper, pp. 1302–1309, 4–6 July 2011
12. Spengler, M.: Dynamik von Eisenbahnbrücken unter Hochgeschwindigkeitsverkehr (in German). PhD thesis, Institut für Massivbau, TU Darmstadt, Heft 19 (2010)
13. Salcher, P., Adam, C.: Simplified assessment of high-speed train induced bridge vibrations considering shear effects. Proc. Appl. Math. Mech. **12**, 197–198 (2012)

14. Timoshenko, S., Young, D.H.: Vibration Problems in Engineering, 3rd edn. Van Nostrand Reinhold, New York (1955)
15. Chopra, A.: Dynamics of Structures: Theory and Applications to Earthquake Engineering, 3rd edn. Prentice Hall, Upper Saddle River (2007)

Speed-Gradient Control of Mechanical Systems with Constraints

M.S. Ananyevskiy and A.L. Fradkov

Abstract State-of-the-art and some applications of the Speed-Gradient method to control of complex systems is presented. A universal speed-gradient method and speed-gradient method for control problems with phase constraints is proposed. Some analytical results are obtained. The application of proposed methods is illustrated by two examples: the selective energy control problem of two pendulums and the average energy control problem of quantum diatomic molecule. Computer simulation results confirm fast convergence rate of algorithms.

1 Problem Formulation

Consider a nonlinear time-varying system

$$\dot{x} = f(x, u, t), \quad x \in R^n, \quad u \in R^m, \quad t \in R, \quad x(0) = x_0, \tag{1}$$

with control goal

$$J\left(u(\cdot)\right) = \limsup_{t \to +\infty} Q\left(x(t, u(\cdot), x_0), t\right), \quad J\left(u(\cdot)\right) \to \min, \tag{2}$$

and constraints

$$\forall t \geqslant 0: \quad B_k\left(x(t, u(\cdot), x_0), t\right) > 0, \quad k = 1, \ldots, \mu, \tag{3}$$

M.S. Ananyevskiy (✉) · A.L. Fradkov
Institute of Problems of Mechanical Engineering, RAS, Bolshoj pr. 61, V.O., Saint-Petersburg 199178, Russia
e-mail: msaipme@yandex.ru; fradkov@mail.ru

Alexander K. Belyaev et al. (eds.), *Mechanics and Model-Based Control of Advanced Engineering Systems*, DOI 10.1007/978-3-7091-1571-8_3,
© Springer-Verlag Wien 2014

here x—state, u—control, t—time, x_0—initial condition, $Q(\cdot)$, $B_k(\cdot) : R^n \times R \to R$—some functions, $x(t, u(\cdot), x_0)$—solutions of the system (1) with control $u(\cdot)$ and initial condition x_0.

2 Speed-Gradient Method

In order to design control algorithm the scalar function $w(x, u, t)$ is calculated that is the speed of changing $Q(x, t)$ along trajectories $x(t)$ of (1)

$$w(x, u, t) = \frac{\partial Q(x, t)}{\partial t} + \frac{\partial Q}{\partial x} f(x, u, t). \tag{4}$$

Then it is needed to evaluate the gradient of $w(x, u, t)$ with respect to input variables

$$\nabla_u w(x, u, t) = \nabla_u \frac{\partial Q}{\partial x} f(x, u, t). \tag{5}$$

Finally the algorithm of changing $u(t)$ is determinated according to the differential equation (differential form)

$$\dot{u} = -\Gamma \nabla_u w(x, u, t), \quad u(0) = u_0 \tag{6}$$

or to the algebraic equation (finite form)

$$u = u_0 - \Gamma \nabla_u w(x, u, t), \tag{7}$$

where $\Gamma = \Gamma^T > 0$ is the positive definite gain matrix, u_0 is some initial value of control algorithm. It can be also introduced a speed-pseudogradient algorithm

$$u = u_0 - \Gamma \psi(x, u, t), \tag{8}$$

where $\psi(x, u, t)$ satisfies the pseudogradient condition

$$\psi(x, u, t)^T \nabla_u w(x, u, t) \geqslant 0. \tag{9}$$

The algorithm (6) is called speed-gradient algorithm [1], since it suggests to change $u(t)$ proportionally to the gradient of the speed of changing $Q\big(x(t), t\big)$.

2.1 Universal Speed-Gradient Method

Consider the Taylor approximation for $w(t) = w\big(x(t), u(t), t\big)$, where $\big(x(t), u(t)\big)$ is a trajectory of the system (1)

$$w(t+\tau) = w\big(x(t), u(t), t\big) + \\ + \left(\frac{\partial w}{\partial t}\big(x(t), u(t), t\big) + \frac{\partial w}{\partial x} f\big(x(t), u(t), t\big) + \frac{\partial w}{\partial u}\dot{u}\right)\tau + o(\tau). \tag{10}$$

If $w(x,u,t)$ is non-positive, then (with some additional assumptions) according to a La-Salle principle the control goal (2) is fulfilled.

Consider

$$\dot{u} = -\gamma(x,u,t)\psi(x,u,t), \tag{11}$$

where $\psi : R^n \times R^m \times R \to R^m$, $\gamma : R^n \times R^m \times R \to R$ and

$$\gamma(x,u,t)\frac{\partial w}{\partial u}\psi(x,u,t) > \frac{\partial w(x,u,t)}{\partial t} + \frac{\partial w}{\partial x} f(x,u,t). \tag{12}$$

For example it can be used $\psi(\cdot) = \nabla_u w(\cdot)$ and

$$\gamma(x,u,t) = \frac{\eta(x,u,t) + \lambda\sqrt{\eta(x,u,t)^2 + \zeta(x,u,t)^2}}{\zeta(x,u,t)}, \quad \lambda > 0, \tag{13}$$

with

$$\eta(x,u,t) = \frac{\partial w(x,u,t)}{\partial t} + \frac{\partial w}{\partial x} f(x,u,t), \quad \zeta(x,u,t) = \frac{\partial w}{\partial u}\psi(x,u,t). \tag{14}$$

For affine systems the same algorithm was proposed by Sontag in 1989 [2].

The control algorithm (11) with the inequality (12) we named the "Universal speed-gradient method".

Theorem 1. *Let the following assumptions be valid:*

1. $w\big(x_*(t), u_*(t), t\big)$ *is a twice continuously differentiable function along the trajectories* $\big(x_*(t), u_*(t)\big)$ *of system* (1), (11)*;*
2. *The function* $Q(x,t)$ *is nonnegative, uniformly continuous in any set of the form* $\{(x,t) : ||x|| < \beta, t \geqslant 0\}$ *and radially unbounded;*
3. *For initial condition the inequality* $w\big(x(0), u(0), 0\big) \leqslant 0$ *is true;*
4. *Inequality* (12) *is true for all* $(x,u,t) : w(x,u,t) = 0,\ Q(x,t) \neq 0$*;*
5. *Control* (11) *is a continuous in* (x,u) *function;*

then any solution $\big(x(t), u(t)\big)$ *of* (1), (11) *is bounded and* $\lim_{t\to+\infty} \frac{d}{dt} Q\big(x_*(t), t\big) = 0$.

Proof. The Taylor approximation (10) is true according to assumption 1. According to assumptions 3, 4 inequality $\dot{Q}\big(x(t), t\big) < 0$ is true for all $(x,u,t) : Q(x,t) \neq 0$. Consequently, from assumption 2 follows that any solution $\big(x(t), u(t)\big)$ of (1) and (11) is bounded and $\lim_{t\to+\infty} \frac{d}{dt} Q\big(x_*(t), t\big) = 0$. □

2.2 Universal Speed-Gradient Method with Constraints

Consider the derivative of constraints (3) along the trajectories of the system (1)

$$\frac{\partial B_k}{\partial t}(x,u,t) + \frac{\partial B_k}{\partial x} f(x,u,t) + \frac{\partial B_k}{\partial u}\dot{u} > 0\,, \quad k = 1,\dots,\mu. \tag{15}$$

According to (11) consider the following inequalities

$$\gamma(x,u,t)\frac{\partial B_k}{\partial u}\psi(x,u,t) > -\frac{\partial B_k}{\partial t}(x,u,t) - \frac{\partial B_k}{\partial x} f(x,u,t), \quad k = 1,\dots,\mu. \tag{16}$$

The control algorithm (11) with the inequalities (12) and (16) we named the "Universal speed-gradient method with constraints".

Theorem 2. *Let the following assumptions be valid:*

1. *$w\left(x_*(t), u_*(t), t\right)$ is a twice continuously differentiable function along the trajectories $\left(x_*(t), u_*(t)\right)$ of system* (1), (11)*;*
2. *The function $Q(x,t)$ is nonnegative, uniformly continuous in any set of the form $\{(x,t) : ||x|| < \beta, t \geqslant 0\}$ and radially unbounded;*
3. *For initial condition inequalities $w\left(x(0), u(0), 0\right) \leqslant 0$ and $g\left(x(0), u(0), 0\right) > 0$ are true;*
4. *Inequalities* (12), (16) *are true for all $(x,u,t) : w(x,u,t) = 0$, $Q(x,t) \neq 0$;*
5. *Control* (11) *is a continuous in (x,u) function;*

then any solution $\left(x(t), u(t)\right)$ of (1), (11) *is bounded, the constraint* (3) *fulfilled and* $\lim\limits_{t\to+\infty} \frac{d}{dt} Q\left(x_*(t), t\right) = 0$.

Proof. The Taylor approximation (10) is true according to assumption 1. According to assumptions 3, 4 inequality $\dot{Q}\left(x(t), t\right) < 0$ is true for all $(x,u,t) : Q(x,t) \neq 0$. Consequently, from assumption 2 follows that any solution $\left(x(t), u(t)\right)$ of (1) and (11) is bounded and $\lim\limits_{t\to+\infty} \frac{d}{dt} Q\left(x_*(t), t\right) = 0$. From the assumptions 3, 4 follows that constraints are fulfilled for any solution $\left(x(t), u(t)\right)$ of (1) and (11). □

3 Two Pendulums Example

Consider the system of two pendulums (Fig. 1) with a single control input

$$\begin{cases} \dot{q}_k = \dfrac{1}{ml^2} p_k\,, \\ \dot{p}_k = -mgl\sin q_k + ul\cos q_k\,, \quad k = 1,2, \end{cases} \tag{17}$$

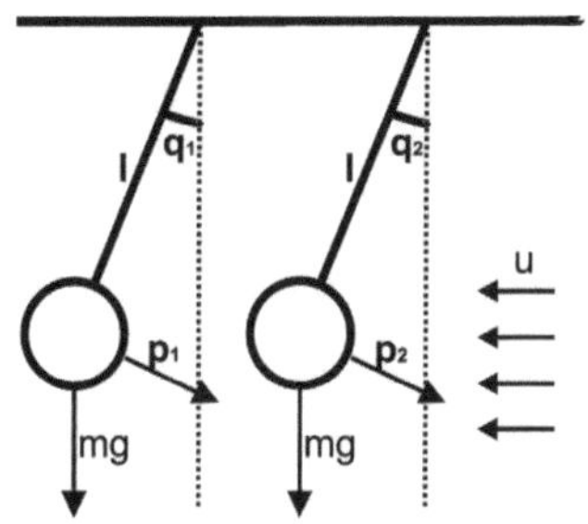

Fig. 1 Two pendulums with a single control input u

the Hamiltonians of pendulums are the following

$$H_0^k(p_k, q_k) = \frac{1}{ml^2} p_k^2 + mgl(1 - \cos q_k), \qquad k = 1, 2. \tag{18}$$

Consider the control goal

$$\lim_{t \to +\infty} H_0^1\big(p_1(t), q_1(t)\big) = E_1. \tag{19}$$

with phase constraints

$$H_0^2\big(p_2(t), q_2(t)\big) < E_2\,, \quad t \geqslant 0. \tag{20}$$

According to the Speed-gradient approach we obtained the following control function

$$u(p,q) = -\Gamma\left(\frac{p_1 \cos q_1}{ml}\Big(H_0^1(p_1, q_1) - E_1\Big) + \alpha \frac{p_2 \cos q_2}{ml\,\big(H_0^2(p_2, q_2) - E_2\big)^2}\right). \tag{21}$$

To demonstrate the ability of the controller to achieve the control goal and to fulfill the phase constraints we carried out computer simulation. The following value of system parameters and initial conditions were chosen: $m = 1$, $l = 1$, $g = 10$, $q_1(0) = 0$, $q_2(0) = 0.05$, $p_1(0) = 0$, $p_2(0) = 0$. Energy goal value for the first pendulum was taken $E_1 = 8$, energy constraint for the second one was taken $E_2 = 5$. Algorithm parameters were: $\Gamma = 0.015$, $\alpha = 10$. Time for simulating was 80 s. Simulations shows that proposed algorithm solve the control problem: energy of the first pendulum converged to the goal value E_1 and the energy of the second was constrained by E_2. The simulating results are presented in (Figs. 2 and 3). The complete analysis of this control system was presented in [3].

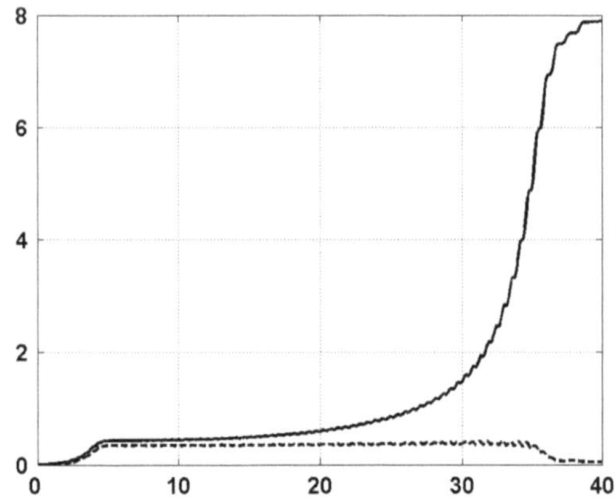

Fig. 2 Energy of pendulums. *Solid line* corresponds to the energy of the first pendulum $H_0^1\left(q_1(t), p_1(t)\right)$, *dash line*—to the energy of the second one $H_0^2\left(q_2(t), p_2(t)\right)$

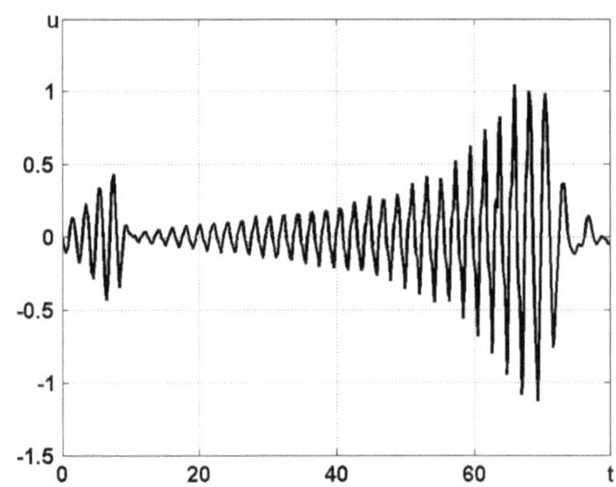

Fig. 3 Control function: $u(t) = u\left(q_1(t), q_2(t), p_1(t), p_2(t)\right)$

4 Molecular Example

Consider a quantum model for diatomic molecule, described by the Schrodinger equation with control [4–6]

$$i\hbar\frac{\partial\Psi(r,t)}{\partial t} = H_0\Psi(r,t) + f(u)H_1\Psi(r,t), \tag{22}$$

where

$$H_0 = -\frac{\hbar^2}{2M}\frac{\partial^2}{\partial r^2} + V(r)\,, \quad H_1 = A\mu(r), \tag{23}$$

and Morse potential

$$V(r) = D\left(exp\left(-\alpha\frac{r-r_0}{r_0}\right) - 1\right)^2 - D, \tag{24}$$

here $i = \sqrt{-1}$, $\hbar = 1$—Planck constant, $\Psi(t, r)$—wave function, r—distance between nuclei of the molecule, M—reduce mass of the molecule, α—parameter of a nonlinearity, $\mu(r)$—molecular dipole momentum, D—dissociation energy, r_0—distance of equilibrium, u—control function of electromagnetic field, $f(\cdot)$—some function. All parameters are in the atomic Hartree unit system.

The problem is to design the control function $u(t)$ to stabilize the average energy on the goal value:

$$\lim_{t\to+\infty} \phi(t)^* H_0 \phi(t) = E_* \, . \tag{25}$$

All the following calculations are made for a finite-level approximation obtained by a Bubnov-Galerkin method.

According to a speed-gradient method the following goal function is introduced

$$Q(\phi) = (\phi^* H_0 \phi - E_*)^2 \tag{26}$$

and $w(\phi, u, t)$ is calculated

$$w(\phi, u, t) = \dot{Q}(\phi) = 2\frac{i}{\hbar}(\phi^* H_0 \phi - E_*)\phi^*[H_1, H_0]\phi f(u) \, . \tag{27}$$

A Tailor approximation for $w(t) = w\big(\phi(t), u(t), t\big)$ is the following

$$w(t+\tau) = g_0(\phi, u) + \big(g_1(\phi, u) + g_2(\phi, u)\dot{u}\big)\,\tau + o(\tau) \, , \tag{28}$$

where

$$g_0(\phi, u) = \frac{i}{\hbar}(\phi^* H_0 \phi - E_*)\phi^*[H_1, H_0]\phi f(u) \, , \tag{29}$$

$$g_1(\phi, u) = \frac{i}{\hbar} f(u) \frac{d}{dt}\big((\phi^* H_0 \phi - E_*)\phi^*[H_1, H_0]\phi\big) \, , \tag{30}$$

$$g_2(\phi, u) = \frac{i}{\hbar}(\phi^* H_0 \phi - E_*)\phi^*[H_1, H_0]\phi \frac{d}{du} f(u) \, . \tag{31}$$

According to a universal speed-gradient method the following algorithm was obtained

$$\dot{u} = sat\left(\frac{-g_2(\phi, u)^2 - g_1(\phi, u)}{g_2(\phi, u)}\right), \qquad u(0) = 0. \tag{32}$$

For computer simulation we used the parameters of iodine molecule $J^{127}J^{127}$: $M = 114{,}842$, $\alpha = 4.954$, $D = 0.0572$, $r_0 = 5.0366$, with control function in the following form $f(u) = 0.02\sin(u)$. Computer simulations shows that the energy converged to the goal value. The simulation results presented in Figs. 4 and 5.

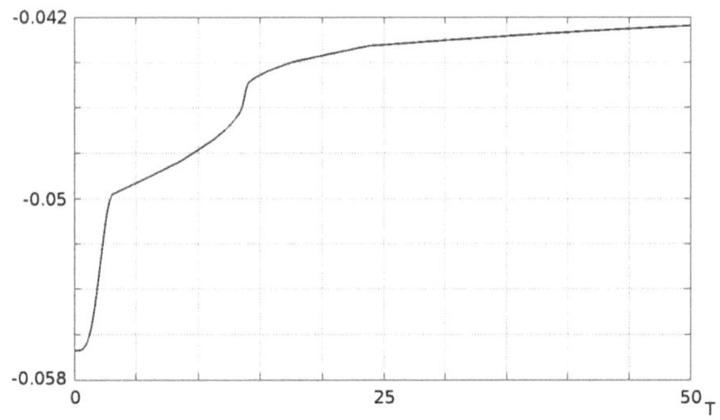

Fig. 4 Average energy: $\phi(t)^* H_0 \phi(t)$, duration: 50 fs (10^{-15} s), goal energy value is $E_* = 0.043$

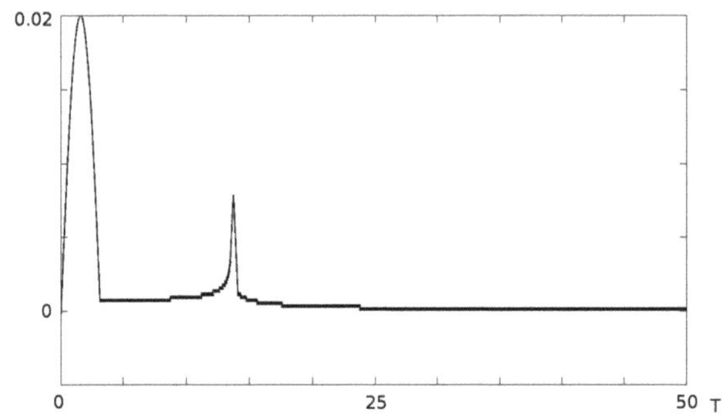

Fig. 5 Control function $f\left(u(t)\right) = A \sin\left(u(t)\right)$, duration: 50 fs ($10^{-15}$ s)

5 Conclusion

A new version of speed-gradient method is proposed that generates "universal" control algorithms both for differential and for finite form. Efficiency of this "universal" method is illustrated by computer simulation for energy control of quantum diatomic molecule. A speed-gradient method for control problems with phase constraints is also proposed and its efficiency is illustrated by computer simulation for selective energy control of two pendulums.

Acknowledgements The work is supported by Russian Foundation for Basic research (projects 12-01-31354, 11-08-01218).

References

1. Fradkov, A.L.: Speed-gradient scheme and its applications in adaptive control. Autom. Remote Control **40**, 1333–1342 (1979)
2. Sontag, E.D.: A "universal" construction of a Artstein's theorem on nonlinear stabilization. Syst. Control Lett. **13**, 117–123 (1989)
3. Ananyevskiy, M.S., Fradkov, A.L., Nijmeijer, H.: Control of mechanical systems with constraints: two pendulums case study. In: Proceedings of the 17th IFAC World Congress on Automatic Control, Seoul (2008)

4. Goggin, M.E., Milonni, P.W.: Driven Morse oscillator: classical chaos, quantum theory and photodissociation. Phys. Rev. Lett. **68** (1988)
5. Anan'evskii, M.S., Fradkov, A.L.: Control of the observables in the finite-level quantum systems. Autom. Remote Control **66**(5), 734–745 (2005)
6. Anan'evskii, M.S.: Selective control of the observables in the ensemble of quantum mechanical molecular systems. Autom. Remote Control **68**(8), 1322–1332 (2007)

About Coupling of the Block Elements

V.A. Babeshko, O.V. Evdokimova, and O.M. Babeshko

Abstract The block element method gives the representations of the solutions of the partial differential equations more exactly than other numerical methods such as finite and boundary element methods. However it is achieved at expenses of more complicated structure and algorithm of the block element method. This is the reason for limitation of wider use of this method. The block element method provides one with the analytical representation of solution of partial differential equations and opens the way to extract the new regularities of the natural and mechanical process. The main operations for the block element method are the factorization and coupling of the block elements. It is necessary to select small block elements to be coupled to the considered domain. This paper addresses coupling of the block elements having a spherical boundary.

1 Introduction

Let us consider the problem of selection of small block elements which have to be connected to a bounded convex region with smooth boundary to be covered by blocks. For this aim it is necessary to develop the appropriate block elements. First of all we construct the block elements for the ball region.

Following the paper [1,2], we demonstrate the block element method algorithm. We constructed here the block elements for the boundary-value problem in the ball region Ω_0 with boundary $\partial\Omega_0$ of radius a for the Helmholtz differential equation in the form of

V.A. Babeshko (✉) · O.M. Babeshko
Department of Science, Kuban State University, 149 Stavropolskia st. 350040 Krasnodar, Russia
e-mail: babeshko41@mail.ru

O.V. Evdokimova
Southern Center of the RAS, 41 Chehov avenue, 344006 Rostov-Don, Russia
e-mail: evdokimoba.olga@mail.ru

Alexander K. Belyaev et al. (eds.), *Mechanics and Model-Based Control of Advanced Engineering Systems*, DOI 10.1007/978-3-7091-1571-8_4,
© Springer-Verlag Wien 2014

$$Q(\partial x_1, \partial x_2, \partial x_3)\varphi = \left[\partial^2 x_1 + \partial^2 x_2 + \partial^2 x_3 + k^2\right]\psi(x_1, x_2, x_3) = 0 \qquad (1)$$

It is shown in [1, 2] that the pseudo-differential equations for the block element enable us to consider all possible variants of boundary conditions in terms of θ, φ, r for the partial differential equation. For this purpose, we considered both the Dirichlet and Neumann boundary conditions.

In the spherical system of coordinate's θ, φ, r Eq. (1) for the ball has the form

$$(\Delta + k^2)\psi = 0\,,$$

$$\Delta = \frac{1}{r^2}\frac{\partial}{\partial r}\left(r^2\frac{\partial}{\partial r}\right) + \frac{1}{r^2}\frac{1}{\sin\theta}\frac{\partial}{\partial\theta}\left(\sin\theta\frac{\partial}{\partial\theta}\right) + \frac{1}{r^2\sin^2\theta}\frac{\partial^2}{\partial\varphi^2}, \qquad (2)$$

$$r, \theta, \varphi \in \Omega_1$$

The solutions of the boundary-value problems (2) are found in the spaces of slowly increasing generalized functions $\mathbf{H}_s$. For investigating this equation by the differential factorization method, we introduce the Fourier-Bessel transform and write down the spherical functions in the following form

$$\mathbf{B}_2(l, m) = \int_0^{\pi}\int_0^{2\pi} g(\theta, \varphi) Y_l^{m-}(\theta, \varphi)\sin\theta d\theta d\varphi = G(l, m)$$

$$\mathbf{B}_2^{-1}(\theta, \varphi)G = \sum_{l=0}^{\infty}\sum_{m=-l}^{l} G(l, m) Y_l^{m+}(\theta, \varphi) = g(\theta, \varphi)$$

$$\mathbf{B}_3(\lambda, l, m)g = \int_0^{\infty}\int_0^{\pi}\int_0^{2\pi} g(r, \theta, \varphi) J_{l+\frac{1}{2}}(\lambda r) Y_l^{m-}(\theta, \varphi)\sin\theta d\theta d\varphi r dr = G(\lambda, l, m)$$

$$\mathbf{B}_3^{-1}(r, \theta, \varphi)G = \sum_{l=0}^{\infty}\sum_{m=-l}^{l}\int_0^{\infty} G(\lambda, l, m) J_{l+\frac{1}{2}}(\lambda r) Y_l^{m+}(\theta, \varphi)\lambda d\lambda = g(r, \theta, \varphi) \qquad (3)$$

Here $J_\nu(\lambda r)$ is the Bessel function, and $Y_l^m(\theta, \varphi)$ is the spherical function,

$$Y_l^{m\pm}(\theta, \varphi) = \frac{1}{2}\sqrt{\frac{2l+1}{\pi}\frac{(l-|m|)!}{(l+|m|)!}}\, P_l^{|m|}(\cos\theta) e^{\pm im\varphi}$$

Applying transforms (3) to Eq. (2), we construct the external form [1, 2]

$$\omega = Pb^2\sin\theta d\theta \wedge d\varphi + Qbdr \wedge d\theta + Rb\sin\theta d\varphi \wedge dr$$

where P, Q, R are some functions. We carry out the transition to the functional equation in the form [1, 2]

$$K(\lambda)\Psi(l,m,\lambda) = \int\limits_{\partial\Omega_0} \omega\,, \qquad K(\lambda) = \lambda^2 - k^2. \tag{4}$$

In the case of a ball, we have

$$(\lambda^2 - k^2)\Psi(l,m,\lambda) = L_{lm}(\lambda),$$
$$L_{lm}(\lambda) = a^2\psi'_{lm}(a)T_{lm}(\lambda,a) - a^2\psi_{lm}(a)T'_{lm}(\lambda,a),$$
$$\psi_{lm}(r) = \mathbf{B}_2(l,m)\psi(r,\theta,\varphi)\,, \quad T'_{lm}(\lambda,r) = \frac{1}{\sqrt{r}}J_{l+\frac{1}{2}}(\lambda r).$$

In order to ensure the automorphism and obtain the pseudo-differential equation, we write down the solution of the boundary-value problem as follows

$$\psi(r,\theta,\varphi) = \mathbf{B}_3^{-1}(r,\theta,\varphi)\frac{L_{lm}(\lambda)}{(\lambda^2 - k^2)}. \tag{5}$$

The automorphism requirement implies satisfaction of the equality [1, 2]

$$\psi(r,\theta,\varphi) = 0, \quad r > a.$$

As a result of transformations for the simple problem under consideration we obtain a pseudo-differential equation that degenerates into algebraic one in the form of

$$L_{lm}(k) = 0. \tag{6}$$

In complex spatial problems, this equation turns out to be pseudo-differential.

On this example we can observe the difference of the generalized factorization from the simple one: although the characteristic equation $K(\lambda)$ has two roots, Eq. (6) should be satisfied only for one root. A similar problem considered by simple factorization in a layer would require satisfying Eq. (6) for both roots.

Using pseudo-differential equation (6), we consider the formulation of the boundary-value problems for Eq. (2). For example, in the case of Dirichlet conditions for the boundary $\partial\Omega$ in the form of

$$\psi(a,\theta,\varphi) = \psi_0(a,\theta,\varphi) \tag{7}$$

the solution of pseudo-differential equation (6) is obtained in the form of

$$\psi'_{lm}(a) = \frac{\psi_{lm0}(a)T'_{lm}(k,a)}{T_{lm}(k,a)}\,, \qquad \text{where} \quad \psi_{lm0}(a) = \mathbf{B}_2(l,m)\psi_0(a,\theta,\varphi).$$

In the complex spatial problems, we obtain either integral or integro-differential equation instead of algebraic one. For example, in [3] we applied the integral factorization method.

Introducing this relation into Eq. (5) and carrying out the necessary calculations, we obtain for $r \to a$ the following result $\psi(a,\theta,\varphi) \to \psi_0(a,\theta,\varphi)$. By using the same algorithm, we have solved the problem with the Neumann boundary condition. In this case, instead of boundary condition (7), the derivative is prescribed on the boundary; i.e.

$$\psi'(a,\theta,\varphi) = \psi_1(a,\theta,\varphi). \tag{8}$$

The solution of the pseudo-differential equation has the following form

$$\psi_{lm}(b) = \frac{\psi'_{lm1}(b)T_{lm}(k,b)}{T'_{lm}(k,b)}, \quad \text{where} \quad \psi_{lm1}(a) = \mathbf{B}_2(l,m)\psi_1(a,\theta,\varphi).$$

Inserting this relation into Eq. (5) and carrying out the transformations, we again obtain that the boundary conditions are satisfied in the Dirichlet problem; i.e., $\psi'(r,\theta,\varphi) \to \psi_1(a,\theta,\varphi)$, $r \to a$ however for the classical component of the solution.

2 The Nonplanar Boundary Block Elements

In order to construct the block elements for the convex domain with smooth boundary it is necessary to use the block elements of the complicated forms calculated in [4]. Let us consider two spheres of the radii a and b, $a < b$ in the Cartesian system of coordinates with the centers on Oz axis, the spheres occupying the regions Ω_1 and Ω_2 of the space Ω, respectively. Let the distance between the centers be h. We designate the region obtained by the intersection of the spheres as Ω_3; i.e.

$$\Omega_3 = \Omega_1 \cap \Omega_2\,, \quad \Omega_4 = \Omega_2 - \Omega_3\,, \quad h < a < b < h + a. \tag{9}$$

Let us construct the block element in this region for the inner boundary-value problem for the Helmholtz differential equation in the form (2).

Following the algorithm of the differential factorization method [4] we introduce two spherical systems of coordinates with the origins in the centers of spheres on the axis z of the original Cartesian system of coordinates with the parameters r, φ, z and ρ, γ, ξ in the regions Ω_1 and Ω_2, respectively. These systems of coordinates provide the tangential stratification of the boundary for the chosen manifold with the edge. Applying transformation (3) to Eq. (2), we construct the outer form [4], which in the coordinates r, φ, z takes the form

$$\omega = g\left\langle \left[\frac{\partial \psi}{\partial \theta} - \psi \frac{\partial P_l^{|m|}(\cos\theta)}{\partial \theta} \left\{ P_l^{|m|}(\cos\theta) \right\}^{-1} \right] r \sin\theta d\varphi \wedge dr - \right.$$

$$- \left[\frac{\partial \psi}{\partial \varphi} - im\psi \right] r d\theta \wedge dr +$$

$$\left. + \left[\frac{\partial \psi}{\partial r} - \psi \frac{\partial r^{-\frac{1}{2}} J_{l+\frac{1}{2}}(\lambda r)}{\partial r} \left\{ r^{-\frac{1}{2}} J_{l+\frac{1}{2}}(\lambda r) \right\}^{-1} \right] r^2 \sin\theta d\theta \wedge d\varphi \right\rangle, \tag{10}$$

$$g(\theta, \varphi, r) = Y_l^{m+}(\theta, \varphi) r^{-\frac{1}{2}} J_{l+\frac{1}{2}}(\lambda r).$$

A similar form is obtained in terms of the system ρ, γ, ξ.

We implement the transition to the functional equation which can be represented in the form (4).

Let us designate the boundary of the body under consideration as $\partial\Omega_3 = \partial\Omega_{10} \cup \partial\Omega_{20}$ where $\partial\Omega_{10}$ represents the surface of the sphere of radius a of boundary $\partial\Omega_3$, while $\partial\Omega_{20}$ is the surface of the sphere of radius b of the indicated boundary.

Applying the conventional algorithm of construction of the pseudo-differential equations [1, 2] we use the introduced local spherical systems of coordinates for each boundary in Eq. (5) taking that

$$\int_{\partial\Omega_3} \omega = \int_{\partial\Omega_{10}} \omega_{2m-1} + \int_{\partial\Omega_{20}} \omega_{2m}, \quad m = 1, 2, 3, 4, \quad \partial\Omega_{30} = \partial\Omega_2 - \partial\Omega_{20}, \tag{11}$$

where $\partial\Omega_{30}$ is the addition to the boundary $\partial\Omega_{20}$ in $\partial\Omega_2$.

It should be noted that in what follows we consider the outer and inner boundary-value problems for Eq. (3) in various regions which are formed by the boundaries $\partial\Omega_{10}$, $\partial\Omega_{20}$ and $\partial\Omega_{30}$. For the sake of brevity, the solutions on these boundaries are designated as ψ_1 and ψ_2, respectively, although their numerical values for various boundary-value problems are different.

Now, we construct the block element for the boundary-value problem in the region (9), i.e., in addition to the intersection $\Omega_1 \cap \Omega_2$ in Ω_2, where the inequality means that the centers of spheres are outside of the region Ω_4. For this case, the values of the outer form on the boundaries have the form

$$\omega_1 = g_1 \left(r^{-\frac{1}{2}} H_{l+\frac{1}{2}}^{(1)}(\lambda r) \frac{\partial \psi_1}{\partial r} - \psi_1 \frac{\partial r^{-\frac{1}{2}} H_{l+\frac{1}{2}}^{(1)}(\lambda r)}{\partial r} \right) a^2 \sin\theta d\theta \wedge d\varphi,$$

$$\omega_2 = g_2 \left(f_2^p \frac{\partial \psi_2}{\partial \rho} - \psi_2 \frac{\partial f_2^p}{\partial \rho} \right) b^2 \sin\gamma d\gamma \wedge d\sigma,$$

$$\omega_3 = g_1 \left(f_1^r \frac{\partial \psi_1}{\partial r} - \psi_1 \frac{\partial f_1^r}{\partial r} \right) a^2 \sin\theta d\theta \wedge d\varphi,$$

$$\omega_4 = g_2 \left(\rho^{-\frac{1}{2}} J_{l+\frac{1}{2}}(\lambda\rho) \frac{\partial \psi_2}{\partial \rho} - \psi_2 \frac{\partial \rho^{-\frac{1}{2}} J_{l+\frac{1}{2}}(\lambda\rho)}{\partial \rho} \right) b^2 \sin\gamma d\gamma \wedge d\sigma.$$

The automorphism requirement results in the following pseudo-differential equations:

$$\mathbf{B}_{21}^{-1}(\theta,\varphi)\left[\left(r^{-\frac{1}{2}} H_{l+\frac{1}{2}}^{(1)}(kr) \right)^{-1} \times \right.$$

$$\times \left\{ \int_0^{2\pi} \int_{\pi}^{\pi-\theta_0} g_1 \left\langle r^{-\frac{1}{2}} H_{l+\frac{1}{2}}^{(1)}(kr) \frac{\partial \psi_1}{\partial r} - \psi_1 \frac{\partial r^{-\frac{1}{2}} H_{l+\frac{1}{2}}^{(1)}(kr)}{\partial r} \right\rangle a^2 \sin\theta d\theta d\varphi + \right.$$

$$\left. \left. + \int_0^{2\pi} \int_{\gamma_0}^{\pi} g_2(\gamma,\sigma,l,m) \left\langle f_2(\rho,\gamma,l,k) \frac{\partial \psi_2}{\partial \rho} - \psi_2 \frac{f_2(\rho,\gamma,l,k)}{\partial \rho} \right\rangle b^2 \sin\gamma d\gamma d\sigma \right\} \right] = 0,$$

$$r = a\,, \quad \rho = b\,, \qquad \theta,\varphi \in \partial\Omega_{10},$$

$$\mathbf{B}_{21}^{-1}(\gamma,\sigma)\left[\left(\rho^{-\frac{1}{2}} J_{l+\frac{1}{2}}(kr) \right)^{-1} \times \right.$$

$$\times \left\{ \int_0^{2\pi} \int_{\gamma_0}^{\pi} g_1 \left\langle \rho^{-\frac{1}{2}} J_{l+\frac{1}{2}}(kr) \frac{\partial \psi_2}{\partial \rho} - \psi_2 \frac{\partial r^{-\frac{1}{2}} J_{l+\frac{1}{2}}(kr)}{\partial \rho} \right\rangle b^2 \sin\gamma d\gamma d\sigma + \right.$$

$$\left. \left. + \int_0^{2\pi} \int_{\pi}^{\pi-\theta_0} g_1(\theta,\varphi,s,n) \left\langle f_1(r,\theta,s,k) \frac{\partial \psi_1}{\partial r} - \psi_1 \frac{f_1(r,\theta,s,k)}{\partial r} \right\rangle a^2 \sin\theta d\theta d\varphi \right\} \right] = 0,$$

$$\gamma,\sigma \in \partial\Omega_{20},$$

$$\theta_0 = \arccos\frac{b^2 - a^2 - h^2}{2ah}, \quad \gamma_0 = \arccos\frac{a^2 - b^2 - h^2}{2bh}. \tag{12}$$

The pseudo-differential equations taken in one of these forms enable us to formulate an arbitrary number of possible boundary-value problems for the Helmholtz equation. The above form is intended for the both Dirichlet and Neumann boundary-value problems.

In order to couple two block elements and create a new join block element it is necessary to put the boundary conditions (7) and (8) of the first block element into the pseudo-differential equations (12) of the second block element on the boundary of their contact $\partial\Omega_{10}$. This boundary contains parameter a in Eq. (12). The boundary conditions obtained from these equations are introduced in the representation of outer forms. After this procedure, the general representation of solutions in all considered cases is given by the relation

$$\psi(r,\theta,\varphi) = \mathbf{B}_3^{-1}(r,\theta,\varphi)K^{-1}(\lambda,k)\int\limits_{\partial\Omega}\omega\,, \quad r,\theta,\varphi \in \Omega_k\,, \quad \partial\Omega = \partial\Omega_0 \cup \Omega_{10} \cup \Omega_{20}. \tag{13}$$

3 Conclusions

By extending the presented way of coupling the block elements it is possible to cover the arbitrary convex region with smooth boundary by the constructed blocks. This method is used to consider the outer boundary problems when the restricted convex region with smooth boundary is cut from the space. It should be taking into account that the calculation of integral (13) demands the special selection of the way for its integration. The principle of radiation of the energy must be applied. For the outer problems, the contour of integration in Eq. (12) should be deformed in an appropriate way for the operator $\mathbf{B}_3^{-1}(r,\varphi,z)$ when going around the material pole. The described results are transferred by the presented algorithm to the cases of sets of partial differential equations, for example, in the theory of elasticity, which proves to be more complicated [5] only from a technical perspective.

This approach can be applied for the arbitrary block structures. By block structures, we mean materials occupying bounded, semi-bounded, or unbounded domains which are called contacting blocks. It is assumed that each block in a block structure has its own specific behavior in response to physical fields of a various nature. It is also assumed that these fields are described by boundary-value problems for systems of coupled partial differential equations with constant coefficients. Media of this type are typical of the earth's crust, structural materials under complex physical-mechanical conditions, non-material structures of various types and electronic materials. A similar structure is also typical for various materials, including those created by combining only nanoscale components or macro- and nanoscale components.

The absence of considerable constraints on boundary value problems describing the properties of individual blocks suggests that these block structures can have a wide variety of properties. In the general case, the concept of a block requires the boundary of the domain a boundary value problem including multiply connected domains to be unchanged and piecewise smooth. Each block can be bounded or unbounded and can involve coupled processes related to solid and fluid mechanics and electromagnetic, diffusion, thermal, acoustic and other processes.

Block structures are more general objects than piecewise homogeneous structures in which the physical parameters of the medium are assumed to change in jumps in the transition from one block to another with the preservation of the medium material. The latter property means that certain coefficients in the differential equations of a boundary value problem undergo jump variations in the transition from one block to another with the type of the boundary value problem being preserved.

Block structures have a wider range of properties than piecewise homogeneous structures. This follows from the variety of blocks' properties, their shapes, and the character of interblock interactions and also results from the interaction of physical fields, some of which are produced or transformed by blocks.

Short information about application of the block element method is presented in what follows, cf. [6].

1. Reduction of the differential equation to a functional equation by applying the Fourier transforms. The three-dimensional Fourier transform is applied to the system of the differential equation to reduce it to a functional equation. The components of the vector of exterior forms are introduced, namely vectors of an arbitrary coordinate system lying in the coverings of the tangent bundle of the body surface are introduced. In a Cartesian coordinate system, we used the tangent vectors of an arbitrary element of a covering.
2. Fulfillment of prescribed boundary conditions. To achieve this, the solution and its normal derivatives on boundary taken from the boundary conditions are introduced into the representations of the exterior forms. The tangent derivatives are not taken into account. The exterior forms contain the solution and its derivatives on boundary. The functions or normal derivatives on the boundary are found by fitting and inverting the nonsingular matrix from boundary conditions and are introduced into the corresponding representations of exterior forms. The remaining functions or normal derivatives have to be found from the pseudo-differential equations obtained by transformations of the functional equations. The following steps are to be performed to determine the remaining unknowns in the representation of the solution.
3. Factorization of the matrix function coefficient in the functional equation. This representation implies that the elements of coefficient in the functional equation are rational functions only with singularities in the restricted region. For the factorization to be realized we need to apply the method suggested in [6].
4. Reduction of the functional equation to a system of pseudo-differential equations. In order to obtain the required pseudo-differential equations, it is necessary to require that the corresponding Leray's residue forms vanish. Calculating these residue forms in the neighborhood of the local coordinate system we obtain the sought-for relationships.
5. Derivation of a representation of the solution to the boundary-value problem. Introducing the determined components of solution of the pseudo-differential equations into the vector of exterior forms and applying the three-dimensional Fourier transform of the solution we obtain the boundary value problem.

Acknowledgements The work was supported by the Russian Ministry of Education and Science, Agreement No. 14.B37.21.0646 at 20.08.2012.

References

1. Babeshko, V.A., Evdokimova, O.V., Babeshko, O.M.: Block elements with a spherical boundary. Dokl. Phys. **55**(10), 510–513 (2010)
2. Evdokimova, O.V., Babeshko, O.M., Babeshko, V.A.: On the differential factorization method in inhomogeneous problems. Dokl. Math. **77**(1), 140–142 (2008)
3. Babeshko, V.A., Evdokimova, O.V., Babeshko, O.M.: The integral factorization method in mixed problems for anisotropic media. Dokl. Phys. **54**(6), 285–289 (2009)
4. Babeshko, V.A., Evdokimova, O.V., Babeshko, O.M.: Block elements with a nonplanar boundary. Dokl. Phys. **57**(6), 245–249 (2012)
5. Vorovich, I.I., Babeshko, V.A.: Combined Dynamic Problems of Elasticity Theory in Unbounded Domains. Nauka, Moscow (1979). (in Russian)
6. Babeshko, V.A., Babeshko, O.M.: On integral and differential factorization methods. Dokl. Math. **74**(2), 762–766 (2006)

Fractional Derivatives Appearing in Some Dynamic Problems

Alexander K. Belyaev

Abstract Three types of suspension of a semi-infinite Bernoulli-Euler beam and a fluid-conveying pipe are considered. It is shown that the environment in the form of a semi-infinite Bernoulli-Euler beam or a fluid-conveying pipe is taken into account by adding a fractional derivative into the suspension equation. The eigenvector expansion method based upon transformation of the derived equation into a set of four semi-differential equations is utilised for solving the equations with fractional derivatives. A simple expression for the critical velocity of the fluid in the pipe is obtained. If this value is exceeded, both the pipe and its suspension become unstable.

1 Introduction

The intent of the paper is to show that the governing equation for simple mechanical systems may contain fractional derivatives. We consider three types of oscillator to which a semi-infinite Bernoulli-Euler beam is attached. It is shown that if the consideration is limited only to the oscillator, then the environment (i.e. the semi-infinite Bernoulli-Euler beam) adds a fractional derivative into the oscillator equation. Another system governed by a differential equation with fractional derivative is the suspension of the fluid-conveying-pipe. The eigenvector expansion method based upon transformation of the equation into a set of four semi-differential equations is utilised for solving the obtained differential equation with fractional derivatives.

Alexander K. Belyaev (✉)
Institute for Problems in Mechanical Engineering, RAS, V.O. Bolshoy pr. 61, 199178, Saint-Petersburg, Russia

Saint-Petersburg State Polytechnic University, Polytechnicheskaya 29, 195251, Saint-Petersburg, Russia
e-mail: vice.ipme@gmail.com

Alexander K. Belyaev et al. (eds.), *Mechanics and Model-Based Control of Advanced Engineering Systems*, DOI 10.1007/978-3-7091-1571-8_5,
© Springer-Verlag Wien 2014

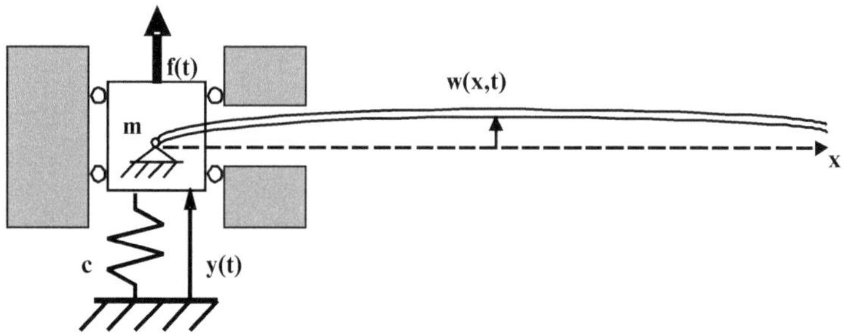

Fig. 1 Schematics of the first model

2 Three Mechanical Systems

Consider a Single-Degree-Of-Freedom system and a semi-infinite Bernoulli-Euler beam $x > 0$, which is attached to mass m at $x = 0$, cf. Fig. 1.

The mass m is allowed to perform only vertical displacements and governed by the equation

$$m\frac{\mathrm{d}^2 y}{\mathrm{d}t^2} = -cy + Q\Big|_{x=0} + f(t)\ . \tag{1}$$

where y is the absolute displacement of the mass m, $f(t)$ is an external driving force, t is time and $Q\big|_{x=0}$ is the shear force in the beam acting on the mass m.

The equation of the beam bending is as follows

$$EI\frac{\partial^4 w}{\partial x^4} + \rho A\frac{\partial^2 w}{\partial t^2} = 0\ , \quad 0 < x < \infty\ , \tag{2}$$

where $w(x,t)$ is the absolute displacement, EI is the bending stiffness of the beam, ρ is the mass density and A is the cross-sectional area. The condition of coupling of mass m and the beam is given by

$$y(t) = w(0,t)\ . \tag{3}$$

The zero initial conditions are assumed, then the Laplace transformation gives

$$mp^2\bar{y}(p) + c\bar{y}(p) = \bar{Q}(p)\Big|_{x=0} + \bar{f}(p)\ , \tag{4}$$

$$EI\tfrac{\mathrm{d}^4\bar{w}}{\mathrm{d}x^4} + \rho A p^2\bar{w} = 0\ , \quad 0 < x < \infty\ . \tag{5}$$

The solution of Eq. (5) bounded at infinity is as follows

$$\bar{w}(x,p) = A_2\exp(\lambda_2 x) + A_4\exp(\lambda_4 x)\ . \tag{6}$$

where the wave numbers are

$$\lambda_2 = -(1+i)\beta\sqrt{p}\,, \quad \lambda_4 = -(1-i)\beta\sqrt{p}\,, \quad \beta = \sqrt[4]{\rho A/4EI}$$

for $\sqrt{p} > 0$. The bending moment in the beam vanishes at $x = 0$

$$\bar{M}\bigg|_{x=0} = -EI\frac{\mathrm{d}^2\bar{w}}{\mathrm{d}x^2}\bigg|_{x=0} = 2EIi\beta^2 p(A_2 - A_4) = 0\,.$$

Hence, $A_2 = A_4$ and the shear force $\bar{Q}\big|_{x=0}$ to be substituted into Eq. (4) is

$$\bar{Q}\bigg|_{x=0} = -EI\frac{\mathrm{d}^3\bar{w}}{\mathrm{d}x^3}\bigg|_{x=0} = -EI\left[\lambda_2^3 A_2 + \lambda_4^3 A_4\right] = -4EI\beta^3 p\sqrt{p}\,A_2\,. \tag{7}$$

As follows from Eq. (6) $\bar{w}(0, p) = \bar{y}(p) = 2A_2$, that allows one to establish the following relationship between the shear force $\bar{Q}\big|_{x=0}$ and the beam displacement $\bar{w}(0, p)$

$$\bar{Q}\bigg|_{x=0} = -2EI\beta^3 p\sqrt{p}\,\bar{y}(p)\,. \tag{8}$$

Inserting the latter equation into Eq. (4) yields

$$mp^2\bar{y}(p) + 2EI\beta^3 p\sqrt{p}\,\bar{y}(p) + c\bar{y}(p) = \bar{f}(p)\,. \tag{9}$$

Since the trivial initial conditions were assumed, Eq. (9) corresponds to the following ordinary differential equation for displacement $y(t)$

$$m\frac{\mathrm{d}^2 y}{\mathrm{d}t^2} + 2EI\beta^3\frac{\mathrm{d}^{3/2} y}{\mathrm{d}t^{3/2}} + cy = f(t)\,. \tag{10}$$

As seen from Eq. (10), the dynamics of mass m is governed by a single differential equation with a fractional derivative of the order $3/2$.

Another mechanical system governed by the differential equation with a fractional derivative is obtained from the above system provided that the semi-infinite Bernoulli-Euler beam is clamped to a rigid mass m rather than it is simply supported. In this case the governing equation is as follows (the derivation is omitted as it is fully analogous to the previous one)

$$m\frac{\mathrm{d}^2 y}{\mathrm{d}t^2} + 4EI\beta^3\frac{\mathrm{d}^{3/2} y}{\mathrm{d}t^{3/2}} + cy = f(t)\,. \tag{11}$$

The third mechanical system consists of a disk attached to an angular spring and a semi-infinite Bernoulli-Euler beam, Fig. 2.

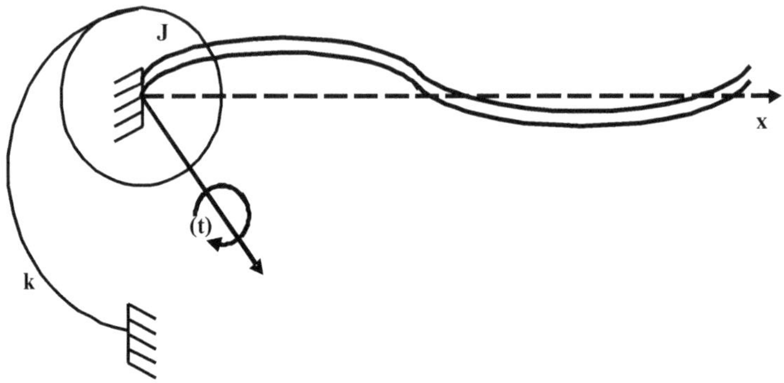

Fig. 2 Schematics of the third model

The beam is supposed to be clamped to the disc in such a way that the angle of rotation of the disk and that of the beam at $x = 0$ coincide. The differential equation of the disk is given as

$$J\frac{\mathrm{d}^2\varphi}{\mathrm{d}t^2} + 2\beta EI\frac{\mathrm{d}^{1/2}\varphi}{\mathrm{d}t^{1/2}} + k\varphi = m(t)\ . \tag{12}$$

where J is the moment of the mass inertia, k is the angular stiffness of the spring and $m(t)$ is the external driving moment. We omit the derivation however one can easily perform it by analogy with the above one. Again, a differential equation with a fractional derivative is obtained however, in contrast to Eqs. (10) and (11), the governing equation contains the fractional derivative of the order $1/2$.

3 Mechanical System with a Pipeline Conveying Fluid

We consider now a pipeline conveying a heavy fluid. The pipe is assumed to perform bending vibration in the plane xz. The suspension of the pipe is assumed to be modeled by a spring of stiffness c and a dashpot b. The mass of the suspension is m and the velocity of the fluid is denoted by v, see Fig. 3.

The governing equation for the pipe bending vibration is as follows

$$EI\frac{\partial^4 w}{\partial x^4} = -(\rho A)_p\frac{\partial^2 w}{\partial t^2} - (\rho A)_f\, a_z\ . \tag{13}$$

Here the subscripts p and f refer to the pipe and fluid respectively and a_z denotes the fluid acceleration in direction z . Using the rule of determining the material derivative we obtain

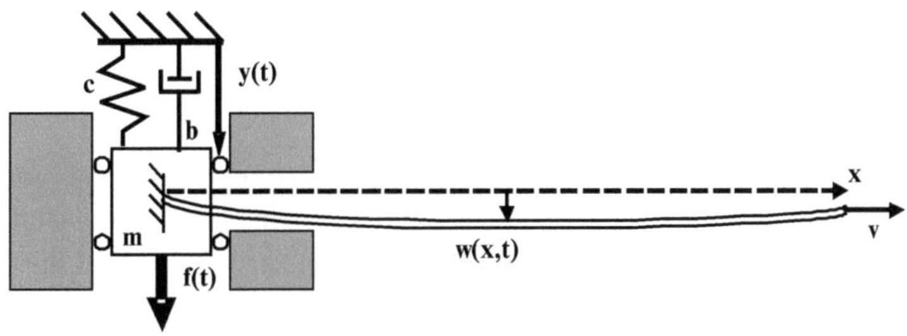

Fig. 3 Schematics of the suspended pipe conveying fluid

$$a_z = \frac{\partial^2 w}{\partial t^2} + 2v\frac{\partial^2 w}{\partial x\,\partial t} + v^2\frac{\partial^2 w}{\partial x^2}\,. \tag{14}$$

Substituting Eq. (14) into Eq. (13) yields

$$EI\frac{\partial^4 w}{\partial x^4} + (\rho A)_f\left(\frac{\partial^2 w}{\partial t^2} + 2v\frac{\partial^2 w}{\partial x\,\partial t} + v^2\frac{\partial^2 w}{\partial x^2}\right) = 0\,, \tag{15}$$

where the inertia term of the pipe is dropped under the assumption $(\rho A)_p \ll (\rho A)_f$ which implies that the distributed mass of the pipe is much smaller than that of the conveyed fluid.

Assuming zero initial conditions for the beam and applying the Laplace transformation we obtain the following ordinary differential equation in the Laplace domain

$$EI\bar{w}^{\,IV} + (\rho A)_f v^2\bar{w}^{\,II} + 2v(\rho A)_f p\bar{w}^{\,I} + (\rho A)_f p^2\bar{w} = 0\,. \tag{16}$$

The eigenvalues $\lambda_n, n = 1, 2, 3, 4$ are now the solutions of the equation

$$\lambda^4 + \beta^2\left(2\lambda\sqrt{\eta} + \beta p\right)^2 = 0\,,\quad \eta = \frac{\beta^2 v^2}{4}\,,\quad \beta = \sqrt[4]{\frac{(\rho A)_f}{EI}} > 0\,. \tag{17}$$

Repeating the derivation of the second part of the paper we arrive at the following equation for the Laplace transform of the displacement y

$$mp^2\bar{y}(p) + 4EI\beta^3 p^{3/2}\bar{w}(0,p) + p\left[b - (\rho A)_f v\right]\bar{w}(0,p) + c\bar{y}(p) = f(p) \tag{18}$$

which corresponds to the following ordinary differential equation with a fractional derivative of order $3/2$

$$m\frac{\mathrm{d}^2 y}{\mathrm{d}t^2} + 4EI\beta^3\frac{\mathrm{d}^{3/2} y}{\mathrm{d}t^{3/2}} + \left[b - (\rho A)_f v\right]\frac{\mathrm{d}y}{\mathrm{d}t} + cy = f(t)\,. \tag{19}$$

This equation governs the motion of the pipe suspension.

4 Eigenvector Expansion Method for Solving Differential Equation with Fractional Derivative

Equations (10)–(12), (19) are ordinary differential equations of second order with the derivatives of the order $1/2$ or $3/2$. We take Eq. (19) in the case $f(t) = 0$ and solve it by means of the eigenvector expansion method suggested in [1] for differential equations with fractional derivatives. To this end, we introduce a non-dimensional time $\tau = kt$, $k = \sqrt{c/m}$ and two non-dimensional system parameters

$$\delta = 2\frac{EI\beta^3}{m\sqrt{k}}, \qquad \epsilon = \frac{b - (\rho A)_f v}{mk} \tag{20}$$

Then we can set Eq. (19) in the following form

$$D^2 y + 4\delta D^{3/2} y + \varepsilon D y + y = 0\,, \quad D = \frac{\mathrm{d}}{\mathrm{d}\tau} \tag{21}$$

This equation can be represented in the normal form of four semi-differential equations by means of the substitution

$$z_1 = D^{3/2} y(t)\,, \quad z_2 = Dy(t)\,, \quad z_3 = D^{1/2} y(t)\,, \quad z_4 = y(t) \tag{22}$$

which allow us to rewrite the latter equation in the matrix form

$$\{\mathbf{A}\} D^{1/2} \{\mathbf{z}\} = \{\mathbf{B}\}\{\mathbf{z}\}$$

where $\{\mathbf{z}\}$ denotes the column composed of z_n, $n = 1,2,3,4$ in Eq. (22). Applying the standard methods of linear algebra yields the eigenvectors $\{\Psi\}$ and eigenvalues λ_j

$$\{\mathbf{A}\}\{\Psi\}_j = \lambda_j \{\mathbf{B}\}\{\Psi\}_j \tag{23}$$

where the eigenvectors are orthonormalized, i.e.

$$\{\Psi\}_i^T \{\mathbf{B}\}\{\Psi\}_j = \delta_{ij}\,, \quad \{\mathbf{A}\}\{\Psi\}_j = \lambda_j \delta_{ij}\,. \tag{24}$$

Let us notice at this place that the eigenvalues λ_j of the matrix equation (23) have nothing in common with the eigenvalues of the differential equation (21). Namely the eigenvalues λ_j are solutions of equation $\lambda^4 + a\lambda + b = 0$ and are given by

$$\lambda_1 = \bar{\lambda}_2 = p + iq\,, \quad \lambda_{3,4} = -p \pm is \tag{25}$$

where

$$p=\sqrt{\kappa}\,,\quad q=\sqrt{\kappa+\frac{\delta}{2\sqrt{\kappa}}}\,,\qquad s=\sqrt{\kappa-\frac{\delta}{2\sqrt{\kappa}}}\,,$$
$$\kappa=\frac{2^{1/3}}{4}\left[\left(\delta^2+\sqrt{\delta^4-\frac{16}{27}}\right)^{1/3}+\left(\delta^2-\sqrt{\delta^4-\frac{16}{27}}\right)^{1/3}\right] \tag{26}$$

By means of the substitution $\{\mathbf{z}\} = \{\mathbf{h}\}\{\Psi\}$ where matrix $\{\Psi\}$ is built from the eigenvectors columns $\{\Psi\}_j$ we arrive at the system of four uncoupled semi-differential equations

$$D^{1/2}h_j-\lambda_j h_j(t)=0\,,\quad j=1,2,3,4 \tag{27}$$

Solving these equations with the help of Laplace transformation, applying the inverse Laplace transformation and satisfying the initial conditions, we obtain the sought-for result. We refer the reader to [1] for detail.

Obtaining a closed form solution assumes the well-known property of the Laplace transformation, namely the Laplace transform $L[\ldots]$ of a fractional derivative of order α of function $\varphi(t)$ is as follows

$$L\big[D^{\alpha}\varphi(t)\big]=p^{\alpha}L\big[\varphi(t)\big]-C \tag{28}$$

It follows from the formal definition of a fractional derivative of order α which is given by

$$D^{\alpha}\left\{\varphi(t)\right\}=\frac{\mathrm{d}}{\mathrm{d}t}\left\{\frac{1}{\Gamma(1-\alpha)}\int\limits_0^t(t-\tau)^{-\alpha}\varphi(t)\,d\tau\right\} \tag{29}$$

see [2]. Here C is the constant determined by the following condition

$$C=D^{\alpha-1}\,\varphi(t)\Big|_{t=0} \tag{30}$$

It is worth mentioning that the value of C is not necessarily equal to zero even for the zero initial conditions for the system. There exists a seeming discrepancy between the number of initial conditions in the system (two initial conditions in the initial-value problem) and the number of the integration constants in the system (27) of four uncoupled semi-differential equations (four integration constants). This discrepancy is easily removed since the general expressions for the displacement and velocity contains some the functions which are unbounded at $t\to 0$. The requirement that these functions must vanish provide us with two additional conditions, see [1] for detail.

5 Critical Velocity of the Fluid in the Pipe

We now proceed to analysis of stability of the suspension. Numerical analysis of free vibration of the pipe, i.e. under the assumption $f(t) = 0$, shows that the stability border is described by the condition $\varepsilon = \delta$. Since parameter ε depends on the velocity of fluid from this condition one obtains the critical velocity of the flow

$$v_{crit} = \frac{b}{\rho A} + \sqrt[4]{\frac{1}{8}\frac{EI}{\rho A}\frac{c}{m}} \tag{31}$$

If this value is exceeded, i.e. $v > v_{crit}$, then the suspension and hence the pipe are unstable.

6 Conclusion

It is shown that some mechanical systems are governed by differential equations with fractional derivatives. The eigenvector expansion method is used for solving the obtained equations with fractional derivatives and deriving closed-form solutions. For example, this closed form solution is appropriate for obtaining simple formula for the critical velocity for systems conveying fluids.

Acknowledgements The work was supported by the joint project of the Russian Foundation for Basic Research and the National Science Council, Taiwan, grant 12-01-92000 HHC_a.

References

1. Suarez, L.E., Shokooh, A.: An eigenvector expansion method for the solution of motion containing fractional derivatives. ASME J. Appl. Mech. **64**, 629–635 (1997)
2. Oldham, K.B., Spanier, J.: Fractional Calculus. Academic, New York (1974)

Hydroelastic Stability of Single and Coaxial Cylindrical Shells Interacting with Axial and Rotational Fluid Flows

S.A. Bochkarev and V.P. Matveenko

Abstract In this paper, the hydroelastic stability of single elastic and coaxial cylindrical shells of revolution subject to compressible fluid flows having axial and tangential velocity components are analyzed numerically. The behavior of flowing and rotating fluid is described in the framework of the potential theory. The behavior of elastic shells is investigated based on the model of the classical shell theory. The results of numerical experiments, which were carried out to analyze the shell stability for various boundary conditions, geometrical dimensions and different values of the width of the inter-shell space, have been discussed.

1 Introduction

Single and coaxial shells of revolution are the integral parts of many technological applications and while in operation can interact with the axial and rotational fluid flows occurring simultaneously. There are a lot of papers in the literature [1], in which the authors based on the numerical and experimental investigations have come to a conclusion that the axial flow of a fluid as well as its rotation exerts a destabilizing effect. However, as far as we know there have been practically no investigations dealing with their combined action on the stability boundary. In this paper we discuss a numerical method of solving this problem. The numerical experiments made for this study allow us to determine a relationship between the dynamic behavior of cantilevered single and coaxial shells and their linear dimensions as well as the width of the annular space between the inner and outer shells.

S.A. Bochkarev (✉) · V.P. Matveenko
Institute of Continuous Media Mechanics, RAS, Acad. Korolev Str 1, Perm, 614013, Russia
e-mail: bochkarev@icmm.ru; mvp@icmm.ru

Alexander K. Belyaev et al. (eds.), *Mechanics and Model-Based Control of Advanced Engineering Systems*, DOI 10.1007/978-3-7091-1571-8_6,
© Springer-Verlag Wien 2014

2 Statement of the Problem and Constitutive Relations

Let us consider two elastic coaxial cylindrical shells of length L (Fig. 1). The inner shell has radius a, and the outer shell has radius b. The shells are subject to two flows of ideal compressible fluids: one occurring inside the inner shell and the other – in the annular gap between the shells. A single shell in the problem formulation considered in this paper can be treated as a particular case. The axial velocity of the internal flow, its angular velocity, specific density and sound speed are denoted by U_i, Ω_i, ϱ_f^i and c_i, respectively. The corresponding parameters of the annular flow are denoted by the same symbols, in which the subscript i is replaced by o. It is necessary to find such a combination of the axial and angular velocity components of the fluid flow, at which the elastic body will lose stability.

The motion of an ideal compressible fluid in the case of potential flow is described by the wave equation, which for the internal flow, occupying volume V_j^i, can be written in the cylindrical coordinate (r, θ, x) as [2]

$$\nabla^2\phi_i - \frac{1}{c_i^2}\left[\frac{\partial}{\partial t} + U_i\frac{\partial}{\partial x}\right]^2 \phi_i = \\ = \frac{2\Omega_i}{c_i}\left(M_i\frac{\partial^2\phi_i}{\partial x\partial\theta} + \frac{1}{c_i}\frac{\partial^2\phi_i}{\partial\theta\partial t}\right) + \frac{\Omega_i^2}{c_i^2}\left(\frac{\partial^2\phi_i}{\partial\theta^2} - r\frac{\partial\phi_i}{\partial r}\right), \tag{1}$$

where ϕ is the perturbation velocity potential, $M_i = U_i/c_i$ is the Mach number. The pressure exerted by the internal fluid flow p_i on the interface between the inner shell and the fluid S_σ^i is calculated by the linearized Bernoulli formula

$$p_i = -\varrho_f^i\left(\frac{\partial\phi_i}{\partial t} + U_i\frac{\partial\phi_i}{\partial s} + \Omega_i\frac{\partial\phi_i}{\partial\theta}\right). \tag{2}$$

The equations for the annular flow are similar to, (1)–(2) with the only difference that the subscript i is replaced by o. The interface between the inner shell and internal flow S_σ^i must satisfy the impermeability condition

$$\frac{\partial\phi_i}{\partial r} = \frac{\partial w_i}{\partial t} + U_i\frac{\partial w_i}{\partial s} + \Omega_i\frac{\partial w_i}{\partial\theta}. \tag{3}$$

The conditions imposed on the inner shell-annular flow interface S_σ^{io} and the outer shell-annular flow interface S_σ^o are written as

$$\frac{\partial\phi_o}{\partial r} = \frac{\partial w_i}{\partial t} + U_o\frac{\partial w_i}{\partial s} + \Omega_o\frac{\partial w_i}{\partial\theta}, \tag{4}$$

$$\frac{\partial\phi_o}{\partial r} = \frac{\partial w_o}{\partial t} + U_o\frac{\partial w_o}{\partial s} + \Omega_o\frac{\partial w_o}{\partial\theta}. \tag{5}$$

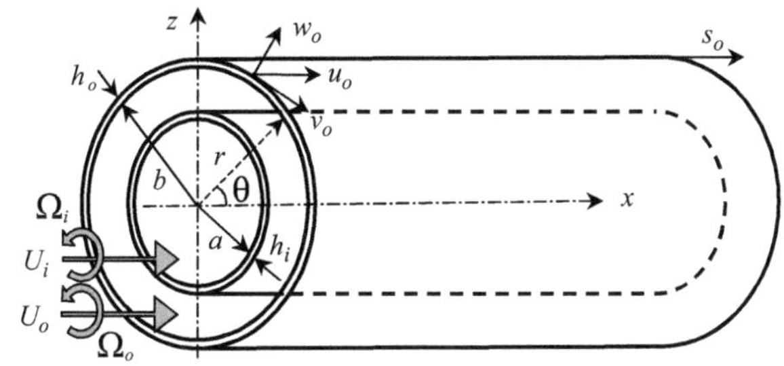

Fig. 1 Computational scheme

Here w_i and w_o are the normal components of the displacement vector of the inner and outer shells. The inlet and outlet perturbation velocity potentials are subject to the following boundary conditions

$$x = 0\colon \phi_i = \phi_o = 0\,, \qquad x = L\colon \frac{\partial \phi_i}{\partial x} = \frac{\partial \phi_o}{\partial x} = 0\,. \tag{6}$$

For numerical implementation of the problem based on the semi-analytical version of the finite element method the equations for perturbation velocity potential (1) together with the boundary conditions (3)–(6) should be transformed using the Bubnov-Galerkin method [3].

The model equations of shells considered in this paper are derived by accepting the Kirchhoff-Love hypothesis, according to which the components of the strain vector of the middle surface and the curvature and torsion changes written in the coordinate system (s, θ, z) are given as [4]

$$\begin{aligned}
&\varepsilon_1^i = \frac{\partial u_i}{\partial s}, \quad \varepsilon_2^i = \frac{1}{a}\left(\frac{\partial v_i}{\partial \theta} + w_i\right), \quad \varepsilon_{12}^i = \frac{1}{a}\frac{\partial u_i}{\partial \theta} + \frac{\partial v_i}{\partial s}, \\
&\kappa_1^i = -\frac{\partial^2 w_i}{\partial s^2}, \quad \kappa_2^i = \frac{1}{a^2}\left(\frac{\partial v_i}{\partial \theta} - \frac{\partial^2 w_i}{\partial \theta^2}\right), \quad \kappa_{12}^i = \frac{1}{a}\left(\frac{\partial v_i}{\partial s} - \frac{\partial^2 w_i}{\partial s \partial \theta}\right).
\end{aligned} \tag{7}$$

Here u_i and v_i are the meridional and circumferential components of the displacement vector of the inner shell.

The physical equations relating the vector of generalized forces and moments $\mathbf{T}_i$ to the vector of the generalized strains $\boldsymbol{\varepsilon}_i = \left\{\varepsilon_1^i, \varepsilon_2^i, \varepsilon_{12}^i, \kappa_1^i, \kappa_2^i, 2\kappa_{12}^i\right\}^{\mathrm{T}}$ can be represented in the matrix form

$$\mathbf{T}_i = \left\{T_{11}^i, T_{22}^i, T_{12}^i, M_{11}^i, M_{22}^i, M_{12}^i\right\}^{\mathrm{T}} = \mathsf{D}_i \boldsymbol{\varepsilon}_i, \tag{8}$$

where the non-zero matrix elements D_i for an isotropic material are conventionally defined in terms of the elasticity modulus E_i and Poisson's ratio ν_i.

The mathematical description of the dynamic behavior of the shells is based on the virtual displacement principle, which for the inner shell can be written in the matrix form as

$$\int_{S_i} \delta\boldsymbol{\varepsilon}_i^{\mathrm{T}} \mathbf{T}_i dS + \int_{S_i} \delta\mathbf{d}_i^{\mathrm{T}} \varrho_0^i \ddot{\mathbf{d}}_i dS - \int_{S_i} \delta\mathbf{d}_i^{\mathrm{T}} \mathbf{P}_i dS = 0. \tag{9}$$

Here $\mathbf{d}_i$ and $\mathbf{P}_i = \left\{0\ 0\ p_i|_{r=a} - p_o|_{r=a}\right\}^{\mathrm{T}}$ are the vectors of the generalized displacements and surface loads, $\rho_0^i = \int_{h_i} \varrho_s^i dz$, ϱ_s^i is the specific density of the material of the inner shell, h_i is the thickness of the inner shell. An analogous equation (where i is replaced by o) is written for the outer shell, for which $\mathbf{P}_o = \{0\ 0\ p_o|_{r=b}\}^{\mathrm{T}}$.

Applying the standard finite element procedures and representing the perturbed motion of the shell and the fluid as $(\mathbf{d}_i, \boldsymbol{\phi}_i, \mathbf{d}_o, \boldsymbol{\phi}_o) = (\mathbf{q}_i, \mathbf{f}_i, \mathbf{q}_o, \mathbf{f}_o)\exp(\mathrm{i}\lambda t)$, we obtains the systems of equations, which can be combined into one expression

$$\left(\mathsf{K} + \mathsf{A} - \lambda^2\mathsf{M} + \mathrm{i}\lambda\mathsf{C}\right)\{\mathbf{q}_i, \mathbf{f}_i, \mathbf{q}_o, \mathbf{f}_o\}^{\mathrm{T}} = 0, \tag{10}$$

where

$$\mathsf{K} = \mathrm{diag}\left\{\mathsf{K}_s^i, \mathsf{K}_f^i + \mathsf{K}_f^{\omega i}, \mathsf{K}_s^o, \mathsf{K}_f^o + \mathsf{K}_f^{\omega o}\right\},$$

$$\mathsf{M} = \mathrm{diag}\left\{\mathsf{M}_s^i, \mathsf{M}_f^i, \mathsf{M}_s^o, \mathsf{M}_f^o\right\},$$

$$\mathsf{C} = \begin{bmatrix} 0 & \mathsf{C}_{sf}^i & 0 & \mathsf{C}_{sf}^o \\ \mathsf{C}_{fs}^i & \mathsf{C}_f^{ci} + \mathsf{C}_f^{\omega i} & 0 & 0 \\ 0 & 0 & 0 & \mathsf{C}_{sf}^o \\ \mathsf{C}_{fs}^o & 0 & \mathsf{C}_{fs}^o & \mathsf{C}_f^{co} + \mathsf{C}_f^{\omega o} \end{bmatrix},$$

$$\mathsf{A} = \begin{bmatrix} 0 & \mathsf{A}_{sf}^i + \mathsf{A}_{sf}^{\omega i} & 0 & \mathsf{A}_{sf}^o + \mathsf{A}_{sf}^{\omega o} \\ \mathsf{A}_{fs}^i + \mathsf{A}_{fs}^{\omega i} & \mathsf{A}_f^{ci} + \mathsf{A}_f^{\omega i} & 0 & 0 \\ 0 & 0 & 0 & \mathsf{A}_{sf}^o + \mathsf{A}_{sf}^{\omega o} \\ \mathsf{A}_{fs}^o + \mathsf{A}_{fs}^{\omega o} & 0 & \mathsf{A}_{fs}^o + \mathsf{A}_{fs}^{\omega o} & \mathsf{A}_f^{co} + \mathsf{A}_f^{\omega o} \end{bmatrix},$$

$$\mathsf{K}_s^i = \sum_{m_s^i} \int_{S_s^i} \mathsf{B}_i^{\mathrm{T}} \mathsf{D}_i \mathsf{B}_i dS,\ \mathsf{M}_s^i = \sum_{m_s^i} \int_{S_s^i} \mathsf{N}_i^{\mathrm{T}} \varrho_0^i \mathsf{N}_i dS,\ \mathsf{A}_{sf}^i = \sum_{m_s^i} \int_{S_\sigma^i} \varrho_f^i U_i \bar{\mathsf{N}}_i^{\mathrm{T}} \frac{\partial \mathsf{F}_i}{\partial s} dS,$$

$$\mathsf{K}_f^i = \sum_{m_f^i} \int_{V_f^i} \left(\frac{\partial \mathsf{F}_i^{\mathrm{T}}}{\partial r} \frac{\partial \mathsf{F}_i}{\partial r} + \frac{1}{r^2} \frac{\partial \mathsf{F}_i^{\mathrm{T}}}{\partial \theta} \frac{\partial \mathsf{F}_i}{\partial \theta} + \frac{\partial \mathsf{F}_i^{\mathrm{T}}}{\partial x} \frac{\partial \mathsf{F}_i}{\partial x} \right) dV,\ \mathsf{M}_f^i = \sum_{m_f^i} \int_{V_f^i} \frac{1}{c_i^2} \mathsf{F}_i^{\mathrm{T}} \mathsf{F}_i dV,$$

$$\mathsf{C}_{fs}^{i} = -\sum_{m_s^i} \int_{S_\sigma^i} \mathsf{F}_i^{\mathrm{T}} \bar{\mathsf{N}}_i dS,\ \mathsf{C}_f^{ci} = \sum_{m_f^i} \int_{V_f^i} \frac{2U_i}{c_i^2} \frac{\partial \mathsf{F}_i^{\mathrm{T}}}{\partial x} \mathsf{F}_i dV,\ \mathsf{A}_f^{ci} = -\sum_{m_f^i} \int_{V_f^i} M_i^2 \frac{\partial \mathsf{F}_i^{\mathrm{T}}}{\partial s} \frac{\partial \mathsf{F}_i}{\partial s} dV,$$

$$\mathsf{A}_{fs}^{i} = -\sum_{m_s^i} \int_{S_\sigma^i} U_i \mathsf{F}_i^{\mathrm{T}} \frac{\partial \bar{\mathsf{N}}_i}{\partial s} dS,\ \mathsf{C}_{sf}^{i} = \sum_{m_s^i} \int_{S_\sigma^i} \varrho_f^i \bar{\mathsf{N}}_i^{\mathrm{T}} \mathsf{F}_i dS,\ \mathsf{A}_{fs}^{\omega i} = \sum_{m_f^i} \int_{S_\sigma^i} \Omega_i \frac{\partial \bar{\mathsf{N}}_i}{\partial \theta} \mathsf{F}_i dS,$$

$$\mathsf{K}_f^{\omega i} = \sum_{m_f^i} \int_{V_f^i} \frac{\Omega_i^2}{c_i^2} \left[\frac{\partial^2 \mathsf{F}_i^{\mathrm{T}}}{\partial \theta^2} \mathsf{F}_i - r \frac{\partial \mathsf{F}_i^{\mathrm{T}}}{\partial r} \mathsf{F}_i \right] dV,\ \mathsf{C}_f^{\omega i} = -\sum_{m_f^i} \int_{V_f^i} \frac{2\Omega_i}{c_i^2} \frac{\partial \mathsf{F}_i^{\mathrm{T}}}{\partial \theta} \mathsf{F}_i dV,$$

$$\mathsf{A}_{sf}^{\omega i} = \sum_{m_s^i} \int_{S_\sigma^i} \varrho_f^i \Omega_i \bar{\mathsf{N}}_i^{\mathrm{T}} \frac{\partial \mathsf{F}_i}{\partial \theta} dS,\ \mathsf{A}_f^{\omega i} = \sum_{m_f^i} \int_{V_f^i} 2\Omega_i \frac{M_i}{c_i} \frac{\partial^2 \mathsf{F}_i^{\mathrm{T}}}{\partial x \partial \theta} \mathsf{F}_i dS.$$

Here B_i is the matrix relating the strain vector $\boldsymbol{\varepsilon}_i$ to the vector of nodal displacements of the shell-type finite element; m_f^i and m_s^i is the number of finite elements used to decompose the fluid domain V_f^i and the inner shell domain V_s^i; F_i, N_i, $\bar{\mathsf{N}}_i$ are the shape functions for the perturbation velocity potential of the internal flow, the shell-type element and the normal component of the inner shell displacement vector; $\mathbf{q}_i, \mathbf{f}_i, \mathbf{q}_o, \mathbf{f}_o$ are some functions of the coordinates; $\mathrm{i}^2 = -1$; $\lambda = \lambda_1 + \mathrm{i}\lambda_2$ is the characteristic number. Missing matrices can be obtained by replacing the index i by the index o.

Problem solving is reduced to the computation and analysis of the eigenvalues of system (10). Complex eigenvalues are calculated by the algorithm based on the Müller method [5]. To maximize the computational efficiency of the algorithm, the degree of freedom of system (10) was renumbered using the reverse Cuthill-McKee algorithm [6].

3 Results of Computations

The computations discussed below were made for different values of the width of the annular gap between the outer and inner shells, which is defined by the following relation: $k = (b - a)/a$. Here we present several examples of numerical simulation for a cylindrical shell ($E = 2 \times 10^{11}$ N/m^2, $\nu = 0.29$, $\varrho_s = 7{,}812$ kg/m^3, $R = 1$ m, $h = 0.01$ m) and a system of coaxial shells ($E_i = E_o = 2 \times 10^{11}$ N/m^2, $\nu_i = \nu_o = 0.3$, $\varrho_s^i = \varrho_s^o = 7{,}800$ kg/m^3, $L = 1$ m, $b = 0.1$ m, $h_i = h_o = 5 \times 10^{-4}$ m), simply supported ($v = w = 0$) at both ends ($x = 0, L$) or supported as a cantilever. The shells are subject to a compressible fluid $\varrho_f^i = \varrho_f^o = 10^3$ kg/m^3, $c_i = c_o = 1{,}500$ m/s. In all calculations we used 40 elements for each shell and 1,000 and 1,600 elements for the fluid in a single shell and coaxial shell, respectively. In the latter case the number of elements was defined by the width of the space between the shells.

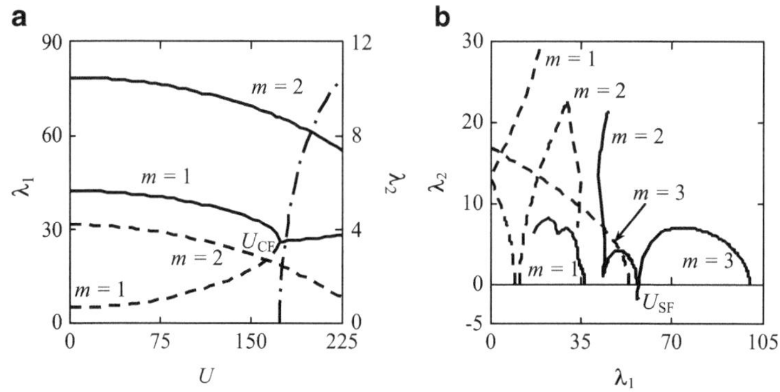

Fig. 2 The real and imaginary parts (**a**) and loci (**b**) of eigenvalues (Hz) versus axial velocity component of the rotating flow U: (**a**) simply supported shell; (**b**) cantilevered shell

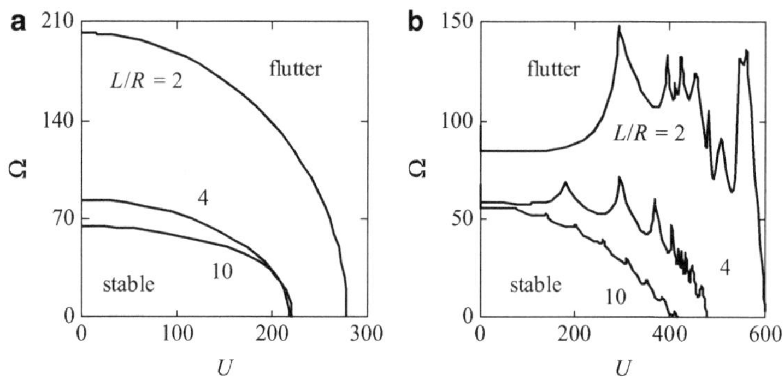

Fig. 3 Stability diagram for simply supported (**a**) and cantilevered (**b**) shells having different linear dimensions L/R

The numerical experiments have shown that under the combined action of simultaneous axial and rotational fluid flows the shells lose stability. The type of stability loss is determined by the boundary conditions. In particular, the dependence of the eigenvalues λ(Hz) on the axial velocity of the fluid U(m/s) was obtained for single shells of revolution, in which the axial and rotational flows with the angular velocity $\Omega = 50$ rad/s (Fig. 2) occur concurrently. In the figure, the dashed lines denote eigenvalues, corresponding to the backward waves, and solid lines denote eigenvalues, corresponding to the forward waves. The results of computation show that for shells simply supported (Fig. 2a) or clamped at both ends, the loss of stability occurs in the form of a coupled-mode flutter, since at the axial flow velocity U_{CF} the real parts of the forward and backward waves of the first mode ($m = 1$) coalesce. For cantilevered shells (Fig. 2b) the loss of stability occurs in the form of a single-mode flutter, at which the imaginary part of the third mode for the fluid velocity U_{SF} becomes negative.

Figure 3 shows the stability diagrams obtained in the case of combined action of both velocity components for single shells under different boundary conditions

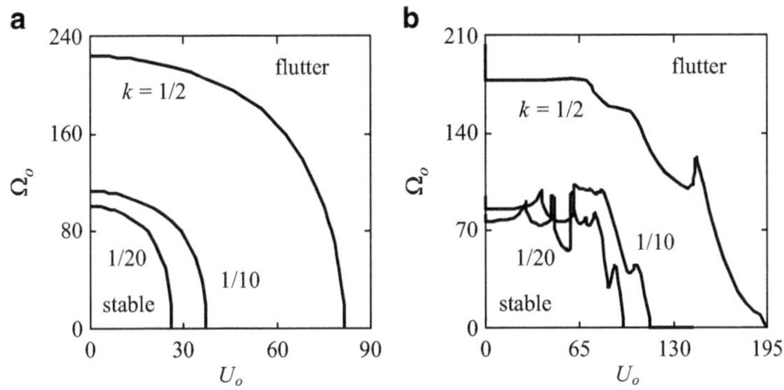

Fig. 4 Stability diagram for simply supported (**a**) and cantilevered (**b**) coaxial shells at different values of the annular gap width k

and having different linear dimensions L/R (the number of the harmonic in the circumferential direction is $j = 4$). From the results shown in Fig. 3b it can be concluded that for cantilevered shells the axial velocity component has a stabilizing effect, which strongly depends on the linear dimensions of the system – the smaller dimensions the higher is the stability boundary. With increasing L/R the stabilizing effect of the axial velocity vanishes.

Figure 4 shows the stability boundaries obtained for coaxial shells for different variants of the boundary conditions and different values of the annular gap width ($j = 3$). In this case, for cantilevered shells the axial fluid flow also exerts a stabilizing effect, whereas for shells with other boundary conditions it has only a destabilizing effect. The strength of the stabilizing effect depends on the width of the annular gap – the smaller the width, the higher is the stability boundary.

4 Conclusion

In this paper, a mathematical statement of the loss-of-stability problem and a finite element algorithm for its numerical simulation have been proposed to study the dynamic behavior of single elastic and coaxial cylindrical shells of revolution subject to compressible fluid flows having axial and tangential velocity components. Numerical calculations have shown that for both examined configurations the stability boundary, which has been determined by assigning a fixed value to one of the velocities and exhaustive searching for the critical value of the other velocity, depends on the type of the boundary conditions and linear dimensions. A combined action of both velocity components essentially affects the character of the dynamic behavior of elastic bodies. This effect is more pronounced in the case of cantilevered shells. Moreover, the stabilizing action of the axial flow involves two other effects – a jump-wise change in the critical value of one of the velocity components at a minimum value of the other velocity component and non-monotonic dependence of one velocity component on the other. Such a behavior is caused either by different responses of the cantilevered shell to the axial and tangential velocity components acting separately or by the existence of hydrodynamic damping, which plays a decisive role in the dynamic behavior of the system.

Acknowledgements This study is supported by the Russian Foundation for Basic Research (grant 12-01-00323).

References

1. Païdoussis, M.P.: Fluid-Structure Interactions: Slender Structures and Axial Flow, vol. 2. Elsevier, London (2004)
2. Ilgamov, M.A.: Oscillations of Elastic Shells Containing Liquid and Gas. Nauka, Moscow (1969). (in Russian)
3. Bochkarev, S.A., Matveenko, V.P.: Numerical study of the influence of boundary conditions on the dynamic behavior of a cylindrical shell conveying a fluid. Mech. Solids **43**(3), 477–486 (2008)
4. Biderman, V.L.: The Mechanics of Thin-Walled Structures. Mashinostroyeniye, Moscow (1977). (in Russian)
5. Matveenko, V.P.: On an algorithm of solving the problem on natural vibrations of elastic bodies by the finite element method. In: Boundary-Value Problems of the Elasticity and Viscoelasticity Theory, pp. 20–24. UNTs Akad Nauk SSSR, Sverdlovsk (1980). (in Russian)
6. George, A., Liu, J.W.H.: Computer Solution of Large Sparse Positive Definite Systems. Prentice-Hall, Englewood Cliffs (1981)

Flexible Robots: Modelling and Simulation

Hartmut Bremer

Abstract Modelling and simulation needs adequate procedures. These can be divided into the analytical and the synthetic methods the basis of which is the *Central Equation of Dynamics* (1904/1988), derived from *Lagrange's Principle* (1764). The analytical methods (Lagrange [17]; Maggi, Principii di Stereodinamica. Ulrico Hoepli, Milano, 1903; Hamel, Zeitschr Ang Math u Phys 50:1–57, 1904) thereby do not really meet the engineering needs. But the Central Equation also leads to the synthetical method(s) which in its most general representation is the *Projection Equation* (1988/2003). Along with a Ritz series expansion it leads to an Order-n-Formalism for flexible robots as the most powerful procedure, approximating also real time demands (2008/2011). The historical data show a continuous development over the decades which makes mechanics an enjoyable and inspiring science and reveals classical mechanics as modern as can be. Applications at the end of the paper demonstrate once more its success.

1 The Aim of Modelling

Flexible Robots are characterized by structural elasticity along with flexible gears. They are of course operated by control. Modelling thus pursues two aims:

- Representation of the physical world
- Control plant representation

The first models physical reality, or nature, because a (computer) simulation only makes sense if it represents the robot in its natural surroundings. Simulation saves investigation time and costs. The second point means to reduce the considerations

---

H. Bremer (✉)
Institute for Robotics, Johannes Kepler University Linz, Linz, Austria
e-mail: hartmut.bremer@jku.at

Alexander K. Belyaev et al. (eds.), *Mechanics and Model-Based Control of Advanced Engineering Systems*, DOI 10.1007/978-3-7091-1571-8_7,
© Springer-Verlag Wien 2014

to the essential and important motions, i.e. those motions which are mainly to be influenced. The simplest control concept which aims the desired target is always the best one. Of course, both aims go hand in hand. Nevertheless, at the end of all theoretical investigations nature itself, as a resolute and incorruptible arbiter, evaluates the results. Experiments are unavoidable.

We are going to focus on the first topic since the second one is easily obtained from it. Its ingredients are rigid and flexible bodies, undergoing fast motions, along with the use of non-holonomic (velocity-) variables and, if need be, non-holonomic constraints. But even in case of purely holonomic constraints the use of non-holonomic variables can simplify the mathematical representation significantly [4].

Most of the theoretical background has already been reported [5, 11]. The presented paper may thus be seen a kind of a historical overview, or – *cum granu salis* – an advertisement for our philosophy. Its great success will be outlined at the end of the paper.

2 Basics

2.1 A Powerful Tool: Lagrange's Principle

It was in the 1980s when the one and best method in multi body dynamics leaped into view – the man who mastered motion! –, simultaneously unmasking his predecessors (namely Newton and Lagrange) as black magicians.

> ...among the most obscure concepts [$\cdots$] are those marked by the word "virtual". ... They are the closest thing in dynamics to black magic, and they are the heart of Lagrange's approach.

Radetzki [26]. This makes curious, thus let us have a look to one of Lagrange's (black?) magic contributions: "Investigations on the moon's libration" [20] – nearly 250 years old.

Lagrange himself did not draw pictures because his aim was analysis, in contrary to the geometric methods of his time. However, a simple sketch is helpful to see what is going on (Fig. 1).

On page 9 we find

$$(A) \quad \frac{1}{dt^2}\int \alpha(d^2X\delta X + d^2Y\delta Y + d^2Z\delta Z) + T\int \frac{\alpha\delta R}{R^2} + S\int \frac{\alpha\delta R'}{R'^2} = 0. \tag{1}$$

Today, we would call the mass element dm, collect the components X, Y, Z in a vector $\mathbf{r}$ and refer the variations $\delta R, \delta R'$ to those of $\mathbf{r}$ ($\Rightarrow \delta R = (\partial R/\partial \mathbf{r})\delta \mathbf{r}, \delta R' = (\partial R'/\partial \mathbf{r})\delta \mathbf{r}$), coming out with

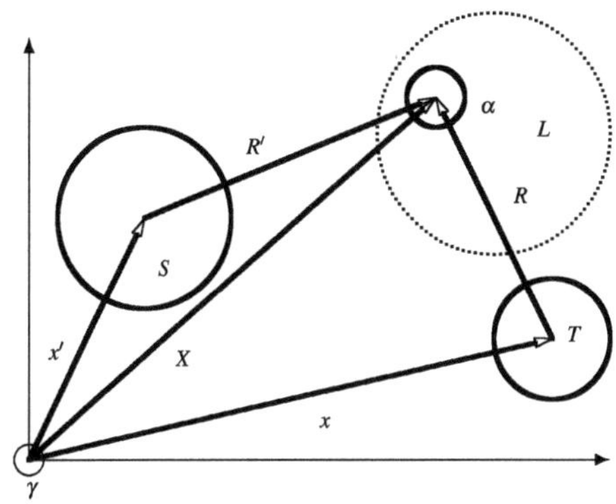

Fig. 1 Lagrange's Problem. γ: first point of Aries characterizing the inertial frame (Lagrange shifted the frame into the gravity center of the moon which is, however, out of interest here). The three bodies (S: "soleil" (sun mass), L: "lune" (moon mass), T: "terre" (earth mass)) are drawn perspectively along with its position vectors x', X, x. The distances of the considered moon particle α from the gravity centers of the sun and the earth, resp., are represented by R', R

$$(A) \qquad \int_{(S)} (\ddot{\mathbf{r}}dm - d\mathbf{f}^e)^T \delta\mathbf{r} = 0 \tag{2}$$

where $d\mathbf{f}^e$ is the resultant of all impressed forces which are exerted on the considered particle. (It is remarkable that Lagrange formulated his δ-method already in 1755 [21] (printed 1762 [19]) when he was 19 years old – very much appreciated by 29 years older Leonhard Euler (letter dated April 24, 1756 [10])). Three pages later Lagrange continues:

...one obtains always an equation similar to equation (A), and

- the whole difficulty consists of nothing but to find the analytical expression of the forces which one assumes to act ...,
- to insert the lowest possible number of undetermined variables...,
- and one obtains at once as many particular equations as needed for the solution of the problem.

This means, in our words: insert force laws, insert minimal coordinates $\mathbf{q}$ and you obtain the equations of motion.

Force laws! This is the breakthrough [6]! It freed him from the restrictive burden of conservative systems which he had started with: As stated in the introduction, he first tried to elaborate a general procedure from the well established principle of least action. Now he turns the argumentation the other way round (see e.g. his Analytical Mechanics from 1788 [22]) and shows the principle of least action, the mass center principle, the energy conservation principle, even finite rotations, to be a consequence of his "general formula" as he calls equation (A), nowadays referred to as the Lagrange-Principle or d'Alembert's Principle. However, the latter does not meet the root of the matter. D'Alembert considered the "lost forces" (w.r.t. motion) to counterbalance [1], and it is almost tedious to find these, while Lagrange's concept makes use of the fact that these do not perform work. He expressively refers to d'Alembert as well as to John II Bernoulli (virtual work in statics). But, as Hamel [14, p. 219] puts it: the combination of these two principles which originally do not have anything in common has been performed by Lagrange. Since nowadays nobody doubts that Eq. (2) is one of the most powerful tools in mechanics, it may indeed be considered "magic" (although not black).

2.2 *The Central Equation of Dynamics*

Nevertheless, up to here Eq. (2) represents a "raw form" because integration $\int_{(S)}$ (S: System under consideration) has still to be carried out.

This is quickly done: extract one time derivation which needs correction due to the product rule, $\frac{d}{dt}\int \dot{\mathbf{r}}^T dm\delta\mathbf{r} - \int \dot{\mathbf{r}}^T dm\frac{d}{dt}(\delta\mathbf{r}) = d\mathbf{f}^{eT}\delta\mathbf{r} := \delta W$. Interchange $d\delta\mathbf{r} \rightarrow \delta d\mathbf{r}$ in the second summation term (which is allowed for an inertial representation, see below) to come out with a quadratic form $\int \dot{\mathbf{r}}^T dm\delta\dot{\mathbf{r}} = \frac{1}{2}\delta\int \dot{\mathbf{r}}^T dm\dot{\mathbf{r}} = \delta T$ (T: kinetic energy). Next, formulate the first term as partial derivative w.r.t. that quadratic form, insert $\mathbf{r} = \mathbf{r}(\mathbf{s}) \Rightarrow \delta\mathbf{r} = (\partial\mathbf{r}/\partial\mathbf{s})\delta\mathbf{s}$ – and we are stuck. However, a short calculation reveals the key: $\dot{\mathbf{r}}(\mathbf{s}) = (\partial\mathbf{r}/\partial\mathbf{s})\dot{\mathbf{s}} \Rightarrow (\partial\mathbf{r}/\partial\mathbf{s}) = (\partial\dot{\mathbf{r}}/\partial\dot{\mathbf{s}})$ – just put the dots. One has then a representation involving the chain rule which may be abandoned: $\frac{d}{dt}\int \frac{\partial}{\partial\dot{\mathbf{r}}}(\frac{1}{2}\dot{\mathbf{r}}^T dm\dot{\mathbf{r}})\frac{\partial\dot{\mathbf{r}}}{\partial\dot{\mathbf{s}}}\delta\mathbf{s} = \frac{d}{dt}[\frac{\partial T}{\partial\dot{\mathbf{s}}}\delta\mathbf{s}]$, leading Eq. (2) to

$$\frac{d}{dt}\left[\frac{\partial T}{\partial\dot{\mathbf{s}}}\delta\mathbf{s}\right] - \delta T - \delta W = 0, \quad \dot{\mathbf{s}} = \mathbf{H}(\mathbf{q})\dot{\mathbf{q}}. \tag{3}$$

The "minimal velocities" $\dot{\mathbf{s}}$ (notation selected in comparison to Lagrange's "minimal coordinates" $\mathbf{q}$ – "le plus petit nombre possible de variables indéterminées") represent a regular linear combination of $\dot{\mathbf{q}}$ – being componentwise integrable or not.

When I found the simple relation $(\partial\mathbf{r}/\partial\mathbf{s}) = (\partial\dot{\mathbf{r}}/\partial\dot{\mathbf{s}})$, as a young scientist, I was rather proud of it – until later I noticed (in Hamel's book) that already Helmholtz used that expression [14, p. 246]. Henceforth we call it "Helmholtz's auxiliary equation".

It was in 1984 when we tried to argue against "the man who mastered motion" and his claim for fame – which seems somehow charlatanry, or at least ignorance. However, his "method" had already been spread all over the western hemisphere and the paper was consecutively rejected – until in 1988 (after full 4 years), it was accepted by ZAMM in east Berlin which in that time belonged to the eastern hemisphere. Its title was "On a Central Equation in Dynamics" [i.e. Eq. (3)] from which a considerable body of procedures was developed [3] (including, of course, that one from the "man who mastered motion"). However, proudness is once more out of place. Although Eq. (3) is expressis verbis not found in Hamel's final work and only fragmentary in his habilitation [12] one finds it in [13] (where one would not expect it. We are grateful to our friend John Papastavridis, Georgia Tech., for this information). Even more: For $\mathbf{r} = \mathbf{r}(\mathbf{q})$, one obtains $d\delta\mathbf{r} - \delta d\mathbf{r} = \sum[(\partial^2\mathbf{r}/\partial q_i\partial q_j)-(\partial^2\mathbf{r}/\partial q_j\partial q_i)]+d\delta\mathbf{q}-\delta d\mathbf{q}$ where $d\delta\mathbf{q}-\delta d\mathbf{q}$ vanishes if one uses a variational ansatz for the $\delta\mathbf{q}$. Then, Schwarz's rule remains which is fulfilled for an inertial representation of $\mathbf{r}$, hence $d\delta\mathbf{r} = \delta d\mathbf{r}$. And: there is no need at all to assume $\delta\mathbf{r}$ or $\delta\mathbf{q}$ infinitesimal and the very last touch of mistery (if remained) disappears. But the variational approach is only one possibility because $\mathbf{q}$ (see Lagrange) is free from constraints. Hamel keeps this door open which leads him later to Lie's group theory.

3 Procedures

The analytical methods are obtained from Eq. (3) by direct calculation. In general, T depends on $\mathbf{q}$ (displacement) and $\dot{\mathbf{s}}$ (velocity), hence $\delta T = (\partial T/\partial\mathbf{q})\delta\mathbf{q}+(\partial T/\partial\dot{\mathbf{s}})\delta\dot{\mathbf{s}}$. Since $\delta\mathbf{q} = (\partial\mathbf{q}/\partial\mathbf{s})\delta\mathbf{s}$ one may write $\delta T = (\partial T/\partial\mathbf{s})\delta\mathbf{s} + (\partial T/\partial\dot{\mathbf{s}})\delta\dot{\mathbf{s}}$ yielding, along with $\delta W = \mathbf{Q}^T\delta\mathbf{s}$, (the variational form of) Hamel's equations: $[\frac{d}{dt}\frac{\partial T}{\partial\dot{\mathbf{s}}} - \frac{\partial T}{\partial\mathbf{s}} - \mathbf{Q}^T]\delta\mathbf{s} + \frac{\partial T}{\partial\dot{\mathbf{s}}}[\frac{d\delta\mathbf{s}-\delta d\mathbf{s}}{dt}] = 0$. Here, within the first term, $\delta\mathbf{s}$ appears explicitly while in the second it is still hidden. This minor flaw is, of course, removed by Hamel with his famous coefficients β_i (although he states that its use is not always practical [14, p. 483]). Nowadays, these equations are mostly called the Hamel-Boltzmann-Equations. Hamel is a very polite man when states (p. 481) that these equations have also independently been developed by the Russian Woronetz, the Italian Volterra and in case of constant β's by Poincaré; also Boltzmann came near to it – so to say: he came near, but only near. Although in the 1990s a certain revival of non-holonomic systems can be recognized, the Hamel equations did not really come into play when compared to Lagrange's equations (of the second kind [18]) which are seemingly easier to treat. They are obtained by setting $\dot{\mathbf{s}} \rightarrow \dot{\mathbf{q}} \Rightarrow [\frac{d}{dt}\frac{\partial T}{\partial\dot{\mathbf{q}}} - \frac{\partial T}{\partial\mathbf{q}} - \mathbf{Q}^T]\delta\mathbf{q} = 0$. If $\mathbf{q}$ itself does not undergo constraints (the Lagrangean case), $\delta\mathbf{q}$ may be canceled. Additional constraints may then be added via Lagrangean multipliers which leads to an elegant mathematical description (e.g. [2]) but far from engineering needs – its evaluation requires an almost unbearable effort.

Parametrizing the kinetic energy as $T = \sum(\mathbf{v}^T\mathbf{p} + \boldsymbol{\omega}^T\mathbf{L})_i/2$ (where $\mathbf{p}, \mathbf{L}$: linear and angular momentum, resp., w.r.t. the mass centers, $\mathbf{v}, \boldsymbol{\omega}$: corresponding velocities) leads to (index i suppressed): (1) $\mathbf{v} = (\partial\mathbf{r}/\partial\mathbf{q})\dot{\mathbf{q}} \Rightarrow \mathbf{a} = \frac{d}{dt}(\partial\mathbf{r}/\partial\mathbf{q})\dot{\mathbf{q}} + (\partial\mathbf{r}/\partial\mathbf{q})\ddot{\mathbf{q}}$, (2) $\mathbf{v} = \mathbf{v}(\mathbf{q}, \dot{\mathbf{q}}) \Rightarrow \mathbf{a} = (\partial\mathbf{v}/\partial\mathbf{q})\dot{\mathbf{q}} + (\partial\mathbf{v}/\partial\dot{\mathbf{q}})\ddot{\mathbf{q}}$. "Helmholtz auxiliary equation" reveals the last summation terms in (1) and (2) to be identical, thus $(\partial\mathbf{v}/\partial\mathbf{q}) = \frac{d}{dt}(\partial\mathbf{r}/\partial\mathbf{q})$. The same holds for $\boldsymbol{\omega} = (\partial\boldsymbol{\pi}/\partial\mathbf{q})\dot{\mathbf{q}}$. One has then (3) $\frac{\partial T}{\partial\dot{\mathbf{q}}} = \frac{\partial T}{\partial\mathbf{v}}\frac{\partial\mathbf{v}}{\partial\dot{\mathbf{q}}} + \frac{\partial T}{\partial\boldsymbol{\omega}}\frac{\partial\boldsymbol{\omega}}{\partial\dot{\mathbf{q}}} = \mathbf{p}^T\frac{\partial\mathbf{r}}{\partial\mathbf{q}} + \mathbf{L}^T\frac{\partial\boldsymbol{\pi}}{\partial\mathbf{q}}$ (4) $\frac{\partial T}{\partial\mathbf{q}} = \frac{\partial T}{\partial\mathbf{v}}\frac{\partial\mathbf{v}}{\partial\mathbf{q}} + \frac{\partial T}{\partial\boldsymbol{\omega}}\frac{\partial\boldsymbol{\omega}}{\partial\mathbf{q}} = \mathbf{p}^T\frac{d}{dt}\frac{\partial\mathbf{r}}{\partial\mathbf{q}} + \mathbf{L}^T\frac{d}{dt}\frac{\partial\boldsymbol{\pi}}{\partial\mathbf{q}}$. Lagrange's equations require time derivation of (3) and subtracting (4) which shows that the $\frac{d}{dt}(\ )$-terms vanish, yielding $\sum[\frac{\partial\mathbf{r}}{\partial\mathbf{q}}^T(\dot{\mathbf{p}} - \mathbf{f}^e) + \frac{\partial\boldsymbol{\pi}}{\partial\mathbf{q}}^T(\dot{\mathbf{L}} - \mathbf{M}^e)]_i = 0$ (where $\mathbf{f}^e, \mathbf{M}^e$: impressed force and torque, resp.). This representation is commonly called a "synthetical" procedure because it synthesizes ("puts together") linear and angular momentum balances. [Notice that this representation is strongly related to an inertial representation because of (1): $\mathbf{v} = \frac{d}{dt}\mathbf{r}(\mathbf{q})$]. "Put together" implies the use of two axioms, namely the linear and the angular momentum theorem. It is most likely that Lagrange was not aware of this fact, and the question arises wether we can proceed this way without being incomprehensible or mistaken. The answer is yes when we use the linear momentum theorem for Eq. (2) and additionally assume "Boltzmann's axiom" as a second one in the background. The naming is probably due to Hamel. István Szabó [28] reports that, when Mach postulated Newton's Principles to be sufficient to solve any mechanical problem [23], he did obviously not recognize Newton's deep sigh in his foreword from 1686:

> Utinam caetera Naturae phaenomena ex principiis Mechanicis eodem argumentandi genere derivare liceret [25] – I wish we could derive the rest of the phaenomena of nature by the same kind of reasoning from mechanical principles

(where he refers to the motions of the planets, the comets, the moon, and the sea). This means, in other words, that Newton (and contemporarians) did not yet have the momentum theorem at their disposal to solve general mechanical problems as Mach suggests. Indeed, nearly 50 years should pass until Leonhard Euler published it [7] and, in the sequel, the angular momentum theorem [8]. It took him once more 25 years to come to the conclusion that the linear and the angular momentum have to be considered independent [9]. Szabó continues:

> But he [Mach] also overlooked the insight of his great compatriot and contemporary Ludwig Boltzmann, namely that from (according to Mach: Newton's) momentum theorem it follows the angular momentum theorem only when postulating the symmetry of the stress tensor. My teacher Georg Hamel here talked about the Boltzmann axiom, and when I asked him (as a student) about Mach's citation he showed an indulgent smile.

Thus, having a safe basis we may proceed with $\sum[\frac{\partial \mathbf{r}}{\partial \mathbf{q}}^T(\dot{\mathbf{p}}-\mathbf{f}^e)+\frac{\partial \boldsymbol{\pi}}{\partial \mathbf{q}}^T(\dot{\mathbf{L}}-\mathbf{M}^e)]_i = 0$: Here, $(\partial\mathbf{r}/\partial\mathbf{q})$ can easily be calculated since $\mathbf{r}$ is represented in an inertial frame. But $\boldsymbol{\pi}$ is a quasicoordinate – how to get $(\partial\boldsymbol{\pi}/\partial\mathbf{q})$? It might be extracted from $\partial\mathbf{A}\,\mathbf{A}^T = \partial\tilde{\boldsymbol{\pi}}$ [with the skew-symmetric 3×3 spin tensor $\tilde{(\ )}$, where $\mathbf{A}$: transformation matrix from body fixed representation into an inertial one] as some authors do. This yields an almost cumbersome calculation; the authors therefore recommend a computer code. However, the key is once more the "Helmholtz Auxiliary Equation": just put the dots to come out with $(\partial\mathbf{r}/\partial\mathbf{q}) = (\partial\dot{\mathbf{r}}/\partial\dot{\mathbf{q}}) = (\partial\mathbf{v}/\partial\dot{\mathbf{q}})$, $(\partial\boldsymbol{\pi}/\partial\mathbf{q}) = (\partial\dot{\boldsymbol{\pi}}/\partial\dot{\mathbf{q}}) = (\partial\boldsymbol{\omega}/\partial\dot{\mathbf{q}})$ $\Rightarrow \sum[\frac{\partial \mathbf{v}}{\partial \dot{\mathbf{q}}}^T(\dot{\mathbf{p}}-\mathbf{f}^e)+\frac{\partial \boldsymbol{\omega}}{\partial \dot{\mathbf{q}}}^T(\dot{\mathbf{L}}-\mathbf{M}^e)]_i = 0$. Because the velocities have to be calculated anyhow, the Jacobians $(\partial\mathbf{v}/\partial\dot{\mathbf{q}})$, $(\partial\boldsymbol{\omega}/\partial\dot{\mathbf{q}})$ are nothing but the coefficient matrices w.r.t. $\dot{\mathbf{q}}$ which are then in advance known – there is no additional calculation necessary (not to mention computer codes). The same holds when using $\dot{\mathbf{s}}$ instead of $\dot{\mathbf{q}}$ to relate the results with Eq. (3). The relationship is characterized according to Maggi's procedure [24]: formulate the virtual work by premultiplying with $\delta\mathbf{q}^T = \delta\mathbf{s}^T(\partial\dot{\mathbf{q}}/\partial\dot{\mathbf{s}})^T$ with $\delta\mathbf{s}$ being arbitrary, yielding $\sum[(\partial\mathbf{v}/\partial\dot{\mathbf{s}})^T(\dot{\mathbf{p}}-\mathbf{f}^e)+(\partial\boldsymbol{\omega}/\partial\dot{\mathbf{s}})^T(\dot{\mathbf{L}}-\mathbf{M}^e)]_i = 0$. The use of velocity depending Jacobians is even more comfortable: One may insert an identity matrix in the form $\mathbf{A}_R^T\mathbf{A}_R$ where $\mathbf{A}_R$ transforms to an arbitrary frame: $\sum[(\partial\mathbf{v}/\partial\dot{\mathbf{s}})^T\mathbf{A}_R^T\mathbf{A}_R(\dot{\mathbf{p}}-\mathbf{f}^e)+(\partial\boldsymbol{\omega}/\partial\dot{\mathbf{s}})^T\mathbf{A}_R^T\mathbf{A}_R(\dot{\mathbf{L}}-\mathbf{M}^e)]_i = 0$.

Since $\mathbf{A}_R$ cannot depend on velocities, one has $\sum[(\partial \mathbf{A}_R \mathbf{v}/\partial \dot{\mathbf{s}})^T \mathbf{A}_R(\dot{\mathbf{p}} - \mathbf{f}^e) + (\partial \mathbf{A}_R \boldsymbol{\omega}/\partial \dot{\mathbf{s}})^T \mathbf{A}_R(\dot{\mathbf{L}} - \mathbf{M}^e)]_i = 0$ where $\mathbf{A}_R \mathbf{v} := \mathbf{v}_s, \mathbf{A}_R \boldsymbol{\omega} := \boldsymbol{\omega}_s$ are vector representations in the reference frame. Let $\mathbf{p}$ be calculated by $\mathbf{p} = \mathbf{A}_R^T {}_R\mathbf{p}$ (and $\mathbf{L}$ analogously). One has then $\mathbf{A}_R(\mathbf{A}_R^T {}_R\dot{\mathbf{p}} + \dot{\mathbf{A}}_R^T {}_R\mathbf{p} - \mathbf{f}^e) = {}_R(\dot{\mathbf{p}} + \tilde{\boldsymbol{\omega}}_R \mathbf{p} - \mathbf{f}^e)$ where $\mathbf{A}_R \dot{\mathbf{A}}_R^T = \tilde{\boldsymbol{\omega}}_R$ is the skew-symmetric spin tensor w.r.t. the angular velocity $\boldsymbol{\omega}_R$ of the chosen reference frame (terms in $\mathbf{L}$ analogously). As a result one obtains the most general description

$$\sum_{i=1}^{N} \left[\left(\frac{\partial \mathbf{v}_s}{\partial \dot{\mathbf{s}}} \right)^T \left(\frac{\partial \boldsymbol{\omega}_s}{\partial \dot{\mathbf{s}}} \right)^T \right]_i \begin{pmatrix} (\dot{\mathbf{p}} + \tilde{\omega}_R \mathbf{p} - \mathbf{f}^e) \\ (\dot{\mathbf{L}} + \tilde{\omega}_R \mathbf{L} - \mathbf{M}^e) \end{pmatrix}_i = 0, \tag{4}$$

N : number of (rigid) bodies, all cartesian vectors represented in a reference frame R. Notice: $\boldsymbol{\omega}_s$ is the absolute angular velocity of the considered body while $\boldsymbol{\omega}_R$ is the angular velocity of the chosen reference frame. For instance: $\boldsymbol{\omega}_R = 0$: inertial frame, $\boldsymbol{\omega}_R = \boldsymbol{\omega}_s$: body-fixed frame. We call Eq. (4) the (final form of) the *Projection Equation* [projecting the force-torque-balances (Euler) into the unconstrained directions (Lagrange) in terms of $\dot{\mathbf{s}}$ (Hamel)].

Why an arbitrary reference frame? Let us have a look at an oil mill as had been used by the old greeks (later adopted by the romans and still in use today). Since we are interested in the pressure at the bottom of the mill stone, we use an inertial representation which is followed by a transformation $\mathbf{A}$ for the actual millstone position $\mathbf{A}(\dot{\mathbf{L}} - \mathbf{M}^e) = 0$. We obtain a very simple result where the underlying calculation is troublesome. Don't care – the computer does it? The computer: ruin of science and threat to mankind [29]! Using a reference frame which rotates with the mill itself (ω_M) shrinks the calculation to nearly nothing: $f_M = (J/r)\,\omega_M^2$ (r: millstone radius, J: moment of inertia). There is no computer needed at all.

Why nonholonomic velocities $\dot{\mathbf{s}}$? Clearly, there is no way out when non-holonomic constraints arise. The Projection equation (4) takes non-holonomic constraints in advance into account (while for the analytical methods (Hamel/Maggi) these are inserted after performing calculation). But also in holonomic systems the use of non-holonomic variables may be advantageous. This is for instance the case when dealing with substructures where the describing variables contain the velocities of the origin of the chosen reference frame. These are, for spatial motions, always non-holonomic.

4 Algorithms

4.1 Structurizing the Problem: The Kinematic Chain

A system p which moves with $\dot{\mathbf{y}}_p$ (velocities of reference frame origin) yields, for a contiguous system i, $\dot{\mathbf{y}}_i = \mathbf{T}_{ip}\dot{\mathbf{y}}_p + \mathbf{F}_i\dot{\mathbf{s}}_i$. Here, $\mathbf{T}_{ip}$ calculates the velocities at the coupling point and transforms into the actual reference frame. Thus, $\mathbf{T}_{ip}\dot{\mathbf{y}}_p$

represents the "guidance velocities" of frame i, while $\mathbf{F}_i\dot{\mathbf{s}}_i$ takes the relative velocities of system i into account. Assuming the first subsystem moving w.r.t. the inertial frame, we obtain the "kinematic chain"

$$\begin{pmatrix} \dot{\mathbf{y}}_1 \\ \dot{\mathbf{y}}_2 \\ \vdots \\ \dot{\mathbf{y}}_n \end{pmatrix} = \begin{bmatrix} \mathbf{F}_1 & & & \\ \mathbf{T}_{21}\mathbf{F}_1 & \mathbf{F}_2 & & \\ \vdots & \vdots & \ddots & \\ \mathbf{T}_{n1}\mathbf{F}_1 & \mathbf{T}_{n2}\mathbf{F}_2 & \cdots & \mathbf{F}_n \end{bmatrix} \begin{pmatrix} \dot{\mathbf{s}}_1 \\ \dot{\mathbf{s}}_2 \\ \vdots \\ \dot{\mathbf{s}}_n \end{pmatrix} := \mathbf{F}_{tot}\,\dot{\mathbf{s}} \tag{5}$$

where $\mathbf{T}_{ij} = \mathbf{T}_{ip(i)} \times \cdots \mathbf{T}_{s(j)j}$ (p: predecessor, s: successor).

4.2 Structurizing the Problem: O(n)-Algorithm

If we split the sum (over a total of N bodies) in Eq. (4) into an double sum and introduce, for the kth summation term, variables $\dot{\mathbf{y}}_k$, then we obtain, along with the chain rule of differentiation,

$$\sum_{k=1}^{n}\left(\frac{\partial\dot{\mathbf{y}}_k}{\partial\dot{\mathbf{s}}}\right)^T \sum_{i=1}^{n_k}\left[\left(\frac{\partial\mathbf{v}_s}{\partial\dot{\mathbf{y}}_k}\right)^T(\dot{\mathbf{p}}+\tilde{\boldsymbol{\omega}}_R\,\mathbf{p}-\mathbf{f}^e)+\left(\frac{\partial\boldsymbol{\omega}_s}{\partial\dot{\mathbf{y}}_k}\right)^T(\dot{\mathbf{L}}+\tilde{\boldsymbol{\omega}}_R\,\mathbf{L}-\mathbf{M}^e)\right]_i = 0$$

which represents a conglomerate of n subsystems, each of which consisting of n_k interconnected bodies, being represented by $\dot{\mathbf{y}}_k$. In terms of a matrix representation one obtains, along with Eq. (5),

$$\begin{bmatrix} \mathbf{F}_1^T & \mathbf{F}_1^T\mathbf{T}_{21}^T & \cdots & \mathbf{F}_1^T\mathbf{T}_{n1}^T \\ & \mathbf{F}_2^T & \cdots & \mathbf{F}_2^T\mathbf{T}_{n2}^T \\ & & \ddots & \vdots \\ & & & \mathbf{F}_n^T \end{bmatrix} \begin{bmatrix} \mathbf{M}_1\ddot{\mathbf{y}}_1+\mathbf{G}_1\dot{\mathbf{y}}_1-\mathbf{Q}_1 \\ \mathbf{M}_2\ddot{\mathbf{y}}_2+\mathbf{G}_2\dot{\mathbf{y}}_2-\mathbf{Q}_2 \\ \vdots \\ \mathbf{M}_n\ddot{\mathbf{y}}_n+\mathbf{G}_n\dot{\mathbf{y}}_n-\mathbf{Q}_n \end{bmatrix} = \begin{bmatrix} 0 \\ 0 \\ \vdots \\ 0 \end{bmatrix}. \tag{6}$$

The non-marked elements are zero. Thus, Eq. (6) may be solved in a (generalized) Gaussian sense [$O(n)$-Algorithm, $n = \sum n_i, n_i = Rank(\mathbf{F}_i)$: relative degree of freedom of considered subsystem]. For details see [5].

4.3 Rigid Body Versus Flexible Body Dynamics

The simplest subsystem is the rigid body itself. Hubinger [15] made an impressive animated simulation on the occasion of the 15th birthday of our institute: motion of a chain which undergoes the following time history: free fall, then getting contact (impact and endpoint fixed, corresponding constraint forces see [11]). After 4 s the endpoint is released in one direction leading it to slide until contact gets lost and free motion once more takes place. A runtime comparison w.r.t. the number of links in comparison to the standard method is found in [11]. Standard here means to calculate the motion equations in minimal form and then invert the total mass matrix[1] for every time step. The time saving using the $O(n)$-algorithm is overwhelming.

Once disposing of a very fast algorithm one could think of using rigid body dynamics for elastic bodies, too, as some authors do (e.g. [16]). This procedure is then called "Finite Segmentation Method". But can a chain represent a threat, or a beam? How many links would be needed? A simple experiment with common household yarn gives an answer [5]. Firstly, an ideal threat does no exist in reality, there are always restoring forces. Thus, we have to consider a beam rather than a threat. Secondly, if we compare beam with a chain with spring interconnected links, then we would need about 100 links to ensure a certain accuracy (the deviations start growing already with the second eigenform). One hundred links for just one bending direction is definitely too much. One should not be lazy-minded here and continue with a Ritz approach. Things become clearer – and easier. The method itself remains unaltered.

5 Flexible Robots: Modeling and Simulation

First results have been obtained with ElRob ("Elastic Robot") a description of which can be found in [27]. ElRob, a gift from the Technical University of Munich when or institute was founded in 1995, was then already 8 years old. After a lot of replacements and repairs it is now in well-deserved pension. Its successor ELLA ("Elastic Laboratory Robot") is already at work. New sensors for (absolute) acceleration and angular velocity at the joints are implemented and the dspace computer is replaced with an industrial one which simplifies direct result transfer to industrial application. Robin [27] shows a straight line manoeuver in space: the simulation model for computational testing is according to Eq. (6) including the Ritz coefficients for variables. The control model is purely rigid to obtain simple control laws. In order to avoid elastic vibrations, the performance index includes jerk minimization.

Acknowledgements Support of the Austrian Center of Competence in Mechatronics (ACCM) is gratefully acknowledged

[1] $\mathbf{M} = \mathbf{F}_{tot}^T\, blockdiag\{\mathbf{M}_1 \cdots \mathbf{M}_n\}\mathbf{F}_{tot}$, see Eqs. (5) and (6).

References

1. Alembert, J. de: Traité de Dynamique, 2nd edn. David, Paris (1758). http://fr.wikisource.org
2. Bloch, A.M.: Nonholonomic Mechanics and Control. Springer, New York (2003). http://fr.wikisource.org
3. Bremer, H.: Über eine Zentralgleichung in der Dynamik. ZAMM **68**, 307–311 (1988)
4. Bremer, H.: On the Use of Nonholonomic Variables in Robotics. In: Belyaev, A., Guran, A. (eds.) Selected Topics in Structronics and Mechatronic Systems, pp. 1–48. World Scientific, River Edge (2003)
5. Bremer, H.: Elastic Multibody Dynamics – A Direct Ritz Approach. Springer Science+Business media, B.V. (2008)
6. Bremer. H.: Lagranges "récherches sur la libration de la lune". GAMM-Mitt. **34**(suppl), 1–14 (2011). doi:10.1002/gamm.201110031
7. Euler, L.: Mechanica sive motus scientia analytice exposita. Comm. Acad. Scient. Petropoli, St. Petersburg (1736). http://www.math.dartmouth.edu/~euler/E15
8. Euler, L.: Découverte d'un nouveau principe de la Mécanique. Mem. de l'acad. Sci. Berlin, pp. 185–217 (1750). http://www.math.dartmouth.edu/~euler/E177
9. Euler, L.: Nova methodus motum corporum rigidorum determinandi. Mem. Acad. Sci. Petropoli, St. Petersburg, pp. 208–238 (1775). http://www.math.dartmouth.edu/~euler/E479
10. Euler, L. in Lagrange, J.: Correspondance. Oeuvres de Lagrange, vol. 14, pp. 152–154. Gauthiers-Villars, Paris (1892). http://gdz.sub.uni-goettingen.de
11. Gattringer, H.: Starr-elastische Robotersysteme – Theorie und Anwendungen. Springer, Berlin/Heidelberg (2011)
12. Hamel, G.: Die Lagrange-Eulerschen Gleichungen in der Mechanik. Zeitschr. Ang. Math. u. Phys. **50**, 1–57 (1904)
13. Hamel, G.: Über die virtuellen Verschiebungen in der Mechanik. Math. Ann. **59**, 416–434 (1904)
14. Hamel, G.: Theoretische Mechanik. Springer, Berlin/Heidelberg/New York (1949)
15. Hubinger, S.: Mehrkörpersimulation mit O(n)-Verfahren (2010). http://www.robotik.jku.at/Forschung/Videos
16. Huston, R.L.: Multibody Dynamics. Butterworth-Heinemann, Oxford (1990)
17. Lagrange, theorie de la libration de la lune. Proc. Pruss. Acad. 5–122 (1780)
18. Lagrange, J. de: théorie de la libration de la lune. Oeuvres de Lagrange, vol. 5. Gauthiers-Villars, Paris (1870)
19. Lagrange, J. de: Essai d'une nouvelle méthode pour déterminer les maxima et les minima des formules intégrales indéfinies. Oeuvres de Lagrange, vol. 1, p. 389. Gauthiers-Villars, Paris (1873). http://gdz.sub.uni-goettingen.de
20. Lagrange, J. de: Récherches sur la libration de la lune. Oeuvres de Lagrange, vol. 6. Gauthiers-Villars, Paris (1873)
21. Lagrange, J. de: Correspondance. Oeuvres de Lagrange, vol. 14, pp. 146–151. Gauthiers-Villars, Paris (1892). http://gdz.sub.uni-goettingen.de
22. Lagrange, J. de: Analytische Mechanik (deutsch von H. Servus), 2nd edn. Springer, Berlin (1897)
23. Mach, E.: Die Mechanik in ihrer Entwickelung, 7nd edn., p. 272. F.A. Brockhaus, Wiesbaden (1912)
24. Maggi, G.A.: Principii di Stereodinamica. Ulrico Hoepli, Milano (1903)
25. Newton, I.: Philosophiae Naturalis Principia Mathematica. London (1686). http://en.wikisource.org/wiki/The_Mathematical_Principles_of_Natural_Philosophy_(1846)
26. Radetzki, P: The Man who Mastered Motion. Science, **86**, 52–60 (1986)
27. Robin (the robotic institute, JKU Linz), http://www.robotik.jku.at/Forschung/Videos
28. Szabó, I.: Bemerkungen zur Literatur über die Geschichte der Mechanik. Humanismus und Technik **22**, 3 (1979)
29. Truesdell, C.: An Idiot's Fugitive Essays on Sience. Springer, Berlin (1984)

Structural Monitoring Through Acquisition of Images

Fabio Casciati and Li Jun Wu

Abstract The availability of a suitable data acquisition sensor network is a key implementation issue to link the models with real world structures. Among various kinds of sensors, the class of non-contact sensor represents an endearing direction; indeed they can be easily installed on existing infrastructure in different scenes. Vision-based techniques, which enable dense global measurements of static deformations, as well as dynamic processes, are currently made available by ongoing technology developments. A vision system, which covers a medium range investigation area and takes advantage of fast-developing digital image processing and computer vision technologies, is constructed in this paper to monitor the vibration of a reduced scale frame available in the laboratory. Several markers are placed on the positions of interest. After preprocessing, calibration, segmentation, object representation and recognition, the 2D displacements of the markers are measured. Experiment results show that this tool for local positioning system (LPS) provides a satisfactory performance.

1 Introduction

In several applications of structural monitoring and/or structural health monitoring (SHM), it is extremely challenging to attach to a structure sensors able to quantify the response to environmental conditions and to the geometrical constraints. This is mainly made for checking that the assigned serviceability requirements are satisfied. Non-contact experimental techniques have been developed as an alternative approach [3, 4, 11]. Non-contact devices are also suitable for the experimental analysis of properly reduced scale models: to adopt conventional sensors in such

F. Casciati (✉) · L.J. Wu
Department of Civil and Architectural Engineering, Division of Structural Mechanics,
University of Pavia, Via Ferrata 3, 27100 Pavia, Italy
e-mail: fabio@dipmec.unipv.it; lijun.wu@unipv.it

Alexander K. Belyaev et al. (eds.), *Mechanics and Model-Based Control of Advanced Engineering Systems*, DOI 10.1007/978-3-7091-1571-8_8,

© Springer-Verlag Wien 2014

experiments could result unfeasible since the added sensor's mass can affect the behavior of the models. For this reason, there is a growing interest in developing alternative techniques for measuring movement without allowing contact with the structure [8]. Moreover, non-contact sensor could be more durable since it doesn't endure the same hazard of the structure. Among them vision-based monitoring systems are currently proposed [7].

A vision-based positioning system is adopted in this paper for in-plane SHM and image processing algorithms are applied to optimize its performance. Velocity and displacement measurements are based upon tracking the object motion between sequences of images. Displacement measurements based on vision system provides a good accuracy and a good robustness although a "sight-on-line scene" is required. The implemented system covers a medium range field and takes advantage of well-developed and still developing processors and digital image processing technologies. On one hand, the fast development of processors provides the designer the flexibility to develop their image acquiring and processing system according to their application scene thanks to a great increase of the image processing ability of the hardware. On the other hand, the image processing algorithm and the theory for computer vision have been well-developed in the last several decades. It is worth mentioning that Intel has developed three cross-platform libraries to support real time computer vision: Intel® Integrated Performance Primitives (Intel® IPP), Open Source Computer Vision (OpenCV) and Image Processing Library (IPL) [6]. In the present study the off-the-shelf image processing software Image Pro Plus 6.0 [9] is used. Image-Pro Plus 6.0 performs the image enhancement using powerful color and contrast filters and other spatial and geometric operations. It can trace and count objects manually or automatically and measure object attributes such as: area, perimeter, roundness etc. The collected data can be viewed numerically, statistically or in graphic form (histogram and scatter gram). The features isolate an area of interest from the rest of the image: it can be extracted by spatial tools, or by using segmentation tools that extract features by color or intensity value. Images from multiple fluorescent probes can be composed.

2 Vision-Based Monitoring Systems

2.1 Hardware and Software

A monochrome camera SV642 (Table 1) is employed to capture the motion of a three-stories frame, which is mounted on a shaking table. It provides 640 by 480 resolutions at 204 frames/s (fps). The sizes (vertical V and horizontal H) of view can be calculated according to the focal length f, the work distance W and the sensor sizes (v and h).

$$\frac{f}{W} = \frac{h}{H} = \frac{v}{V} \tag{1}$$

Table 1 The parameters of the SV642 camera

Pixel	Pixel size	Frame rate	ADC	Interface of camera
$640H \times 480V$	$9.9 \times 9.9\,\mu\text{m}$	204 fps@640×480	8 bits	ShieldedCAT-5 with RJ45 Plugs

The captured image sequences are sent to a PC through a PIXCI digital frame grabber for PCI bus. The software "Image Pro Plus 6.0" is used to process the image sequence. When one measures the displacement based on a vision system, the unit for the displacement of the object is the pixel, which should be multiplied by the scaling factors to obtain the real displacement. Assuming the size of the target is (L_x, L_y) which corresponds to (x pixels, y pixels) on the image, the scaling factors (S_x, S_y) are calculated according to

$$S_x = \frac{L_x}{x}, \; S_y = \frac{L_y}{y}. \tag{2}$$

2.2 *The In-Plane Measurement Method*

Velocity and displacement measurements using images are based upon tracking the object motion between sequences of images. The proposed image processing methodology comprises four steps:

1. Preprocess: it is the process of manipulating an image so that the result is more suitable than the original for a specific application, such as contrast enhancement.
2. Calibration: it is employed to determine the transformation matrix that correlates the image coordinates and their respective actual coordinates. After one ensures that the camera optical axis is perpendicular to the structural plane, there remain a distortion of the image which may be caused by the optical system (radial and tangential lens distortions, perspective distortion) and the electronic system. These distortions were estimated for less than 0.5 pixels on the basis of the observation of the test image and could be reduced to 0.02 pixels by employing a suitable image processing algorithm [10]. In order to obtain the coordinate transformation matrix, which establishes the correlation between the actual coordinates and those in the image, it is necessary to apply an adequate calibration method. It is performed by identifying the (u, v) coordinates of the calibration points selected in the image and inserting the respective actual coordinates (x, y) (2D). This step is carried out by the calibration routine. The calibration can be performed in one of these two ways: (i) extrapolation and (ii) interpolation, as illustrated in Fig. 1a, b. In the first approach, the calibration is performed on the basis of the known dimensions of the structure or through the application of markers with pre-established distances in the structure itself. The advantage of this procedure is that it facilitates the acquisition of the

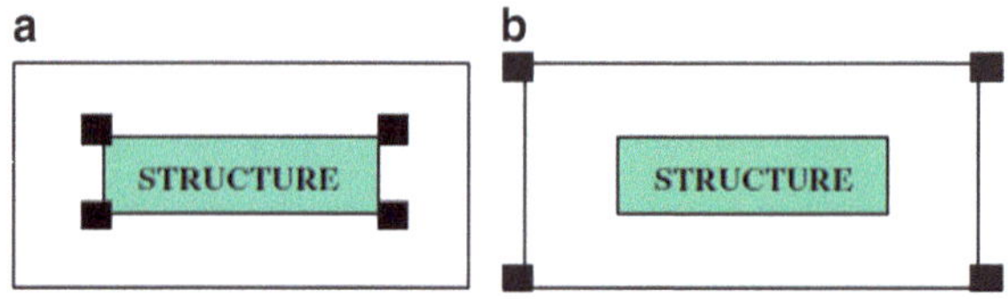

Fig. 1 Forms of calibration: (**a**) extrapolation and (**b**) interpolation

actual data. However, extrapolating the measurements outside the region used for the calibration can result quite imprecise. In the second way, the calibration is performed through markers defining the borders of the region where the movement of the structure occurs. Therefore, any configuration of the structure will always be within the calibration region. Thus, the results obtained through this method are expected to be more precise than those obtained by simple extrapolation. The disadvantage of this second method is that it is necessary to impose markers outside the structure, which in some cases can result difficult [8].

3. Segmentation: it is the process of assigning a label to every pixel in an image, such that pixels with the same label share certain visual characteristics. Thus, it partitions a digital image into multiple segments and simplifies the representation of an image into something that is more meaningful and easier to analyze. It is typically used to locate objects and boundaries (lines, curves, etc.) in images. Several general-purpose algorithms and techniques have been developed for image segmentation, such as thresholding, histogram-based methods, edge detection, watershed transformation method, etc. [5].
4. Object representation, description and recognition. To recognize objects, one must have an internal representation of an object suitable for matching its features to image descriptions which means to describe or represent the object through certain features. There are three kinds of techniques to represent the objects [2]: appearance-based, intensity contour-based, and surface feature-based techniques. Shape represent techniques, as that of appearance-based techniques, include contour-based (perimeter, compactness, eccentricity, etc.) and region-based (area, Euler number, eccentricity, geometric moments, etc.) options [12].

3 Feasibility of Vision-Based Displacement Measurements

In this section, an experiment using SV642 is carried out to analyze the feasibility of a displacement measurement vision-based system. A three-stories frame is mounted on a shaking table which moves at the frequency of 2.4 Hz, with the amplitude of 4 mm. The dimension of each story is $60 \times 30 \times 3$ cm [1]. The first and third stories are braced. The optical axis of the SV642 camera is perpendicular to the front plane of the frame. The sample rate is 120 frames/s and 1,242 frames are recorded (i.e., 10.3 s). Two white markers T1 and T2 are installed at the supported nodes on the second story, while further two markers T3 and T4 are installed on

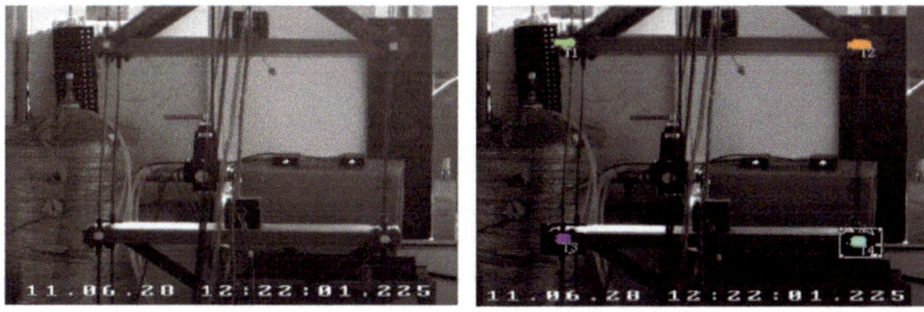

Fig. 2 Photographs accounting for the initial position at rest. On the *right side* four markers are outlined

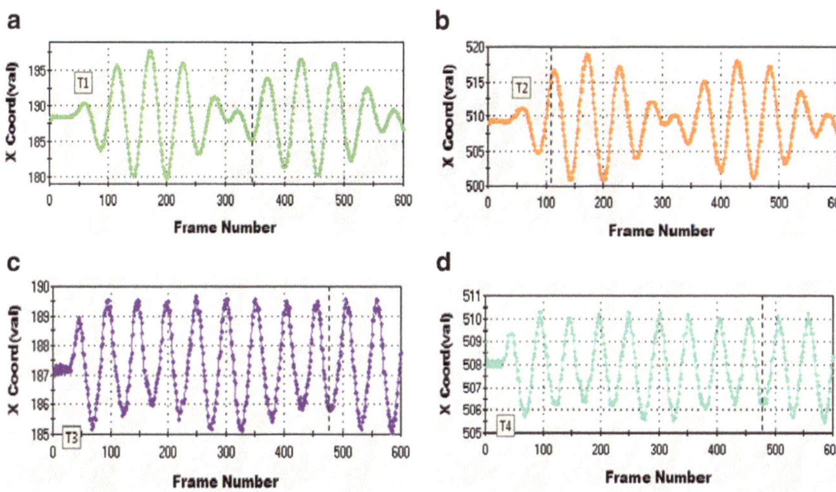

Fig. 3 The track of: (**a**) T1 (*left-top*); (**b**) T2 (*right-top*); (**c**) T3 (*left-bottom*); (**d**) T4 (*right-bottom*); (The unit of vertical coordinate is the pixel)

the first story, as shown in Fig. 2. According to Eq. (2), one gets 1.685 mm/pixel as the scale factor S. Thus the obtained horizontal displacements shown in Fig. 3 are derived and the resolution turns out to be 0.16 mm (0.1 pixels). The movements of T1 and T2 are consistent each with the other, while the movements of T3 and T4 are consistent each with the other. They are quite similar to the shaking table sine wave since the stiffness of the first story is largely increased by the presence of the cross-brace. On the movement of T1 and T2, one can distinguish two mode frequencies: one is 2.4 Hz (the excitation), the other is 0.41 Hz (the system). If one compares the movement between T2 and T4, a 180° phase difference can be discovered. Focus now the attention on the movement of T1 (Fig. 4): one obtains for the maximum displacement of T1 the value 16.85 mm (10 pixels) and the instant when the maximum shift occurred: it corresponds to the digital image n°175 which is shown in Fig. 5. The vertical direction displacements are shown on Figs. 6 and 7.

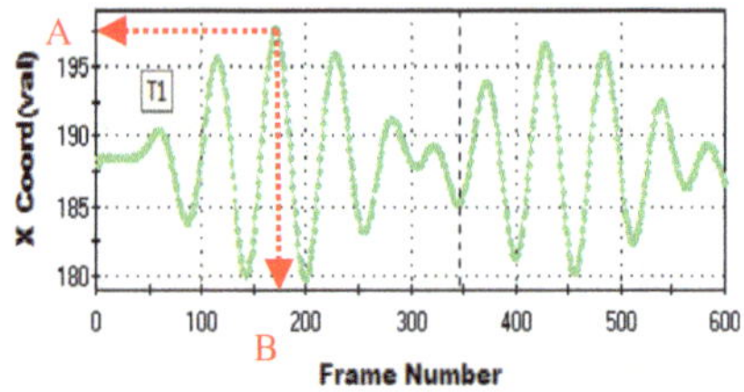

Fig. 4 The maximum displacement of T1

Fig. 5 The instant when maximum shift occurred

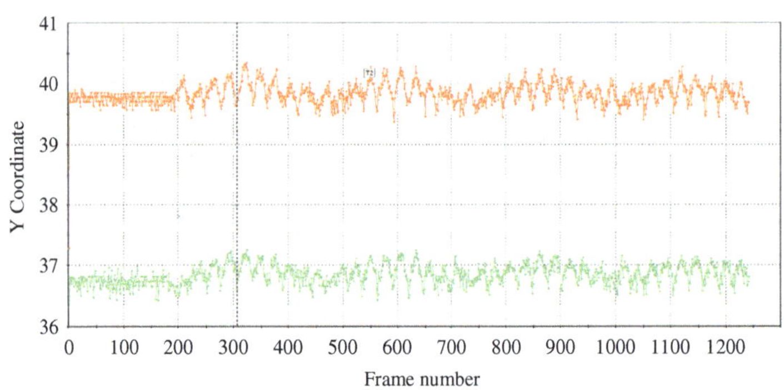

Fig. 6 The vertical displacements of T1 and T2

The vertical displacement of T3 and T4 is similar and is nearly 0 due to the fixing cross-brace. But the vertical displacement of T1 and T2 cannot be ignored. Its peak-peak amplitude is around 0.7 pixels and corresponds to 1.18 mm. One can also distinguish two frequencies from Fig. 6: 4.8 and 0.41 Hz. Figure 8 shows the absolute values of the difference of the horizontal displacement in T1 (and T2) at time t_i and t_{i-1}. Since the movement is uniformly sampled, these values are directly proportional to their velocities. One can see That the velocity of T1 coincides with

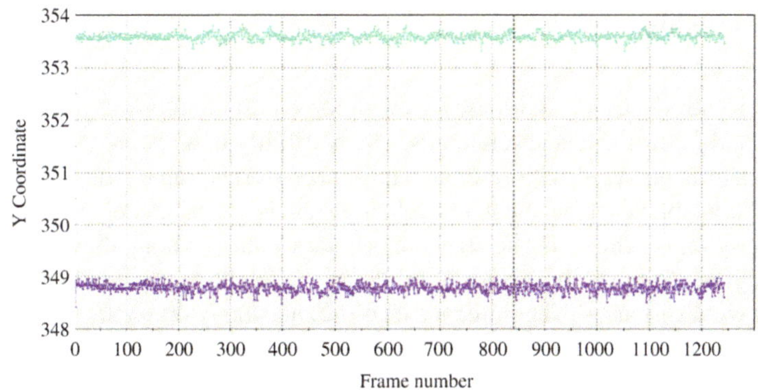

Fig. 7 The vertical displacements of T3 and T4

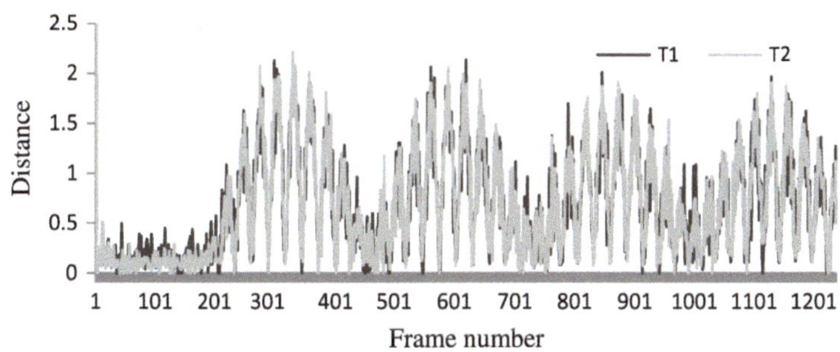

Fig. 8 Absolute values of the difference of the horizontal displacement in T1 (and T2) at time t_i and t_{i-1}

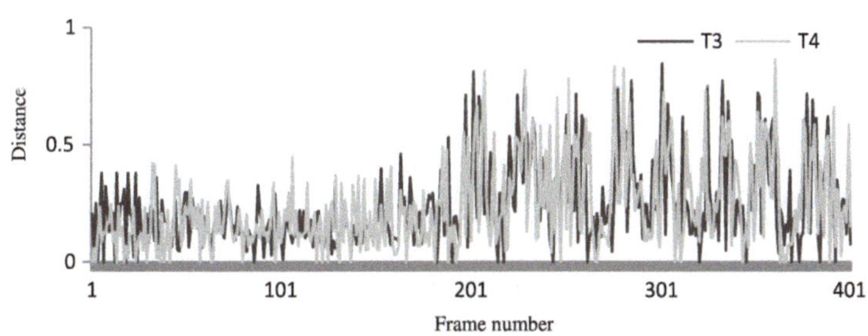

Fig. 9 Absolute values of the difference of the horizontal displacement in T3 (and T4) at time t_i and t_{i-1}

the one of T2. The absolute value of the difference of the horizontal displacement of T3 (and T4) (Fig. 9) is relatively small, since the two markers are rigidly connected to the supporting system. But the absolute distances of T3 (and T4) suffers a serious noise.

4 Conclusions

This paper introduces a vision-based view of structural monitoring. The procedure is illustrated for in-plane measurements. The feasibility of a vision-based measurement system is analysed with reference to a three-stories frame which is mounted on a shaking table. Displacement and velocity signals are adequately inferred and a high resolution is achieved. The errors in accuracy could be magnified by different focal lengths, different flickering illumination, different background colours and different work distances. The influence of different camera orientations is a further aspect to be considered and will be the object of future work.

Acknowledgements This research is supported by a grant from the Athenaeum Research Funds of the University of Pavia (FAR 2011). The research activity summarized in this paper was developed within the framework of the Marie Curie European project SMARTEN.

References

1. Balzi, W.: Monitoraggio strutturale tramite acquisizione di immagini (in Italian). Master thesis, University of Pavia (2011)
2. Campbell, R.J., Flynn, P.J.: A survey of free-form object representation and recognition techniques. Comput. Vis. Image Underst. **81**, 166–210 (2001)
3. Casciati, F., Wu, L.J.: Wireless links for global positioning system receivers. Smart Struct. Syst. **10**(1), 1–14 (2012)
4. Casciati, F., Wu, L.J.: Local positioning accuracy of laser sensors for structural health monitoring. Struct. Control Health Monit. **20**(5), 728–739 (2012)
5. Gonzalez, R.C., Woods, R.E.: Digital Image Processing. Prentice Hall, Upper Saddle River (2002)
6. Intel: Intel® IPP – Open Source Computer Vision Library (OpenCV) FAQ (2012). From http://software.intel.com/en-us/articles/intel-integrated-performance-primitives-intel-ipp-open-source-computer-vision-library-opencv-faq
7. Jahanshahi, M.R., Masri, S.F.: Adaptive vision-based crack detection using 3D scene reconstruction for condition assessment of structures. Autom. Constr. **22**, 567–576 (2012)
8. Jurjo, D.L.B.R., Magluta, C.: Experimental methodology for the dynamic analysis of slender structures based on digital image processing techniques. Mech. Syst. Signal Process. **24**(5), 1369–1382 (2010)
9. mediaCybernetics: Image Pro Plus Software Development Kit (SDK) (2012). From http://www.mediacy.com/index.aspx?page=IPP_SDK
10. Olaszek, P.: Investigation of the dynamic characteristic of bridge structures using a computer vision method. Measurement **25**, 227–236 (1999)
11. Uhl, T., Kohut, P.: Vision based condition assessment of structures. J. Phys. Conf. Ser. **305**, 012043 (2011)
12. Zhang, D.S., Lu, G.J.: Review of shape representation and description techniques. Pattern Recognit. **37**(1), 1–19 (2004)

SMA Passive Elements for Damping in the Stayed Cables of Bridges

Sara Casciati, Antonio Isalgue, Vincenc Torra, and Patrick Terriault

Abstract In this paper, the properties of shape memory alloy (SMA) wires relevant for their application as passive elements to mitigate the vibrations of stayed cables in bridges are discussed. These properties are investigated by carrying out experimental tests on samples of diameter 2.46 mm. Future research will address the scalability issues.

1 Introduction

The hysteresis cycle which characterizes the super-elastic behavior of shape memory alloys (SMA) in austenitic phase offers desirable damping capabilities for passive control solutions. The working principle of SMA dampers relies on a martensitic transformation, i.e., a first-order phase transformation with hysteresis, which occurs between two meta-stable solid phases and involves the release and absorption of latent heat. In literature, the development of SMA passive devices for civil engineering applications is mostly targeted to the mitigation of earthquake induced vibrations [1]. Recent studies in this field are dedicated to the characterization of the functional properties of Copper-based alloys [2–7] and their exploitation into devices for the seismic protection of structures [8–10].

S. Casciati (✉)
Department DICA, University of Catania, Piazza Federico di Svevia, 96100 Siracusa Italy
e-mail: saracasciati@msn.com

A. Isalgue · V. Torra
Applied Physics Department, UPC, E-08034 Barcelona, Catalonia, Spain
e-mail: aisalgue@fa.fnb.upc.edu; vtorra@fa.upc.edu

P. Terriault
Mechanical Engineering Department, ETS, Montreal, QC, Canada
e-mail: patrick.terriault@etsmtl.ca

Alexander K. Belyaev et al. (eds.), *Mechanics and Model-Based Control of Advanced Engineering Systems*, DOI 10.1007/978-3-7091-1571-8_9,
© Springer-Verlag Wien 2014

Fig. 1 An applicative example: the Echingen viaduct, in France (Photographs by V. Torra)

Parallel research activities [11–15] focus on investigating the possibility of using specifically designed SMA dampers to mitigate the vibrations of stayed cables in bridges. The task is to increase the lifetime of the cables under the effects of environmental vibrations due to wind, rain, traffic, etc. In contrast to a seismic context, it requires to consider a large number of oscillations per day, at about 1 Hz frequency or higher and with an order of magnitude not lower than half million cycles (which occur in sequences of several thousands of cycles each separated by pauses of randomly varying duration). When inspecting, for example, the Echingen viaduct shown in Fig. 1, cable oscillations of 10 mm amplitude at a frequency of 5 Hz are observed for low wind speeds (less than 10 km/h). A Nichel-Titanium alloy (55.95 %Ni–44.05 %Ti) with an austenite start temperature (As) of 248 K (−25 °C) is selected for the described application, due to its high damping capacity, the longest durability at low deformations, and the best fatigue resistance.

The results from experimental tests carried out in different facilities (namely, the IFSTTAR institute, in Nantes, France, and the ELSA laboratory, in Varese, Italy) show the capability of reducing to less than one half the vibration amplitudes of sample cables of realistic size by installing specifically built SMA damper prototypes [11–15]. The repeatability of such measurements require that adequate design rules are set for the SMA dampers. The specific conditions of the cable define the appropriate characteristics of the damper, such as the length and the number of the NiTi SMA wires. However, some general criteria can be established in order to guarantee the essential features required for the application of SMA wires as passive elements for damping the vibrations of stayed cables. The definition of such criteria and the experimental investigation of the SMA mechanical properties relevant to the described application represent the ultimate goals of the present work. Further research is needed in order to address the scalability of the obtained results.

2 Definition of SMA Properties as Dampers of Cable Oscillations

The following criteria are established in order to define the thermo-mechanical properties which are necessary to achieve a satisfactory performance of the SMA wires used in dampers of cable oscillations.

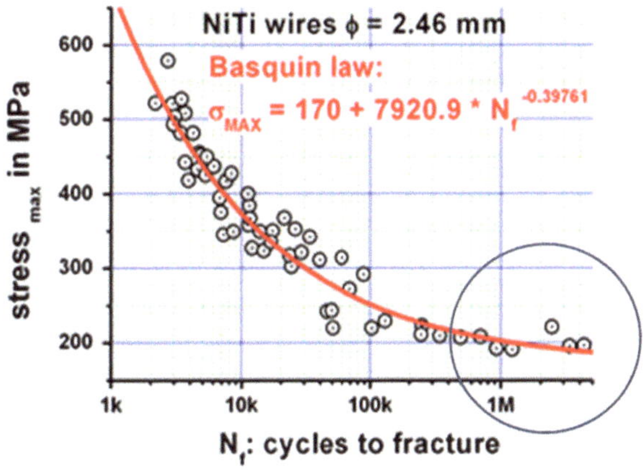

Fig. 2 Number of cycles to rupture versus maximum stress level and fitted Basquin law. The *circled area* indicates the results relevant to the application of SMA wires as dampers of cable oscillations

- The fatigue-fracture life (at a low deformation level near 1 %) must be satisfactory for several days of continuous cycling.
- A correct response of the SMA dampers implies that the progressive deformation which increases the damper length with cycling (i.e., the SMA creep) remains completely suppressed by performing an appropriate preliminary training.
- The capability of dissipating a relevant amount of energy per cycle also at low levels of deformation (below 2.5 %) is verified and quantified for samples of reduced length.
- The temperature effects, for instance, the daily and the summer-winter temperature effects and, also, the self-heating effects of cycling should be either suppressed or well acknowledged and managed.
- The frequency evolution of the hysteresis cycle and the temperature effects must be considered in the estimate of the hysteresis energy, and in assessing the appropriateness of the representative model used for simulation of the SMA behavior.

These aspects will be discussed with reference to the experimental results obtained from NiTi samples of diameter 2.46 mm under cyclic loading-unloading.

Fatigue-fracture life for the considered operational conditions The fatigue-fracture life of the tested specimens is measured in terms of the number of cycles to rupture obtained by carrying out several loading-unloading tests in span control which are characterized by different fixed values of maximum deformation at the end of each cycle. In Fig. 2, the number of cycles to rupture is plotted against the corresponding level of maximum stress. The experimental data can be fitted by a Basquin law, which is also given in Fig. 2. The part of the curve relevant to the application of SMA wires as dampers of cable oscillation is circled in Fig. 2. From the results in Fig. 2, it can be observed that, for strain levels below 1.5 %, the fracture occurs after more than 4.5 million working cycles. Therefore, an adequate fracture life of the SMA wires is achieved when working at low levels of deformation

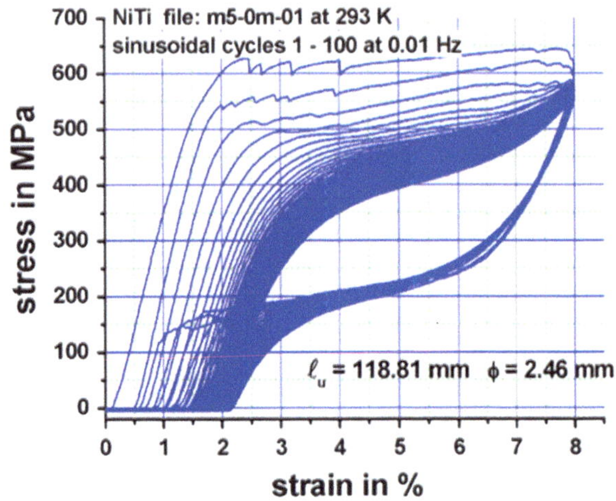

Fig. 3 Preliminary mechanical training: 100 loading-unloading cycles at 0.01 Hz, up to a maximum strain of 8 %

(around 1 %) and for stresses near 200 MPa, such as the ones commonly concerning the dampers of stayed cables.

Preliminary mechanical training and SMA creep suppression The stress-strain relationship in Fig. 3 shows the typical hysteretic cycle that characterizes the super-elastic behavior of SMA elements in austenitic phase. It is obtained by performing 100 loading-unloading cycles, at 0.01 Hz, in span control with a maximum strain level of 8 % at the end of each cycle. The described test is referred to as the training process of the alloy. It is worth noticing that the associated maximum strain level is higher than the operating conditions of the SMA damper during cable oscillations. The need for a preliminary training of the alloy raises from the following remarks. After the sequence of 100 cycles at 0.01 Hz, the shape of the hysteresis loop is altered and its position is lowered and shifted toward the increasing abscissa. In particular, the curve progressively becomes S-shaped, so that a cubic fit is more likely to yield accurate numerical results than a bilinear approximation. Furthermore, the shift of the curve indicates that a residual deformation of the alloy accumulates during the first sequence of 100 cycles, thus leading to an increase of the length of the SMA wire. This phenomenon, known as the SMA creep, stabilizes during successive cycles so that the effect of increasing the length with progressive deformation is suppressed by preliminarily applying the described training to the SMA wire. The above statement is supported by the results plotted in Fig. 4, where the evolution of the SMA creep with the number of working cycles is depicted. The residual deformation at the end of each series of cycles (triangular dots in Fig. 4) can only be partially recovered of a very small amount at the beginning of the successive series of cycles (circular dots in Fig. 4). During the first series of working cycles, the creep increases progressively up to 2.5 %. Then it remains invariant as the series of cycles are repeated. The influence of the cycling frequency is also considered by plotting the results obtained at different frequencies in the range between 0.01 and 16 Hz.

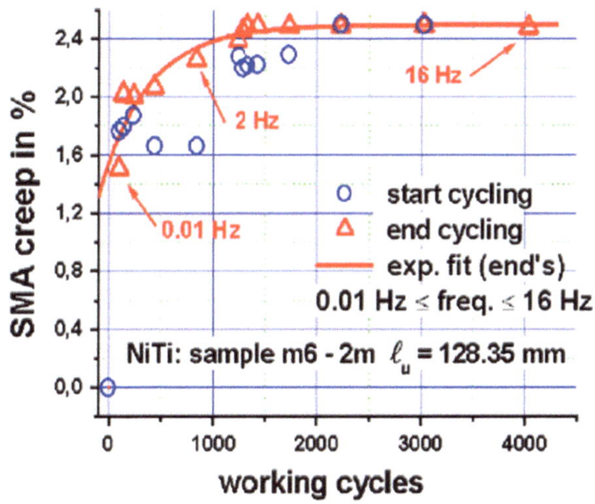

Fig. 4 Saturation of the SMA creep after training

Energy dissipated per cycle at low deformation levels The hysteresis loop characteristic of the stress-strain relationship of SMA wires in austenitic phase is maintained also when working in a strain range below 2.5 %. However, it progressively shrinks as the maximum level of operational strain decreases, as shown in Fig. 5a. Such an evolution is represented also by plotting the energy dissipated per cycle versus the net deformation range (i.e., the maximum strain level subtracted of the initial creep). The experimental data can be fitted by a parabolic curve as shown in Fig. 5b. The obtained results suggest that a reasonable amount of energy can be dissipated during cyclic loading-unloading for deformations as low as 0.4 %.

Temperature and self-heating effects The summer-winter fluctuation effects are investigated by using a thermal chamber to vary the temperature in the range between −15 and 62 °C during the loading-unloading cycles. The cycles are carried out at 0.05 Hz, up to a maximum strain of 5 %. The temperature time history recorded in the thermal chamber is plotted in Fig. 6a. The hysteresis cycles measured at different working temperatures along the given time history are plotted in Fig. 6b. The results suggest that a reduced damping is possible for temperatures as low as −25 °C or −30 °C, at least for "higher" amplitudes of strain up to 5 %. The external temperature effects induce changes in the position of the hysteresis cycle which are governed by the Clausius–Clapeyron thermodynamic equation with a coefficient of 6.3 MPa/K [16]. By decreasing the temperature, part of the sample remains in martensite phase and the size of the hysteresis loop decreases. The hysteresis is then recovered when the temperature returns at 60 °C, as expected.

The temperature changes induced by the self-heating of the sample during a cyclic loading-unloading test up to 1.8 % maximum strain are also investigated.

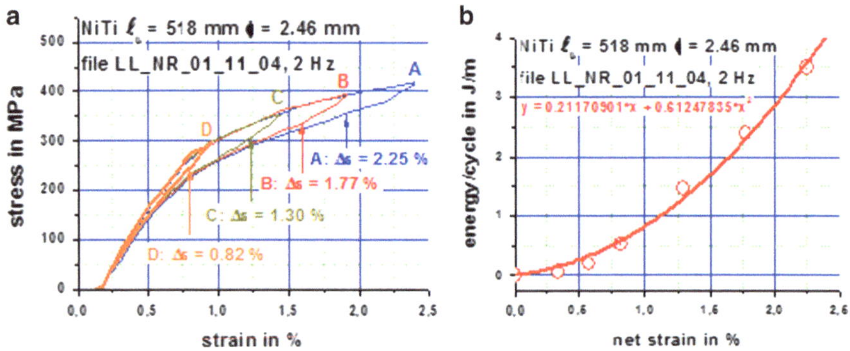

Fig. 5 Damping capabilities at low deformation levels: (**a**) evolution of the hysteresis cycle in the force-deformation relationship as the range of operational strains decreases; (**b**) evolution of the energy dissipated per cycle with the net deformation range

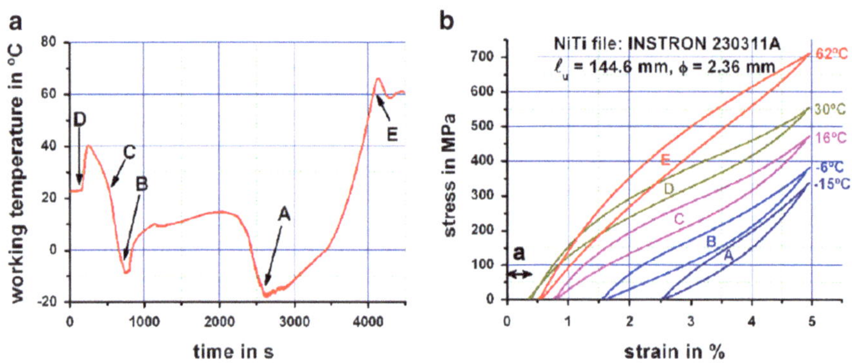

Fig. 6 Summer-winter temperature effects: (**a**) time history of the chamber temperature; (**b**) hysteresis cycles of the stress-strain relationship at different working temperature

Their dependence on the cyclic frequency is illustrated in Fig. 7, where the frequency is changed in the range between 1 and 10 Hz during different time periods of the test. For levels of deformation below 2 %, the self-heating at 10 Hz does not overcome 10 K or 60 MPa, and it reduces to one half for deformations near 1 %. Therefore, when considering the wind effects, the changes in the local temperatures and in the associated stresses are expected to be halved. The values of the force are slightly increased when passing from forced to natural convection (fan "on" and "off" status in Fig. 7, respectively).

Frequency evolution of the hysteresis cycle For low deformation amplitudes (such as 1.5 %), the effect of fast cycles (up to 18 Hz) is represented by a progressive decrease of the energy dissipated per cycle, as shown in Fig. 8. For frequencies overcoming 10 Hz, the MTS machine shows an artifact that "spontaneously" reduces the oscillation amplitude.

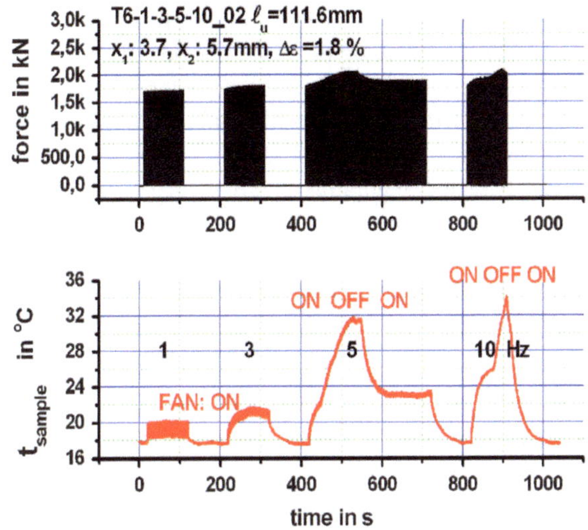

Fig. 7 Self-heating effects during a loading-unloading test with 1.8 % maximum strain and frequency varying from 1 to 10 Hz: time histories of force (*top*) and local temperature (*bottom*) of the sample

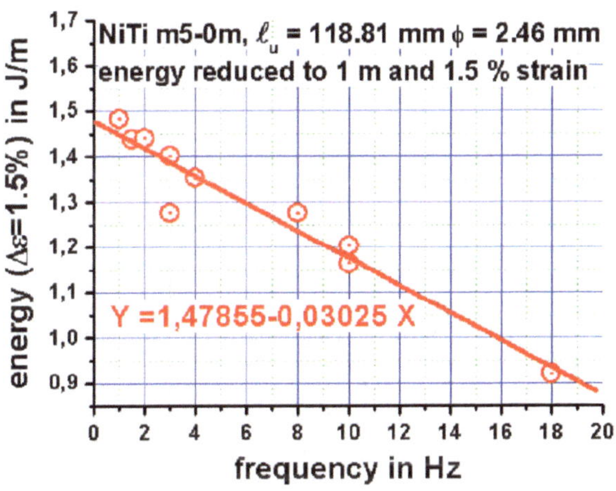

Fig. 8 Cycling frequency effect on the energy dissipated per cycle

3 Conclusions

Based on the presented results, one can conclude that the properties of SMA wires of diameter 2.46 mm are satisfactory for the requirements of damping the oscillations of stayed cables in bridges. Namely, the following summarizing remarks are drawn.

- For a strong storm of 2–4 days, one establishes that the number of working cycles in the damper overcomes two to three millions.
- The task of realizing reliable passive devices is achieved by performing an appropriate training of the SMA wires so that suitable S-shaped hysteresis cycles

characterize the SMA dampers. In this manner, the cycling effects associated to the SMA creep is suppressed to avoid a parasitic increase of the damper length.

- The hysteretic behavior of the NiTi wires and the related capability of dissipating energy during cyclic loading-unloading is observed also at low deformations below 2%.
- A satisfactory damping (with S-shaped cycles) is achieved in a temperature range between −20 and +40 °C, so that the dampers can be used in summer as in winter. The self-heating effects are not relevant at low deformation levels.
- The effects of the cycling frequency in the hysteretic behavior have been considered.

References

1. Ozbulut, O.E., Hurlebaus, S., Desroches, R.: Seismic response control using shape memory alloys: a review. J. Intell. Mater. Syst. Struct. **22**, 1531–1549 (2011)
2. Casciati, F., Casciati, S., Faravelli, L., Marzi, A.: Fatigue damage accumulation in a Cu-based shape memory alloy: preliminary investigation. CMC-Comput. Mater. Contin. **23**(3), 287–306 (2011)
3. Casciati, S., Marzi, A.: Fatigue tests on SMA bars in span control. Eng. Struct. **33**(4), 1232–1239 (2011)
4. Casciati, S., Marzi, A.: Experimental studies on the fatigue life of shape memory alloy bars. Smart Struct. Syst. **6**(1), 73–85 (2010)
5. Casciati, S., Hamdaoui, K.: Experimental and numerical studies toward the implementation of shape memory alloy ties in masonry structures. Smart Struct. Syst. **4**(2), 153–169 (2008)
6. Casciati, S., Faravelli, L.: Structural components in shape memory alloy for localized energy dissipation. Comput. Struct. **86**(3–5), 330–339 (2008)
7. Casciati, F., Casciati, S., Faravelli, L.: Fatigue characterization of a Cu-based shape memory alloy. Proc. Estonian Acad. Sci. Phys. Math. **56**(2), 207–217 (2007)
8. Carreras, G., Casciati, F., Casciati, S., Isalgue, A., Marzi A., Torra, V.: Fatigue laboratory tests toward the design of SMA portico-braces. Smart Struct. Syst. **7**(1), 41–57 (2011)
9. Torra, V., Isalgue, A., Martorell, F., Lovey, F.C., Terriault, P.: Damping in civil engineering using SMA. Part 1: Particular properties of CuAlBe for damping of family houses. Can. Metall. Q. **49**(2), 179–190 (2010)
10. Torra, V., Isalgue, A., Auguet, C., Carreras, G., Casciati, F., Lovey, F.C., Terriault, P.: SMA in mitigation of extreme loads in civil engineering: study of their application in a realistic steel portico. Appl. Mech. Mater. **82**, 278–283 (2011)
11. Torra, V., Isalgue, A., Auguet, C., Casciati, F., Casciati, S., Terriault, P.: SMA dampers for cable vibration: an available solution for oscillation mitigation of stayed cables in bridges. Adv. Sci. Technol. **78**, 92–102 (2013). Online available since 2012/Sep/11 at www.scientific.net. doi:10.4028/www.scientific.net/AST.78.92
12. Isalgue, A., Torra, V., Casciati, F., Casciati, S.: Fatigue of NiTi for dampers and actuators. Adv. Sci. Technol. **83**, 18–27 (2013). Online available since 2012/Sep/11 at www.scientific.net. doi:10.4028/www.scientific.net/AST.83.18
13. Isalgue, A., Torra, V., Casciati, F., Casciati, S.: NiTi wires for dampers and actuators: fatigue. In: Del Grosso, A.E., Basso, P. (eds.) Smart Structures, Proceedings of the 5th European Conference on Structural Control, EACS 2012, Genoa, 18–20 June 2012. Erredi Grafiche Editoriali S.n.c., Genoa (2012). ISBN:978-88-95023-13-7
14. Torra, V., Isalgue, A., Auguet, C., Casciati, F., Casciati, S., Terriault, P.: SMA passive elements for damping in stayed cables: experimental results and simulation. In: Del Grosso, A.E.,

Basso, P. (eds.) Smart Structures, Proceedings of the 5th European Conference on Structural Control, EACS 2012, Genoa, 18–20 June 2012. Erredi Grafiche Editoriali S.n.c., Genoa (2012). ISBN:978-88-95023-13-7

15. Torra, V., Isalgue, A., Auguet, C., Carreras, G., Lovey, F.C., Terriault, P., Dieng, L.: SMA in mitigation of extreme loads in civil engineering: damping actions in stayed cables. Appl. Mech. Mater. **82**, 539–544 (2011)
16. Isalgue, A., Torra, V., Yawny, A., Lovey, F.C.: Metastable effects on martensitic transformation in SMA. Part VI: The Clausius–Clapeyron relationship. J. Therm. Anal. Calorim. **91**(3), 991–998 (2008)

Temperature Effects on the Response of the Bridge "ÖBB Brücke Großhaslau"

Lucia Faravelli, Daniele Bortoluzzi, Thomas B. Messervey, and Ladislav Sasek

Abstract The aim of this paper is the study of the effect of temperature variation on the structural response. In particular, the behavior of a railway bridge located in the south part of Austria is investigated using health monitoring system. Among the large amount of data collected the influence of the temperature variation of the structural response is analyzed.

1 Introduction

The process of detecting, localizing, classifying, and providing a prognosis for damage (i.e. change in material and/or geometric properties, boundary conditions and so on) to engineered structures is referred to as *Structural Health Monitoring-SHM*. This process often involves the observation of a system over time using periodically sampled dynamic response measurements from an array of sensors, the extraction of damage-sensitive features from these measurements, and analysis methods to determine the current state of system health. *SHM* can be used for rapid condition screening and aims to provide, in near real time, reliable information regarding the integrity of the structure. However, in applying such an approach, one must properly account for temperature effects, which might appear as or conceal

L. Faravelli (✉) · D. Bortoluzzi
Department of Civil and Architectural Engineering, Division of Structural Mechanics, University of Pavia, Via Ferrata 3, 27100 Pavia, Italy
e-mail: lucia@dipmec.unipv.it; daniele.bortoluzzi@unipv.it

T.B. Messervey
R2M Solution, Via Monte Sant Agata 16, 95100 Catania, Italy
e-mail: thomasmesservey@r2msolution.com

L. Sasek
Cernokostelecka 1621, 25101 Ricany, Czech Republic
e-mail: ladislav.sasek@safibra.cz

Alexander K. Belyaev et al. (eds.), *Mechanics and Model-Based Control of Advanced Engineering Systems*, DOI 10.1007/978-3-7091-1571-8_10,
© Springer-Verlag Wien 2014

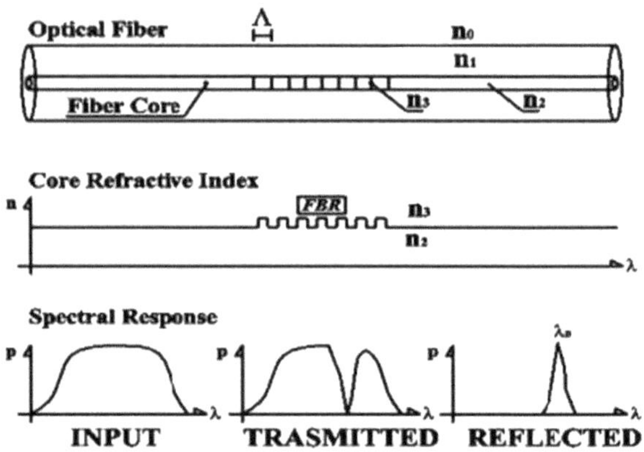

Fig. 1 Scheme of Bragg's grating

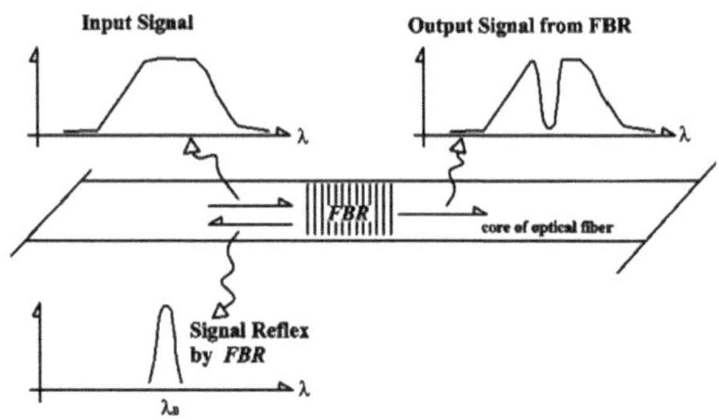

Fig. 2 Data transmission's scheme

damage [1–3]. This paper is dedicated to this topic and uses a case study on the ÖBB Brücke Großhaslau railway bridge to elaborate its discussion.

The *SHM* system installed on the bridge consists of Fiber Bragg Grating (FBG) sensors. The sensors *FBG* are particular spectrum sensors. The peculiarity of these sensors is the ability to modify the local value of the refractive index of the fiber's "core", using in an appropriate way a laser light. In other words, the device behaves as a filter in optical transmission and as selective reflector of the wavelength λ in reflection, as depicted in Fig. 1.

The final consequence due to the passage of a broadband bundle of light along the fiber, is that the photo-etched grating, reflects a specific wavelength called the *Bragg Wavelength* λ_B described by the following equation:

$$\lambda_B = 2n_{eff}\Lambda \quad (nm), \tag{1}$$

where n_{eff} is the refraction index, and Λ the mark distance.

The ability of the *FBG* to select a specific signal from a broadband light spectrum is the ground for the data transmission as shown in Fig. 2. In this way, the various *FBG* sensors filter and reflect back specific frequencies within the frequency package which constitute the signal. In order to understand this, one can

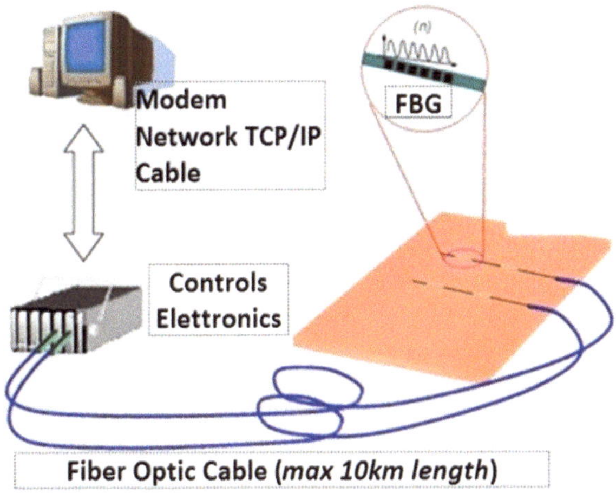

Fig. 3 Scheme of FBG's system monitoring

Fig. 4 Topographic map. Spot (**a**) indicates the site of bridge. Spot (**b**) indicates the site of Glossnitz railway station where is located the operating system control

make the analogy that the set of frequencies are similar to the set of conversations that are running on a telephone line and the *FBG* are appropriate selectors that extract a specific conversation (a specific frequency in our case) by sending it to the recipient where a suitable opto-electrical transducer translates the optical signal into an electrical signal for the final treatment up to our phone [4, 5]. A generic composition of a monitoring system that uses *FBG* sensors is shown in Fig. 3 and summarized as follows:

- Optical fiber with the *FBG* sensors;
- Fiber-optic cable connection between the sensors and the control electronics (max 10 km length);
- Control electronics: the electrical core of the system;
- A PC connected to the control electronics via a local cable, network, Internet or GSM modem, where data are stored and potentially analyzed.

The structure analyzed in this study is a railway bridge called *ÖBB Brücke Großhaslau* located in the south part of Austria (see Fig. 4), built between 2009 and 2010, which is on the railway *ÖBB Martinsberg – Schwarzenau* and crosses

Fig. 5 ÖBB Brücke Großhaslau

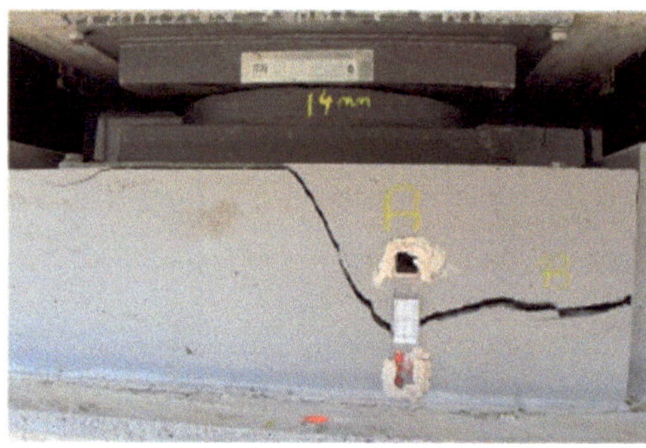

Fig. 6 Cracks on the central support

the road *Zwettler Straße – B36*. It is a single span, reinforced concrete T-beam bridge that is straight and supports dual-lane rail traffic. The dimension of span is about 30 m length and 8 m width. The bottom of the span is about 6.7 m from the street below (see Fig. 5). Just before bridge commissioning in the Spring of 2010, some problems on the structure were detected as a result of visual inspection. This inspection revealed cracks on the central support of the deck (see Fig. 6). They concluded that these problems were related to two principal aspects:

- Unstable supports;
- Soil-structure interaction.

Worsening of the damaged condition was mitigated through the use of elastomeric supports under the deck as shown in Fig. 7. However, due to the damage, the bridge owner in cooperation with *Safibra s.r.o.*, a structural health monitoring company, decided to employ a bridge monitoring system to perform continuous monitoring to check in near real time the health of the structure, and at the same time to create a way to increase the degree of safety for the bridge's users. The *monitoring system* implemented is represented in the following picture (see Fig. 8). Two different kinds of sensors were employed. The first to detect the variation of displacement (see Fig. 9) and the second to detect the variation of temperature. In total, 10 displacement sensors and 4 temperature sensors were used.

Fig. 7 Elastomeric support inserts under the deck

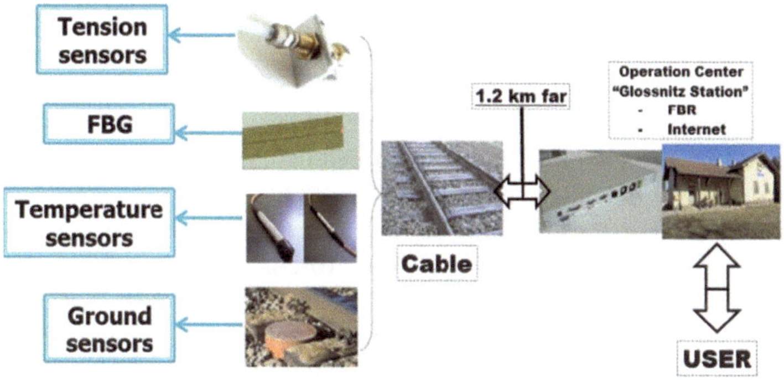

Fig. 8 Monitoring system implemented

Fig. 9 View of the sensor attached to the deck of the bridge

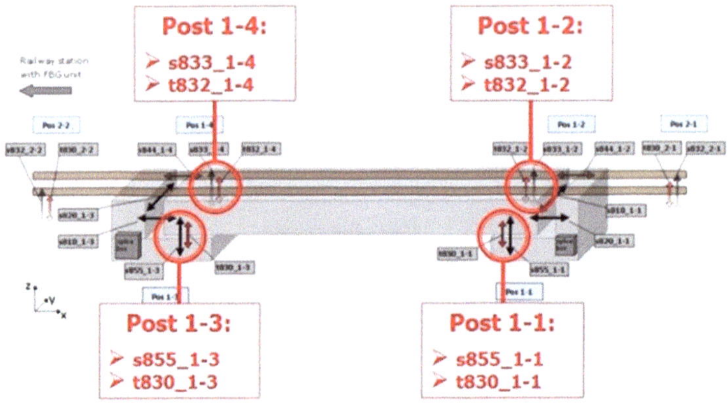

Fig. 10 Map of sensor positions

The sensors are located closer to the support where the cracks appeared. The labels and the positions of the sensors are represented in Fig. 10. Data are registered once per second, stored in a datalogger, and are then sent to a PC situated in an Operation Center where they are also uploaded in a Web page: enabled users can check or downloaded them from a FTP server.

In this study, four sensor pairs (displacement and temperature) are used and are the ones indicated as circled in Fig. 10. To conduct the analysis, data from June to November 2011 are used. For the data displacements, it is noted that relative displacement between two nodes 1 m far each from the other in the original configuration are recorded. The measurement of temperature is expressed in *Celsius* degrees. All the data are managed by *MATLab* and *Microsoft Excel* software. For both displacements and temperature, data representations are provided for seasonal trend; monthly trend and global trend (3D) [6–8]. An example is shown in Fig. 11. For the quantification of the temperature effect for each event (represented by a train crossing) the investigated data are train crossing time, frequency of the first spectrum peak, peak value, and temperature at the time of train crossing. These data can be obtained from periodograms as the one shown in Fig. 12. From these graphs we observed that the event doesn't influence the frequency variation, but the temperature is the main factor responsible for the variations (see Fig. 13).

After this analysis, a numerical (finite element) model of the bridge was developed using *SAP2000* software. *Shell* elements were used to define the mesh. In calibrating the numerical model, attention is focused on the node vertical displacements, rather than on other response parameters. As such, a model very "light" from the computational point of view was obtained. After defining the geometry of the deck, in order to represent the behavior of the bearing system and the bridge abutments, vertical and horizontal springs were used to fix the nodes of the mesh. For soil-structure interaction (ground behavior) Winkler's theory ($k_W = 6\,\text{kg/cm}^3$) was adopted. Regarding the behavior of the bearing device, a vertical stiffness of the springs of 8,000,000 kN/m was used whereas for a horizontal

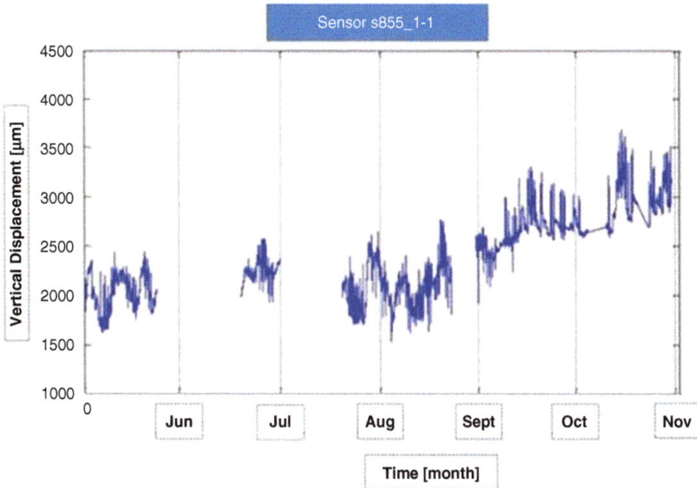

Fig. 11 Vertical displacement along the time monitoring measured by sensor s855-1-1

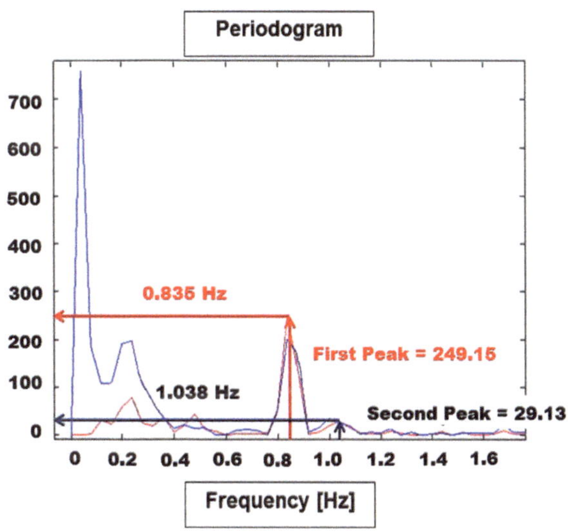

Fig. 12 Periodogram for event of June, 14. Time 15:30:46. Sensor s833-1-2

stiffness a value of 800,000 kN/m was used. This order of magnitude difference was used to represent the greater possibility of the elastomer to deform it-self under shearing loads. A schematic representation of the model is shown in Fig. 14. The model was calibrated using a particular set of displacements and temperature data, measured in field, and extracted on July, 7th 2011. Plotting the data output related to the vertical displacement of *node 20* (corresponding to the sensor s833-1-4) the plot is initially consistent, but delayed with regard to the real behavior measured

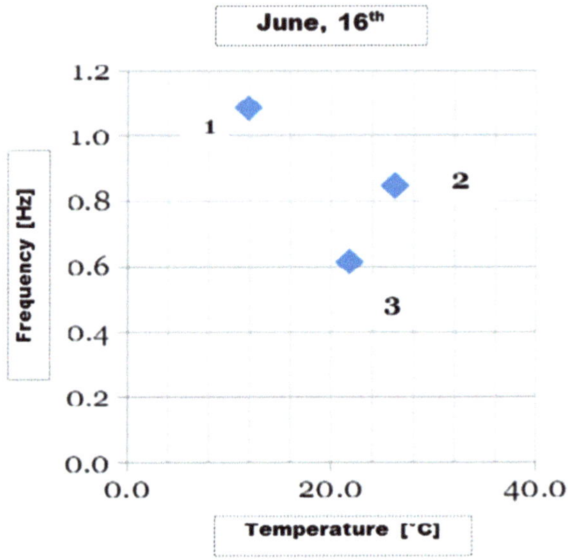

Fig. 13 First frequency peak – June, 16th – sensor s833-1-2 (1 = event from 04 to 06 am; 2 = event from 11 to 14; 3 = event from 15 to 18)

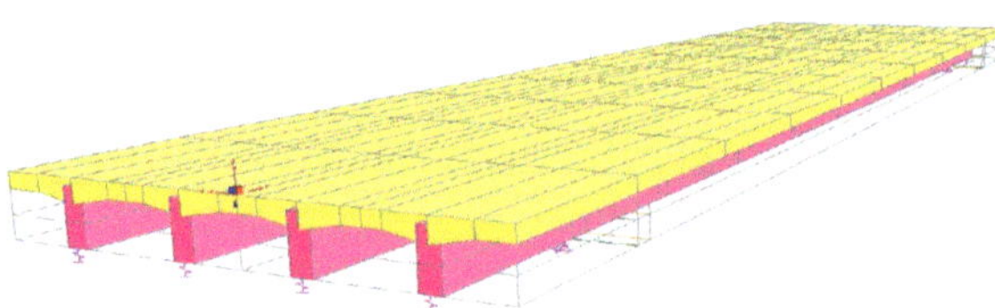

Fig. 14 View of 3D model using *SAP2000*

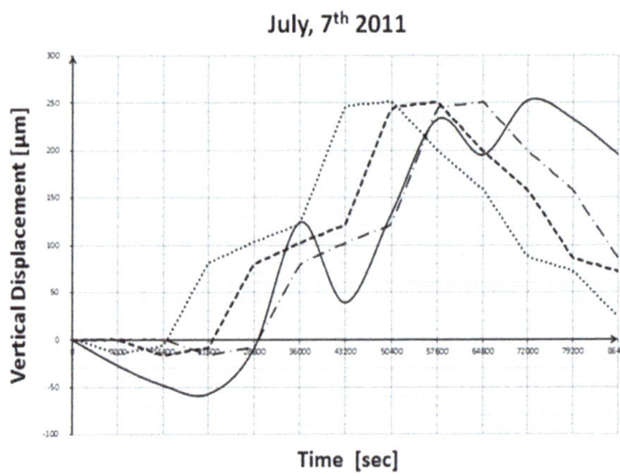

Fig. 15 Comparison vertical displacement versus time. *Solid line* represents the sensor s833-1-4 measurements. *Dotted line* the first attempt (no delay in function load). *Dashed line* a delay of 2 h. *Dashed-dotted line* the final assumption (4 h delay in function load)

by the sensor. This fact was related to the thermal inertia of the concrete. In order to compensate for this problem, and to take into account the thermal inertia of the material in a simple way, the temperature action was delayed by 4 h. In this way, the results of the model matched with the real behavior (see Fig. 15). In order to test

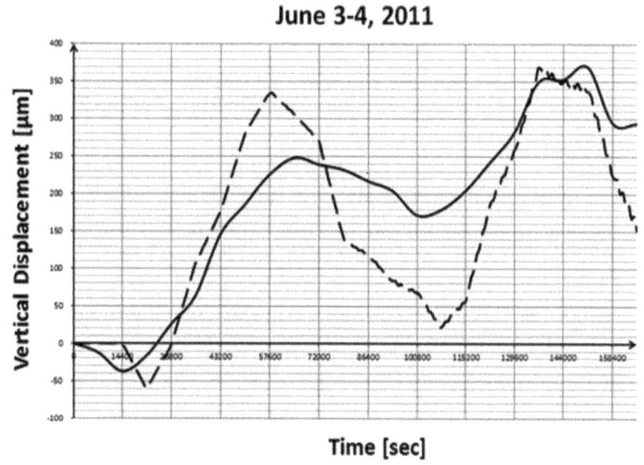

Fig. 16 Comparison vertical displacement versus time. *Solid line* represents the sensor s833-1-4 measurements. *Dashed line* the result of the numerical model (4 h delay in function load)

the goodness of the numerical model it applied to another data set covering June 3rd and 4th. Once again the temperature action was delayed by 4 h. The output result was very good as shown in Fig. 16.

2 Conclusions

In this work, the effect of the temperature variation in the behavior of the Austrian *ÖBB Brücke Großhaslau* bridge has been studied. As a first step, data collected by the sensors located along the bridge were analyzed in order to quantify how and how much temperature could be related to the structural behavior of the bridge. Next, a numerical model was built using the *SAP2000* software. A deliberate decision was taken to construct a simple model that targeted the behavior of interest in order to limit the computational operations. After calibration, the goodness of the model was tested using different sets of data from various sensor measurements. The "output" data were of good quality with respect to both computational time and accuracy (vertical displacement of the controls nodes). By analyzing the output data, it was also possible to understand, quantify, and account for the role of thermal inertia in the structural behavior. After this calibration, it was concluded that the model could be considered good for the prediction of the bridge's behavior under thermal load. As a next development for this study, if would be interesting to take into account other environmental parameters such as the humidity.

Acknowledgements The authors are grateful to the Czech company Safibra for making them the data available. The first author was supported by a research grant from the University of Pavia (FAR 2011).

References

1. Peeters, B., De Roeck, G.: One-year monitoring of the Z24-bridge: environmental effects versus damage events. Earthq. Eng. Struct. Dyn. **30**(2), 149–171 (2001)
2. Kullaa, J.: Eliminating environmental or operational influences in structural health monitoring using the missing data analysis. J. Intell. Mater. Syst. Struct. **20**(11), 1381–1390 (2009)
3. Kullaa, J.: Distinguishing between sensor fault, structural damage, and environmental or operational effects in structural health monitoring. Mech. Syst. Signal Process. **25**(8), 2976–2989 (2011)
4. Inaudi, D., Vurpillot, S.: Monitoring of concrete bridges with long-gage fiber optic sensors. J. Intell. Mater. Syst. Struct. **10**, 280–292 (1999)
5. Glisic, B., Inaudi, D.: Fiber Optic Method for Structural Health Monitoring. Wiley, Chichester (2007)
6. Moro, B.: Sistema Di Monitoraggio Per il Ponte ÖBB Brücke Großhaslau: effetti della temperatura. Master degree thesis, University of Pavia (2012)
7. Orlandoni, M.: Simulazione Numerica Dell'effetto Della Temperatura Sul Ponte ÖBB Brücke Großhaslau: effetti della temperatura, Master Degree Thesis, University of Pavia (2012)
8. www.safibra.cz

Controlled Passage Through Resonance for Two-Rotor Vibration Unit

A.L. Fradkov, D.A. Tomchin, and O.P. Tomchina

Abstract The problem of controlled passage through resonance zone for mechanical systems with several degrees of freedom is studied. Control algorithm design is based on speed-gradient method and estimate for the frequency of the slow motion near resonance (Blekhman frequency). The simulation results for two-rotor vibration units illustrating efficiency of the proposed algorithms and fractal dependence of the passage time on the initial conditions are presented.

1 Introduction

The usage of mechatronic elements sets up broad prospects in designing of modern vibration units. One of the main task for control of mechatronic vibration unit is passing through resonance zone at start up and speed up modes of vibroactuators, when range of operating modes belongs to a post-resonance zone. Such a problem occurs e.g. if the power of a motor is not sufficient for passing through resonance zone due to Sommerfeld effect deeply investigated by I. I. Blekhman.

Historically the first approach to the problem of passing through resonance zone was the so-called "double start" method due to V. V. Gortinskiy et al. [1]. The method is based on the insertion of time relay into motor control circuit for repeatedly switching on and off motor at pre-calculated time instants. Basically,

A.L. Fradkov (✉) · D.A. Tomchin
Institute of Problems in Mechanical Engineering, Bolshoy avenue 61, V.O., Saint-Petersburg 199178, Russia
e-mail: fradkov@mail.ru; kolobok1@yandex.ru

O.P. Tomchina
Saint-Petersburg State Polytechnical University, Polytechnicheskaya str. 29, Saint-Petersburg 195251, Russia
e-mail: otomchina@mail.ru

Alexander K. Belyaev et al. (eds.), *Mechanics and Model-Based Control of Advanced Engineering Systems*, DOI 10.1007/978-3-7091-1571-8_11,
© Springer-Verlag Wien 2014

this and other feedforward (nonfeedback) methods are characterized by difficulties in calculation of switching instants of a motor and sensitivity to inaccuracies of model and to interferences. A prospective approach to the problem is based on feedback control algorithms. Passing through resonance zone control algorithms of mechanical systems with feedback measurements were considered in [2–5]. However the early algorithms [2] did not have enough robustness under uncertain conditions and were hard to design.

For practical implementation it is important to develop reasonably simple passing through resonance zone control algorithms, which have such robustness property: keeping high quality of the controlled system (vibration unit) under variation of parameters and external conditions. A number of such algorithms based on the speed-gradient method, were suggested in works [3–5].

This work is dedicated to extension of the results of [3–5]. Problem statement for control of passage through resonance zone for mechanical systems with several degrees of freedom is adopted from [5]. The control algorithms based on the speed-gradient method for two-rotor vibration units with unbalanced rotors are described. The simulation results illustrating efficiency of the proposed algorithms and fractal dependence of the passage time on the initial conditions are presented.

2 Problem Statement and Approach to Solution

According to [5] the problem of passing through resonance is defined as design of control algorithm providing achievement of the prespecified energy level of the rotating subsystem under limited energy of the supporting body oscillations at a resonance frequency.

Approach to solution of the problem proposed in this paper, is based upon usage of the speed gradient algorithms design [5] and a motion separation into fast and slow components, which occurs near resonance zone. Quantitative analysis of slow "pendular" movements $\psi(t)$ was performed by I. I. Blekhman in [6] where the frequency of an "internal pendulum" was evaluated. It will be further called the "Blekhman frequency".

The idea of control algorithms described below, is in that slow motion $\psi(t)$ is being isolated and then "swinging" starts to obtain rise of energy of a rotating subsystem. To isolate slow motions, low-pass filter is inserted into energy control algorithms. Particularly, if slow component appears in oscillations of angular velocity of a rotor $\dot{\varphi}$, then the control algorithm proposed in [3] is used:

$$u = -\gamma \operatorname{sign}\big((H - H^*)\dot{\psi}\big)\,, \quad T_\psi \dot{\psi} = -\psi + \dot{\varphi}\,, \tag{1}$$

where $H = H(p, q)$ denotes the Hamiltonian (total energy of the system), ψ is the variable of a filter, that corresponds to slow motions of $\psi(t)$, T_ψ is the time constant of a filter. At low damping, slow motions also fade out slowly, what gives control algorithm an opportunity to create suitable conditions to pass through resonance zone, and after that "swinging" can be turned off and then algorithm switches to controlling with constant torque. For a proper work of a filter, it should suppress fast oscillations with frequency ω and pass slow oscillations with ω_B frequency, where ω_B is the Blekhman frequency. That is time constant of a filter T_ψ should be chosen from the inequality

$$T_\psi < 2\pi/\omega_B \,. \tag{2}$$

Algorithms of passing through resonance zone for the two-rotor vibration units are described and studied below.

3 Passing Through Resonance Control Algorithm of Two-Rotor Vibration Unit

Consider the problem of two-rotor vibration unit start-up (spin-up), the unit consists of two rotors, installed on the supporting body elastically connected with fixed basis. Assume that the system dynamics may be considered in the vertical plane (see Fig. 1). Then the equations of dynamics have the following form [7]:

$$\begin{aligned} &m_0\ddot{x}_c - \\ &- m\rho\Big\{\big[\sin(\varphi+\varphi_1)+\sin(\varphi+\varphi_2)\big]\ddot{\varphi}+\big[\cos(\varphi+\varphi_1)+\cos(\varphi+\varphi_2)\big]\dot{\varphi}^2+ \\ &+\sin(\varphi+\varphi_1)\ddot{\varphi}_1+\sin(\varphi+\varphi_2)\ddot{\varphi}_2+\cos(\varphi+\varphi_1)\dot{\varphi}_1^2+\cos(\varphi+\varphi_2)\dot{\varphi}_2^2+ \\ &+2\dot{\varphi}\big(\cos(\varphi+\varphi_1)\dot{\varphi}_1+\cos(\varphi+\varphi_2)\dot{\varphi}_2\big)\Big\}+\beta\dot{x}_c+2c_{01}x_c=0\,; \end{aligned} \tag{3}$$

$$\begin{aligned} &m_0\ddot{y}_c + \\ &+ m\rho\Big\{\big[\cos(\varphi+\varphi_1)+\cos(\varphi+\varphi_2)\big]\ddot{\varphi}-\big[\sin(\varphi+\varphi_1)+\sin(\varphi+\varphi_2)\big]\dot{\varphi}^2+ \\ &+\cos(\varphi+\varphi_1)\ddot{\varphi}_1+\cos(\varphi+\varphi_2)\ddot{\varphi}_2-\sin(\varphi+\varphi_1)\dot{\varphi}_1^2-\sin(\varphi+\varphi_2)\dot{\varphi}_2^2- \\ &-2\dot{\varphi}\big(\sin(\varphi+\varphi_1)\dot{\varphi}_1+\sin(\varphi+\varphi_2)\dot{\varphi}_2\big)\Big\}+\beta\dot{y}_c+2c_{02}y_c+m_0g=0\,; \end{aligned} \tag{4}$$

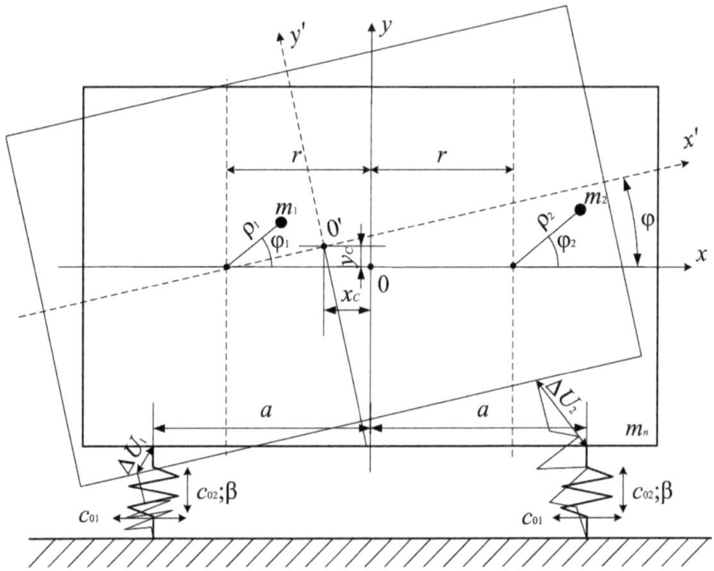

Fig. 1 Two-rotor vibration unit with horizontal supporting body

$$
\begin{aligned}
&-m\rho\Big\{\big[\sin(\varphi+\varphi_1)+\sin(\varphi+\varphi_2)\big]\ddot{x}_c-\big[\cos(\varphi+\varphi_1)+\cos(\varphi+\varphi_2)\big]\ddot{y}_c\Big\}+\\
&\quad+\big[J+J_1+J_2-2m\rho r(\cos\varphi_1-\cos\varphi_2)\big]\ddot{\varphi}+\\
&\quad+\big[J_1-mr\rho\cos\varphi_1\big]\ddot{\varphi}_1+\big[J_2+mr\rho\cos\varphi_2\big]\ddot{\varphi}_2+\\
&\quad+mr\rho\Big\{2\sin\varphi_1\,\dot{\varphi}\dot{\varphi}_1-2\sin\varphi_2\,\dot{\varphi}\dot{\varphi}_2+\sin\varphi_1\,\dot{\varphi}_1^2-\sin\varphi_2\,\dot{\varphi}_2^2\Big\}+\\
&\quad+m\rho g\big[\cos(\varphi+\varphi_1)+\cos(\varphi+\varphi_2)\big]+\beta\dot{\varphi}+c_{03}\varphi=0\,;
\end{aligned}
\tag{5}
$$

$$
\begin{aligned}
&J_1\ddot{\varphi}_1-m\rho\big[\sin(\varphi+\varphi_1)\ddot{x}_c-\cos(\varphi+\varphi_1)\ddot{y}_c\big]+\big[J_1-mr\rho\cos\varphi_1\big]\ddot{\varphi}-\\
&\quad-mr\rho\sin\varphi_1\,\dot{\varphi}^2+mg\rho\cos(\varphi+\varphi_1)+k_c\dot{\varphi}_1=M_1\,;\\
&J_2\ddot{\varphi}_2-m\rho\big[\sin(\varphi+\varphi_2)\ddot{x}_c-\cos(\varphi+\varphi_2)\ddot{y}_c\big]+\big[J_2+mr\rho\cos\varphi_2\big]\ddot{\varphi}+\\
&\quad+mr\rho\sin\varphi_2\,\dot{\varphi}^2+mg\rho\cos(\varphi+\varphi_2)+k_c\dot{\varphi}_2=M_2\,;
\end{aligned}
\tag{6}
$$

Here φ, φ_1, φ_2 are angle of the support and rotation angles of the rotors, respectively, measured from the horizontal position, x_c, y_c are the horizontal and vertical displacement of the supporting body from the equilibrium position, $m_i = m$, $i = 1, 2$ and m_n are the masses of the rotors and supporting body, J_1, J_2 are the inertia moments of the rotors, $\rho_i = \rho$, $i = 1, 2$ are the rotor eccentricities, c_{01}, c_{02} are the horizontal and vertical spring stiffness, g is the gravity acceleration, m_0

is the total mass of the unit, $m_0 = 2m + m_n$, β is the damping coefficient, k_c the friction coefficient in the bearings, $M_i = u_i(t)$ are the motor torques (controlling variables). It is assumed that rotor shafts are orthogonal to the motion of the support.

At the low levels of constant control action $u_i(t) \equiv (-1)^i M_0, i = 1, 2$ at the near resonance zone the rotor angle is "captured", while increase of the control torque leads to passage through resonance zone towards the desired angular velocity. The simulation results for the system (3)–(6) are shown on Fig. 2 with basic system parameters: $J_i = 0.014\,[\mathrm{kg\ m^2}]$, $m = 1.5\,[\mathrm{kg}]$, $m_n = 9\,[\mathrm{kg}]$, $\rho = 0.04\,[\mathrm{m}]$, $k_c = 0.01\,[\mathrm{J/s}]$, $\beta = 5\,[\mathrm{kg/s}]$, $c_{02} = 5{,}300\,[\mathrm{N/m}]$, $c_{01} = 1{,}300\,[\mathrm{N/m}]$ and constant torque $M_0 = 0.65\,[\mathrm{N\ m}]$ (internal curves, "capture") and $M_0 = 0.66\,[\mathrm{N\ m}]$ (external curves, passage).

Synthesis of the control algorithm $u = \mathbf{U}(z)$ is needed for acceleration of the unbalanced rotors, before the system passes through the resonance zone, where $z = [x, \dot{x}, y, \dot{y}, \varphi, \dot{\varphi}, \varphi_1, \dot{\varphi}_1, \varphi_2, \dot{\varphi}_2]^T$ is the state vector of the system. It is assumed that the level of control action is limited and does not allow system to pass through the resonance zone using constant action.

As seen in Fig. 2 the rotational motions separate into fast and slow ones i.e. the proposed approach applies. The following modification of the control algorithm (1) is proposed for passing through resonance zone:

$$\begin{cases} u_i = \begin{cases} (-1)^i M_o\,, & if \quad \gamma = 1\,, \\ (-1)^i M_o\,, & if \quad \gamma = 0 \ \& \ (H - H^*)(\dot{\varphi}_i - \psi_i) > 0\,, \\ 0\,, & else\,, \end{cases} \\ T_\psi \dot{\psi}_i = -\psi_i + \dot{\varphi}_i\,, \quad i = 1, 2 \end{cases} \tag{7}$$

where $H = T + \Pi$, $\psi_i(t)$ are the variables of the filters, $T_\psi > 0$, $T_\psi = \text{const}$ and H^* are the parameters of the algorithm,

$$\gamma(t) = \max_{0 \leqslant \tau < t} \operatorname{sgn}\big(H(\tau) - H^*\big)$$

where $\operatorname{sgn}[z] = 1$ with $z > 0$, $\operatorname{sgn}[z] = 0$ with $z \leqslant 0$. Time constant of the filters T_ψ should satisfy the relation (2). At the same time, the values that are too big lead to decrease in average power of the control signal and to slow down the algorithm.

Efficiency of the proposed control algorithm was studied in the MATLAB environment. The relative simulation error does not exceed 5 %. Calculations were made with the same values of the basic parameters of the system. Then the lowest constant control action providing passing through the resonance is $0.66\,[\mathrm{N\ m}]$.

In the simulation with the proposed algorithm (7), the minimum value of a control torque M_0 to pass through resonance, was calculated.

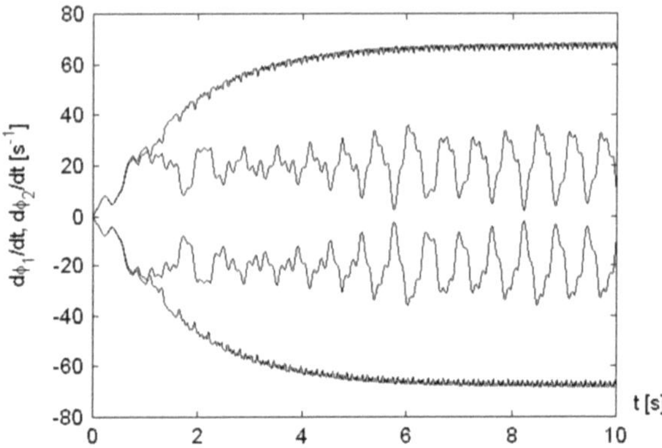

Fig. 2 Constant control action, $u_i(t) \equiv (-1)^i M_0$, $M_0 = 0.65$ [N m] (*internal curves*, "capture") and $M_0 = 0.66$ [N m] (*external curves*, passing)

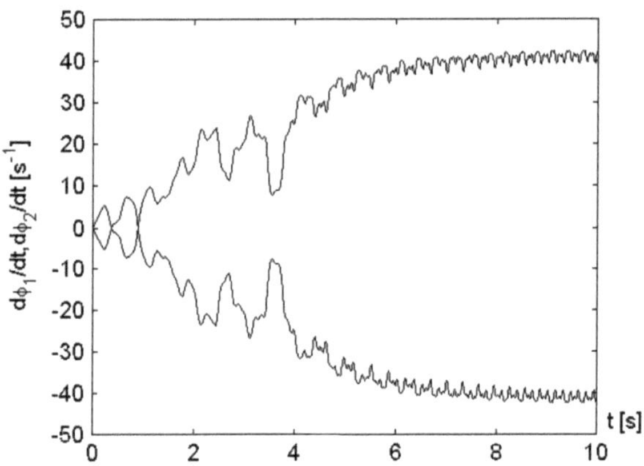

Fig. 3 Passing through resonance with algorithm (7). $M_0 = 0.42$ [N m], $T_\psi = 0.35$ [s]

Simulation results have shown that the minimum value is $M_0 = 0.42$ [N m] (with the precision of 0.01 [N m]), see Fig. 3. In comparison to constant control action, proposed algorithm may reduce the control action level by 1.5 times.

Study of the algorithm with different asymmetric initial positions of rotors [4] has confirmed its efficiency. The graphs are similar to the previous ones, except for the horizontal support deviation, which differs in the initial area. Figure 4 shows the minimum M_0 value change depending on T_ψ in the algorithm (7).

Thus, using the proposed control algorithm, the level of control action required for passing through resonance can be significantly reduced. The algorithm only has two tunable parameters and is simple to use, despite the complicated behavior of the system. The closed control system is low sensitive to an asymmetry of the initial conditions of the unit. However, it is sensitive to the initial conditions themselves. Even more, the dependence of passage time on initial conditions has a fractal shape, see Fig. 5.

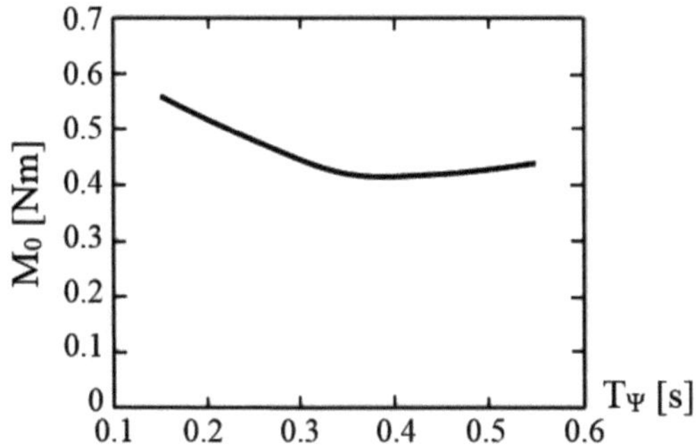

Fig. 4 Algorithm efficiency as a function of the time constant T_ψ

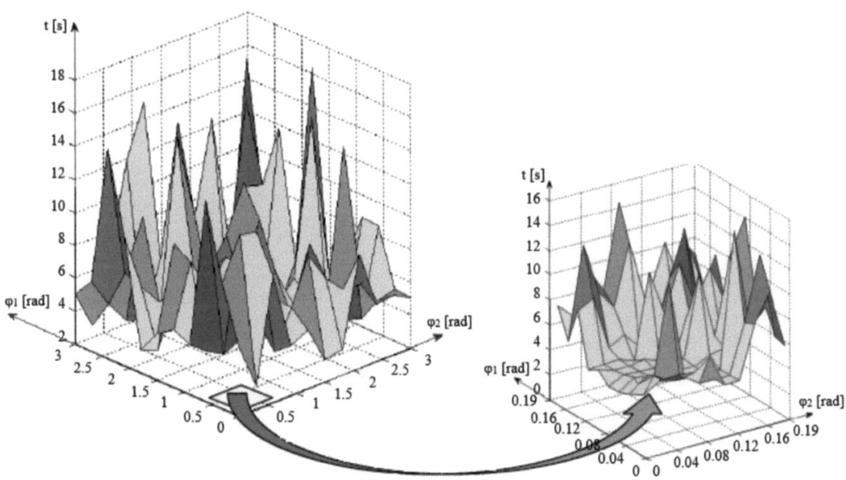

Fig. 5 Time of passing through resonance for different initial conditions

4 Conclusion

This paper systematically shows the approach for synthesis of simple control algorithms of passing through resonance zone in vibration units, proposed in [3–5] and based on "swinging" of slow oscillations of an angular velocity of the rotors. Algorithms are based on the speed-gradient method and the filtering the fast oscillations. The efficiency of the algorithms and fractal dependence of the passage time on the initial conditions is demonstrated by computer simulation.

Acknowledgements The work was supported by the research program 10 (Mechatronics and Robotics) of Russian Academy of Sciences, Russian Foundation for Basic Research (project 11-08-01218) and Russian Federal Program "Cadres" (agreements 8846, 8855).

References

1. Gortinskii, V.V., Savin, A.D., Demskii, A.B., Boriskin, M.A., Alabin, E.A.: A technique for reducing resonance amplitudes during start of vibration machines. Patent of USSR No. 255760, 28 Oct 1969. Bull. No. 33. (in Russian)
2. Malinin, L.N., Pervozvansky, A.A.: Optimization of passage an unbalanced rotor through critical speed. Mashinovedenie (4), 36–41 (1983). (in Russian)
3. Tomchina, O.P.: Passing through resonances in vibratory actuators by speedgradient control and averaging. In: Proceedings of International Conference on "Control of Oscillations and Chaos", St. Petersburg, vol. 1, pp. 138–141. IEEE (1997)
4. Tomchin, D.A., Fradkov, A.L.: Control of passage through a resonance area during the start of a two-rotor vibration machine. J. Mach. Manuf. Reliab. **36**(4), 380–385 (2007)
5. Fradkov, A.L., Tomchina, O.P., Tomchin, D.A.: Controlled passage through resonance in mechanical systems. J. Sound Vib. **330**(6), 1065–1073 (2011)
6. Blekhman, I.I., Indeitsev, D.A., Fradkov, A.L.: Slow motions in systems with inertial excitation of vibrations. J. Mach. Manuf. Reliab. **37**(1), 21–27 (2008)
7. Fradkov, A.L., Tomchina, O.P., Galitskaya, V.A., Gorlatov, D.V.: Integrodifferentiating speed-gradient algorithms for multiple synchronization of vibration units. Nauchno-tekhnicheskii Vestnik of ITMO, St.Petersburg, No. 1, pp. 30–37 (2013)

High-Strength Network Structure of Jungle-Gym Type Polyimide Gels Studied with Scanning Microscopic Light Scattering

Hidemitsu Furukawa, Noriko Tan, Yosuke Watanabe, Jin Gong, M. Hasnat Kabir, Ruri Hidema, Yoshiharu Miyashita, Kazuyuki Horie, and Rikio Yokota

Abstract Latest high-strength gels overcome brittleness due to the inhomogeneities built in their network structure. However, the inhomogeneities still prevent precise characterization of their network structures by scattering methods. A new concept is to take advantage of the ensemble-averaged structure characterization with scanning microscopic light scattering (SMILS), in order to study the network structure and properties of inhomogeneous high-strength gels nondestructively in wide spatio-temporal ranges. In this study, two kinds of the jungle-gym type polyimide gels that have semi-rigid main-chains or rigid main-chains were synthesized in varying the preparing concentration and studied with SMILS. The optimal concentration of polyimide achieved ten times higher Young modulus than before.

H. Furukawa (✉) · Y. Watanabe · J. Gong · M.H. Kabir
Yamagata University, Yonezawa, Yamagata, 992-8510 Japan
e-mail: furukawa@yz.yamagata-u.ac.jp

N. Tan
Tokyo University of Agriculture and Technology, Koganei, Tokyo, 184-8588 Japan

R. Hidema
Kobe University, Kobe, Hyogo, 657-8501 Japan

Y. Miyashita
Ibaraki National College of Technology, Hitachinaka, Ibaraki, 312-8508 Japan

K. Horie
Japan Synchrotron Radiation Research Institute, Sayo, Hyogo, 679-5198 Japan

R. Yokota
Institute of Space and Astronautical Science, Sagamihara, Kanagawa, 229-8510 Japan

Alexander K. Belyaev et al. (eds.), *Mechanics and Model-Based Control of Advanced Engineering Systems*, DOI 10.1007/978-3-7091-1571-8_12,
© Springer-Verlag Wien 2014

1 Introduction

Latest high-strength gels overcome weakness due to incomplete network structure, by different approaches to control the network structure. Okumura and Ito [1] synthesized slide-ring gels (SR gels) having figure-eight cross-links, which are able to slide along polymer chains and effectively relax their tension. Haraguchi and Takeshita [2] synthesized nanocomposite gels (NC gels) containing dispersed sheet-shape clay as a multifunctional crosslinker, which shows a resistance to extension as large as 1,000 %. Gong et al. [3] synthesized double network gels (DN gels), containing 90 % water, which exhibit a high Young's modulus (0.1–0.3 MPa) and large fracture strengths of more than 20 MPa. We also synthesized inter-crosslinking network gels (ICN gels), containing 97 % water and exhibiting more than 60 % increase in the ductility in comparison with normal gels [4, 5]. For each of these gels, the unique mechanisms for the excellent mechanical properties have been studied intensely. Differently from these gels, there is another approach to realize high-strength gels. That is the pursuit of homogeneous and complete network structure. He et al. [6], some of the present authors, developed a novel method of synthesizing polyimide gels with jungle-gym type homogeneous network structure, which exhibit a higher Young's modulus (35–40 MPa). Recently, Sakai et al. [7], synthesized a tetra-PEG gel having ideally homogeneous network structure form tetrahedron-like macromonomer, whose maximum breaking stress is typically 10 MPa. For these homogeneous gels, it is expected that the homogeneity of net-work structure should be important to achieve the highest strength. Thus the relation between the homogenous structure and the strength should be studied. Light scattering is a powerful method to analyze nanometer-scale structure of polymeric system. However, somewhat inhomogeneities still exist even in the homogeneous gels and prevent precise characterization of their network structures by scattering methods. A new concept is to take advantage of the ensemble-averaged structure characterization with scanning microscopic light scattering (SMILS), in order to study the structure and properties of inhomogeneous high-strength gels nondestructively in wide spatio-temporal ranges [8]. Previously, the gelation process of the jungle-gym type polyimide gels was studied with the SMILS [9]. In this study, two kinds of the jungle-gym type polyimide gels that have semi-rigid main-chains or rigid main-chains were synthesized in varying the preparing concentration and studied with SMILS. We found that the optimal concentration of polyimide achieved a quite higher Young modulus (320 MPa).

2 Experimental

2.1 Synthesis of Jungle-Gym Type Polyimide Gels

Materials: Pyromellitic dianhydride (PMDA) was purified by recrystallization from acetic anhydride, and dried at 120 °C in vacuum for 24 h. p-phenylene diamine

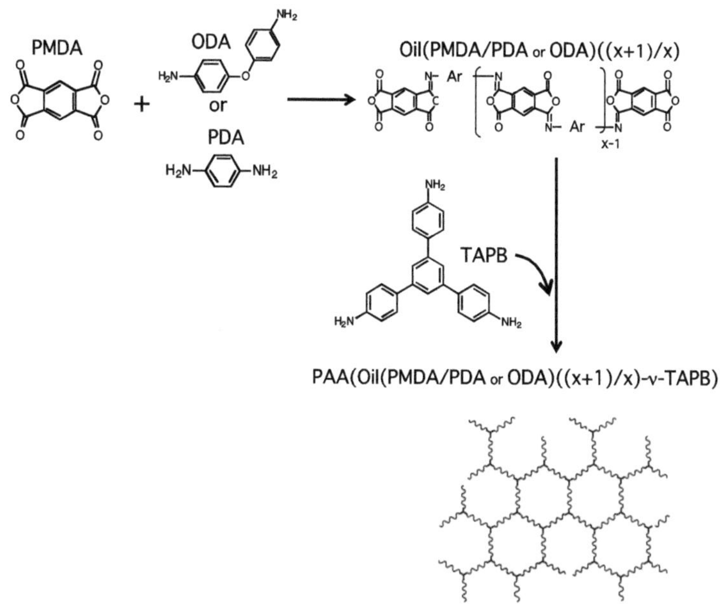

Fig. 1 Preparation scheme of jungle-gym type gels

(PDA) and 4,4-diaminodiphenyl ether (ODA) were purified by recrystallization from ethanol, and dried at 50 °C in vacuum for 24 h. N-methyl-2-pyrrolidone (NMP) and N,N'-dimethylacetoamide (DMAc) and stored with molecular sieves. N,N'-dicyclohexylcarbodiimide (DCC) and the other chemicals were used without further purification. Oligoisoimide macromonomers: The scheme is shown in Fig. 1. Here, the number of repeating unit (the polymerization index) of the macromonomer was controlled by the molar ratio between PMDA and PDA in preparation. If one sets the ratio as (x + 1): x, then the number of repeating unit becomes 2x + 1. Here, we prepared the macromonomer at x = 1, 3, and 18, thus the number became 3, 7, and 37. At first, PMDA was dissolved into NMP, and PDA was added slowly and NMP was added to prepare 10 wt%-solid reaction solution. Immediately, the in-crease of viscosity in the reaction solution was observed. However, the viscosity was decreased after stirring at room temperature for 24 h. The reaction mixture was diluted with 0.1 M LiCl/NMP to prepare 1 wt% solid solution. After stirring for 1 h, 5 wt% DCC/NMP solution was slowly added into the reaction mixture to make the ratio of DCC/PMDA to be 2, and stirred for 24 h. The solution color was changed from light brown to red brown, and became turbid. The solution was poured into 2-propanol. A dark yellow precipitate was collected by vacuum filtration, washed 2-propanol, and finally washed with benzene. The final slurry of benzene and product was freeze-dried and additionally dried at 40 °C in vacuum. This procedure yielded a dark yellow powder of a oligoisimide, OiI(PMDA/PDA)((x + 1)/x). Also, a bright yellow powder of OiI(PMDA/PDA)((x + 1)/x) was similarly synthesized with a solvent DMAc in-stead of NMP. The reaction was checked by FT-IR.

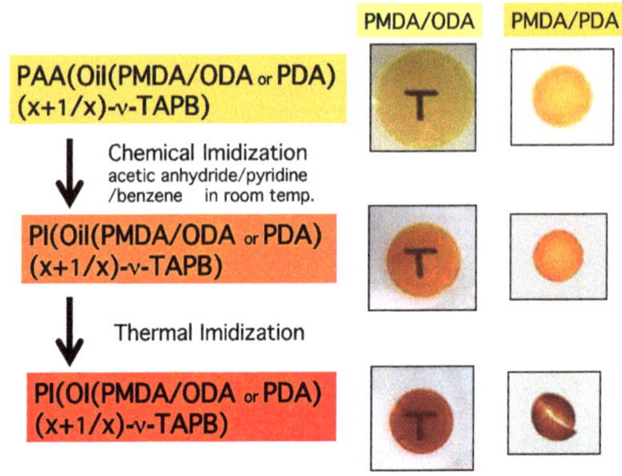

Fig. 2 Appearances of prepared gels

Polyimide gels: In the previous paper [6], a new rigid and symmetric trifunctional amine, 1,3,5-tris(4-aminophenyl)benzene (TAPB) (Fig. 1), was synthesized. In the present study, the TAPB was also used as an end-crosslinker, to prepare jungle-gym type polyimde gels. In order to synthesize appropriate concentration, for example a wt%-solid concentration of polyimide gels in DMAc, the DMAc solution of a wt% oligoisoimide was prepared. Then, the DMAc solution of a wt% TAPB was added to the solution of oligoisoimide with a stoichiometric mole ratio (1:1) between the amino group of oligoisoimide and the terminal acid anhydride group of TAPB. The solution gelled after few minutes in room temperature, and then transparent as-prepared poly(amide acid) (PAA) gels, PAA(OiI(PMDA/PDA)((x + 1)/x)-ν-TAPB) and PAA(OiI(PMDA/ODA)((x + 1)/x)-ν-TAPB), were obtained. The colors of PMDA/PDA and PMDA/ODA gels were light brown and yellow, respectively, as shown in Fig. 2. The swollen PAA gels were dipped for 30 min in chemical imidization solution, which is a mixture of acetic anhydride, pyridine and benzene with 1:1:3 in volume ratio, and then transparent and orange colored gels, PI(OiI(PMDA/PDA)((x + 1)/x)-ν-TAPB) and PI(OiI(PMDA/ODA)((x + 1)/x)-ν-TAPB), were obtained. These gels were repeatedly (about 10 times) washed with a large amount of DMAc, and then the gels were gradually heated toward 200 °C for 4 h. Then transparent polyimide gels, PI(OI(PMDA/PDA)((x + 1)/x)-ν-TAPB) and PI(OiI(PMDA/ODA)((x + 1)/x)-ν-TAPB), were obtained. The colors of these gels become darker than before heating. The reaction was checked by FT-IR, although the results are not shown here.

2.2 Scanning Microscopic Light Scattering (SMILS)

Dynamic light scattering measurement was preformed at 30 °C in the scattering angle range of $\theta = 30°–125°$ with a SMILS apparatus [8]. With this apparatus (its schematic is shown in Fig. 3), many measurements for small scattering volume

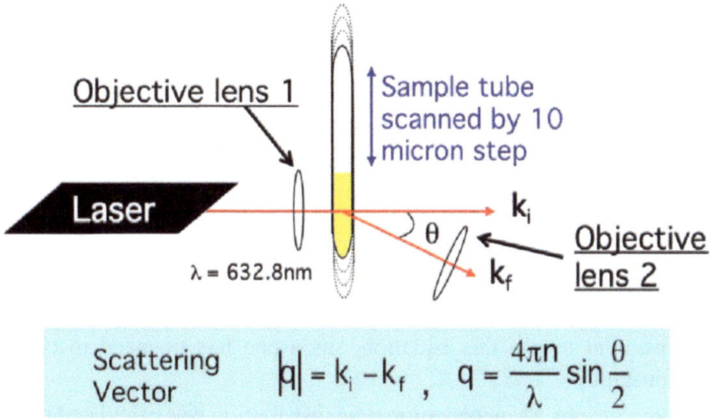

Fig. 3 Schematic of the scattering microscopic light scattering system

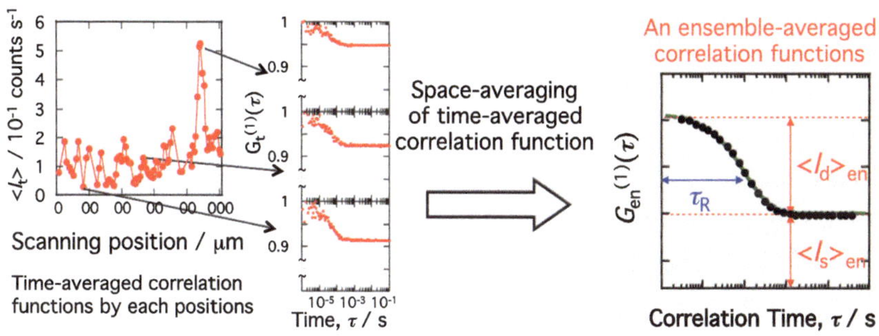

Fig. 4 Scheme to obtain an ensemble-averaged correlation function with the SMILS

can be successively performed at different positions in a sample. Hence, one can rigorously obtain a time- and space-averaged correlation function, i.e., an ensemble-averaged correlation function to overcome the non-ergodicity of inhomogeneous media, as shown in Fig. 4. The photon detection was carried out with a homodyne technique. For each sample, the time-averaged homodyne correlation functions $g_t^{(2)}(\mathbf{q}, \tau)$ were determined at over a 100 points, where $\mathbf{q}$ is scattering vector and τ is correlation time. Then the ensemble-averaged heterodyne correlation function $g_{en}^{(1)}(\mathbf{q}, \tau)$ was calculated by using the following equation [10]

$$g_{en}^{(1)}(\mathbf{q}, \tau) = \frac{\left\langle \langle I \rangle_t \gamma^{-1} \sqrt{1 + g_t^{(2)}(\mathbf{q}, \tau) - g_t^{(2)}(\mathbf{q}, 0)} \right\rangle_{sp}}{\langle I \rangle_{en}}. \qquad (1)$$

where I is the scattering intensity and γ is an apparatus constant. In the present work, since we are interested mainly in the dynamics of fluctuation in gels, we calculated the dynamic components of $g_{en}^{(1)}(\mathbf{q},\tau)$ as

$$\Delta g_{en}^{(1)}(\mathbf{q},\tau) = \frac{g_{en}^{(1)}(\mathbf{q},\tau) - g_{en}^{(1)}(\mathbf{q},\infty)}{g_{en}^{(1)}(\mathbf{q},0) - g_{en}^{(1)}(\mathbf{q},\infty)}$$
$$= \frac{\left\langle \langle I \rangle_t \sqrt{1 + g_t^{(2)}(\mathbf{q},\tau) - g_t^{(2)}(\mathbf{q},0)} \right\rangle_{sp} - \left\langle \langle I \rangle_t \sqrt{2 - g_t^{(2)}(\mathbf{q},0)} \right\rangle_{sp}}{\langle I \rangle_{en} - \left\langle \langle I \rangle_t \sqrt{2 - g_t^{(2)}(\mathbf{q},0)} \right\rangle} \tag{2}$$

It is convenient to use this equation, since one has no need to consider the apparatus constant γ.

To analyze $\Delta g_{en}^{(1)}(\mathbf{q},\tau)$, a relaxation time distribution was calculated numerically with inverse Laplace transform. In general, a monotonically decreasing correlation function can be expressed by the superposition of exponential functions as [8]

$$\Delta g_{en}^{(1)}(\tau) = N \sum_{i=1}^{n} P_{en}\, exp\left(-\tau/\tau_{R,i}\right) \tag{3}$$

where $\tau_{R,i} = \tau_{R,min} \times \left(\tau_{R,max}/\tau_{R,min}\right)^{(i-1)/n}$ $(i = 1, 2, \cdots, n)$ and is a normalized factor defined as $N = \left(1/n\right) log(\tau_{R,max}/\tau_{R,min})$. Thus the nonlinear fitting with Eq. 3 to $\Delta g_{en}^{(1)}(\tau)$ was performed where the analytical condition was set as $log(\tau_{R,min}) = -6$, $log(\tau_{R,min}) = 1$, and $n = 51$.

3 Results and Discussion

In order to characterize the network structure through the analysis of the dynamic fluctuation, the ensemble-averaged correlation function $\Delta g_{en}^{(1)}(\tau)$ was obtained and then the relaxation time distribution $P_{en}(\mathbf{q},\tau_R)$ was determined (not shown here). It was found that one main relaxation peak for each $P_{en}(\mathbf{q},\tau_R)$, which can be assigned to cooperative diffusing (Brownian) motion of nanometer-scale network structure. It is often called gel mode. We tried to analyze the inhomogeneities of the network structure by fitting the following distribution function to $P_{en}(\mathbf{q},\tau_R)$:

$$P_{en}(\mathbf{q},\tau_R) = N\, exp\left[-\frac{(log\tau_R - \langle log\tau_R \rangle)^2}{2\sigma^2}\right] \tag{4}$$

Then we obtained the relaxation time τ_R and determined the cooperative diffusion coefficient D_{coop}, by using the theoretical formula $\tau_R^{-1} = \mathbf{q}^2\, D_{coop}$ [8]. Then, D_{coop} was plotted as a function of the preparing concentration of PAA gels, as shown in Fig. 5.

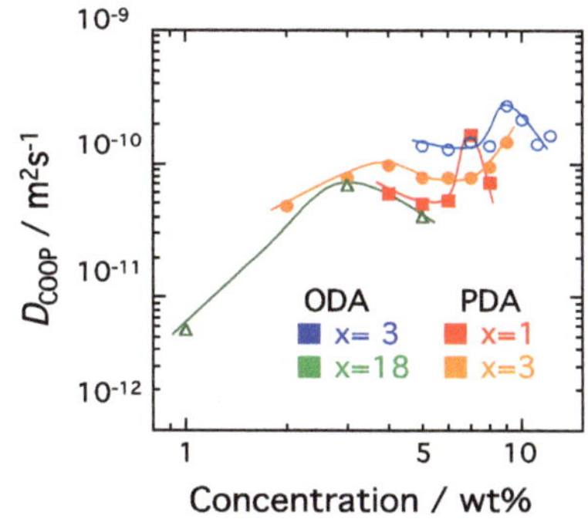

Fig. 5 Cooperative diffusion coefficient of the PAA gels as a function of the preparing concentration

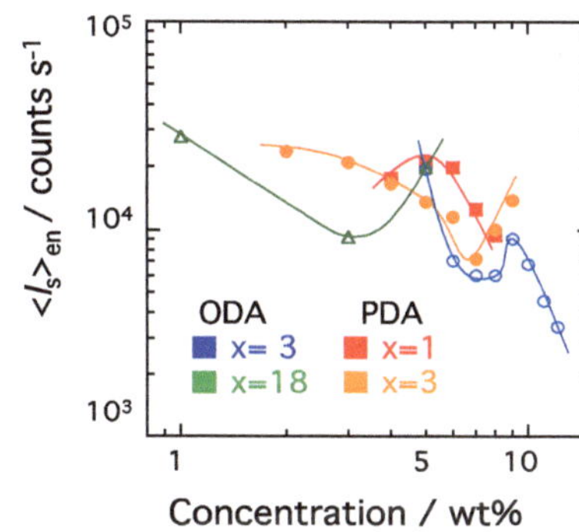

Fig. 6 Static component of ensemble-averaged intensity of the PAA gels as a function of the preparing concentration

We found that D_{coop} has a maximum value and it possibly corresponds to the smallest mesh size due to its homogeneous network structure. For rigid PMDA/PDA gels, 7 and 4 wt% may be optimal concentrations when x = 1 and 3. For semi-rigid PMDA/ODA gels, 9, 7, and 3 wt% may be optimal concentrations when x = 3, 7, and 18 (the date of x = 7 is not shown here). Based on the Stokes-Einstein formula $D_{coop} = k_B T / 6\pi \eta \xi$, the mesh size ξ was estimated as 0.9 nm (x = 3, 9 wt%), 1.5 nm (x = 7, 7 wt%) and 3.5 nm (x = 18, 3 wt%). If we compare these results at x = 3, the optimal concentration for the rigid gels 4 wt% is lower than for the semi-rigid gels 9 wt%. It implies that the main chain of the rigid gels is extended and forms the jungle-gym type network structure in the lower concentration. On the other hand, the main chain of the semi-rigid gels is little shrunken and forms the homogeneous network in the higher concentration. We found that the mesh size determined form $P_{en}(\mathbf{q}, \tau_R)$ and the static scattering intensity also depend on the preparing concentration and indicate the optimal concentration similarly. We also observed the scattering intensities as a function of the preparing concentration of PAA gels, as shown in Fig. 6. We consider the minimum of the static component of the intensity corresponds to the most homogeneous structure. The static intensity shows the similar behavior to the diffusion coefficient, however the detail behavior seems complex and should be studied further in future. Finally, the elastic properties of the polyimide gels swollen in DMAc were measured by Reometrics Solids Analyzer (RSAII, Reometrics Co.) at 0.16–16 Hz and at 20 °C with parallel plates and compress mode. We prepared the PMDA/ODA polyimede gels at an optimal concentration x = 7 and 7 wt% and the samples was cut in the size of 3.2 mm in diameter and 1.8 mm in height. The optimal polyimide gels showed high Young moduli. The modulus of as-prepared poly(amide acid) PAA(OiI(PMDA/ODA)-ν-TAPB) gels, chemically imidized PI(OiI(PMDA/ODA)-ν-TAPB) and thermally full-imidized PI(OI(PMDA/ODA)-ν-TAPB) was 0.32, 13, and 320 MPa, respectively. Comparing to the previous modulus observed for the thermally imidized gels (0.1–35 MPa) [6], it becomes about ten times higher than before. It implies that the modulus of the polyimide gels strongly depend on the preparing concentration and the SMILS is convenient tool to find the optimal preparing concentration nondestructively.

4 Conclusion

Transparent jungle-gym type polyimide gels were developed by using both oligo-isoimide macromonomers and two-step (chemical and thermal) imidizations. The optimum concentration, where as-prepared poly(amide acid) gels have the most homogeneous and jungle-gym type structure, was determined with SMILS.

Acknowledgements This study was supported in part by the Industrial Technology Research Grant Program in 2009 (Project ID: 09A25003a) from the New Energy and Industrial Technology Development Organization (NEDO) of Japan and the Grant-in-Aid for Scientific Research(B) (Project No.: 24360319) from the Japan Society for the Promotion of Science (JSPS).

References

1. Okumura, Y., Ito, K.: The polyrotaxane gel: a topological gel by figure-of-eight cross-links. Adv. Mater. **13**, 485–487 (2001)
2. Haraguchi, K., Takeshita, T.: Nanocomposite hydrogels: a unique organic-inorganic network structure with extraordinary mechanical, optical, and swelling/de-swelling properties. Adv. Mater. **16**, 1120–1124 (2002)
3. Gong, J.P., Katsuyama, Y., Kurokawa, T., Osada, Y.: Double-network hydrogels with extremely high mechanical strength. Adv. Mater. **15**, 1155–1158 (2003)
4. Takada, G., Hidema, R., Furukawa, H.: Ultrahigh ductile gels developed by inter cross-linking network (ICN). J. Solid Mech. Mater. Eng. **6**(2), 169–177 (2012)
5. Takada, G., Hidema, R., Furukawa, H.: Ultrahigh ductile gels having inter-crosslinking network (ICN) structure. e-J. Surf. Sci. Nanotechnol. **10**, 346–350 (2012)
6. He, J., Machida, S., Kishi, H., Horie, K., Furukawa, H., Yokota, R.: Preparation of novel, high-modulus, swollen- or jungle-gym-type polyimide gels end-crosslinked with 1,3,5-tris (4-aminophenyl)benzene. J. Polym. Sci. A Polym. Chem. Ed. **40**, 2501–2512 (2002)
7. Sakai, T., Matsunaga, T., Yamamoto, Y., Ito, C., Yoshida, R., Suzuki, S., Sasaki, N., Shibayama, M., Chung, U.-I.: Design and fabrication of a high-strength hydrogel with ideally homogeneous network structure from tetrahedron-like macromonomers. Macromolecules **41**, 5379–5384 (2008)

8. Furukawa, H., Horie, K., Nozaki, R., Okada, M.: Swelling-induced modulation of static and dynamic fluctuations in polyacrylamide gels observed by scanning microscopic light scattering. Phys. Rev. E **68**, 031406-1-14 (2003)
9. Furukawa, H., Kobayashi, M., Miyashita, Y., Horie, K.: End-crosslinking gelation of poly(amide acid) gels studied with scanning microscopic light scattering. High Perform. Polym. **18**, 837–847 (2006)
10. Furukawa, H., Hirotsu, S.: Dynamic light scattering from static and dynamic fluctuations in inhomogeneous media. J. Phys. Soc. Jpn. **71**, 2873–2880 (2002)

Extension of the Body Force Analogy to Generalized Thermoelasticity

Toshio Furukawa

Abstract The classical body force analogy for static problems of thermoelasticity is extended towards generalized thermoelastic problems including Lord-Shulman's and Green-Lindsay's theories. We consider two dynamic problems, namely a thermal problem without body forces, but with a given distribution of transient sources of heat, and a force problem without sources of heat, but with body forces. Both problems are treated within the two different theories of generalized thermoelasticity. We show that, given suitable boundary and initial conditions, a distribution of body forces can be constructed, such that the dynamic displacements in both problems become equal. This analogy is checked by means of illustrative analytical examples. We also discuss the relations between the stresses and the temperature in both problems.

1 Introduction

In its classical form, the body force analogy can be stated as follows: Consider the static deformation of an isotropic linear thermoelastic body under the action of a given temperature. Then the thermal stresses can be obtained by addition of an imaginary pressure to the isothermal stresses which follow by solving the isothermal governing equations with certain imaginary body forces and surface tractions, for example, see the book by Noda et al. [7] for a contemporary proof. Moreover, the thermal displacements due to the given temperature are identical to the isothermal displacements due to the imaginary body forces and surface tractions. This follows by comparing the thermal boundary value problem in hand with the imaginary isothermal boundary-value problem introduced by the body force

T. Furukawa (✉)
University of the Ryukyus, Nishihara, Okinawa, 903-0213 Japan
e-mail: furukawa@teada.tec.u-ryukyu.ac.jp

Alexander K. Belyaev et al. (eds.), *Mechanics and Model-Based Control of Advanced Engineering Systems*, DOI 10.1007/978-3-7091-1571-8_13,
© Springer-Verlag Wien 2014

analogy. This classical body force analogy cannot be applied, when coupling of deformation and heating must be taken into account in the heat conduction equation, which however is often the case under dynamic conditions. The deformations due to the forces then act as a driving source in the coupled heat conduction equation, such that an additional temperature arises, which enters the stress-strain-temperature relationship. For the coupled form of the heat conduction equation, which contains the time derivative of strain as a source term, see again the book by Noda et al. [7]. Nevertheless, the classical body force analogy can be extended to the coupled theory in the following sense: The thermal displacements arising from an initial-boundary value problem of the coupled theory of thermoelasticity are identical to isothermal dynamic displacements due to suitable imaginary body forces and surface tractions. The value of the latter formulation lies in the fact that results of the coupled theory can be checked by results of the simpler isothermal theory. The drawback, however, of such an imaginary type of analogy is that it cannot be utilized directly in applications, since coupling, if it is of any practical importance, should be taken into account in both, the thermal problem and the force problem. In a recent contribution of Furukawa and Irschik [1], the authors have overcome this drawback by providing a step towards a complete extension of the classical static body force analogy to the coupled theory of dynamic thermoelasticity. Two problems have been considered, namely a thermal problem without body forces, but with a given distribution of transient sources of heat, and a force problem without sources of heat, but with body forces. Both problems were treated within the coupled theory of thermoelasticity, such that temperature has to be taken into account also in the force problem. Within the one-dimensional case, it has been shown that, given suitable boundary and initial conditions, a distribution of body forces can be constructed, such that the dynamic displacements in both problems become equal. Next, we extended this methodology to the theory of generalized thermoelasticity with one relaxation time [2]. It is the scope of the present contribution to extend this methodology to the theory of generalized thermoelasticity with two relaxation times.

2 Basic Equations

We consider two thermoelastic fields, where the boundary conditions and initial conditions for the displacement and temperature of these fields are assumed to be equal. We denote the displacement and temperature in the two problems as $\mathbf{U}$, θ and $\overline{\mathbf{U}}$, $\overline{\theta}$, respectively. The basic equations of the first problem are written as follows.

$$\mathbf{E} = \frac{1}{2}(\nabla\mathbf{U} + \nabla\mathbf{U}^T) \tag{1}$$

$$\operatorname{div}\mathbf{S} + \mathbf{b} = \rho\ddot{\mathbf{U}} \tag{2}$$

$$-\operatorname{div}\mathbf{q} + m\theta_0 \operatorname{tr}\dot{\mathbf{E}} + r = c\,(\dot{\theta} + \delta_{2k}t_0\ddot{\theta}) \qquad (m = (3\lambda + 2\mu)\alpha) \tag{3}$$

$$\mathbf{S} = 2\mu\,\mathbf{E} + \lambda(\operatorname{tr}\mathbf{E})\mathbf{1} + m\,(\theta + \delta_{2k}t_1\dot{\theta})\,\mathbf{1} \tag{4}$$

$$\mathbf{q} + \delta_{1k}t_0\dot{\mathbf{q}} = -k\nabla\theta \tag{5}$$

Here, $\mathbf{S}$ stands for the stress tensor, and $\mathbf{E}$ is the strain tensor. The mechanical parameters ρ and, λ, μ are mass density and Lame's constants, respectively, and thermal properties α, c and k are coefficient of linear thermal expansion, specific heat at constant volume, and thermal conductivity, respectively. We assume that these parameters are independent of temperature θ. Sources of heat are denoted as r and are assumed to be given, $\mathbf{q}$ is heat flux vector, and $\mathbf{b}$ is body force vector. The sign δ_{jk} denotes Kronecker delta whose subscript shows the number of relaxation times, that is, $k = 0$, $k = 1$, and $k = 2$ mean thermomechanical coupled theory, Lord-Shulman's theory [6], and Green-Lindsay's theory [3], respectively. Once and for all we assume that the fields introduced in Eqs. (1)–(5) are sufficiently smooth, in order that our mathematical operations appear to be justified. Hence, we particularly exclude the presence of singular surfaces.

Introduction of Eqs. (5) and (1) into Eq. (3) yields

$$k\,\nabla^2\theta = c(\dot{\theta} + t_0\ddot{\theta}) - m\theta_0(\nabla\dot{\mathbf{U}} + \delta_{1k}t_0\nabla\ddot{\mathbf{U}}) - (r + \delta_{1k}t_0\dot{r}) \tag{6}$$

Introducing Eq. (4) into Eq. (2) and using the relation (1) and assuming that $\mathbf{b} = \mathbf{0}$, we obtain

$$\mu\nabla^2\mathbf{U} + (\lambda + \mu)\nabla(\operatorname{div}\mathbf{U}) + m(\nabla\theta + \delta_{2k}t_1\nabla\dot{\theta}) = \rho\,\ddot{\mathbf{U}} \tag{7}$$

When we use same procedure and an analogous notation, the basic equations of second thermoelastic problem are

$$k\,\nabla^2\overline{\theta} = c(\dot{\overline{\theta}} + t_0\ddot{\overline{\theta}}) - m\theta_0(\nabla\dot{\overline{\mathbf{U}}} + \delta_{1k}t_0\nabla\ddot{\overline{\mathbf{U}}}) \tag{8}$$

$$\mu\nabla^2\overline{\mathbf{U}} + (\lambda + \mu)\nabla(\operatorname{div}\overline{\mathbf{U}}) + m(\nabla\overline{\theta} + \delta_{2k}t_1\nabla\dot{\overline{\theta}}) + \overline{\mathbf{b}} = \rho\,\ddot{\overline{\mathbf{U}}} \tag{9}$$

where we assume that $\overline{r} = 0$. The problem according to Eqs. (6) and (7) then will be denoted as the thermal problem, while the problem stated in Eqs. (8) and (9) will be called the force problem subsequently.

We now introduce new notations

$$\mathbf{U}^* = \mathbf{U} - \overline{\mathbf{U}}, \;\; \Theta^* = \theta - \overline{\theta} \tag{10}$$

Subtracting Eq. (7) from (9) and Eq. (6) from (8) yields the following equations

$$\mu\nabla^2\mathbf{U}^* + (\lambda + \mu)\nabla(\operatorname{div}\mathbf{U}^*) + m(\nabla\Theta^* + \delta_{2k}t_1\nabla\dot{\Theta}) - \overline{\mathbf{b}} = \rho\,\ddot{\mathbf{U}}^* \tag{11}$$

$$k\nabla^2\Theta^* = c(\dot{\Theta}^* + t_0\ddot{\Theta}^*) - m\theta_0\nabla(\dot{\mathbf{U}}^* + \delta_{1k}t_0\ddot{\mathbf{U}}^*) - (r + \delta_{1k}t_0\dot{r}) \tag{12}$$

In order to extend the classical body force analogy to the generalized theory of thermoelasticity, we seek the condition $\mathbf{U}^* = \mathbf{0}$. Hence we take the sources of heat r in Eq. (12) to be given, and we seek for body forces $\overline{\mathbf{b}}$ in Eq. (11), such that the displacements in the thermal and in the force problem become equal.

3 A New Body Force Analogy

The first step in our solution for the body force analogy stated in the previous section is as follows. We assume that

$$\overline{\mathbf{b}} = m(\nabla\Theta^* + \delta_{2k}t_1\nabla\dot{\Theta}^*) \tag{13}$$

Then Eq. (11) reduces to

$$\mu\nabla^2\mathbf{U}^* + (\lambda + \mu)\nabla(\operatorname{div}\mathbf{U}^*) = \rho\,\ddot{\mathbf{U}}^* \tag{14}$$

which indeed has $\mathbf{U}^* = \mathbf{0}$ as a solution. We note that Eq. (13) corresponds to the solution presented by Irschik and Pichler [4, 5] for compensating force induced vibrations, the term Θ^* at the right hand side of Eq. (13) representing the thermal actuation stress necessary for shape control. In the present context, however, the body force $\overline{\mathbf{b}}$ is sought such that the displacements in the thermal and the force problem become equal. An equation for computing the body force $\overline{\mathbf{b}}$ is therefore needed.

The differentiation of Eq. (13) yields

$$\nabla\overline{\mathbf{b}} = m(\nabla^2\Theta^* + \delta_{2k}t_1\nabla^2\dot{\Theta}^*) \tag{15}$$

By use of Eq. (15), Eq. (12) becomes

$$\begin{aligned}\frac{k}{m}\nabla\overline{\mathbf{b}} &= c[\dot{\Theta}^* + \delta_{2k}t_1\ddot{\Theta}^* + t_0(\ddot{\Theta}^* + \delta_{2k}t_1\dddot{\Theta}^*)]\\ &-m\theta_0\nabla[\dot{\mathbf{U}}^* + (\delta_{1k}t_0 + \delta_{2k}t_1)\ddot{\mathbf{U}}^*] - [r + (\delta_{1k}t_0 + \delta_{2k}t_1)\dot{r}]\end{aligned} \tag{16}$$

A further differentiation of this equation and using Eq. (13) causes

$$\begin{aligned}\frac{k}{m}\nabla^2\overline{\mathbf{b}} - \frac{c}{m}(\dot{\overline{\mathbf{b}}} + t_0\ddot{\overline{\mathbf{b}}}) &= -m\theta_0\nabla^2[\dot{\mathbf{U}}^* + (\delta_{1k}t_0 + \delta_{2k}t_1)\ddot{\mathbf{U}}^*]\\ &-\nabla[r + (\delta_{1k}t_0 + \delta_{2k}t_1)\dot{r}]\end{aligned} \tag{17}$$

Hence, when we assume that $\mathbf{U}^* = \mathbf{0}$, we obtain the following partial differential equation for $\overline{\mathbf{b}}$:

$$\nabla^2\overline{\mathbf{b}} - \frac{c}{k}(\dot{\overline{\mathbf{b}}} + t_0\ddot{\overline{\mathbf{b}}}) = -\frac{m}{k}\nabla[r + (\delta_{1k}t_0 + \delta_{2k}t_1)\dot{r}] \tag{18}$$

In other words, when we can guarantee that $\mathbf{U}^* = \mathbf{0}$ is the solution of Eq. (12), then Eq. (18) can be used to compute a body force $\overline{\mathbf{b}}$ which equalizes the displacements of the thermal problem due to the sources of heat r. In order to ensure a trivial solution of Eq. (12), initial and boundary conditions must be taken into account. We state two cases for the mechanical and thermal boundary conditions and initial conditions, which indeed have $\mathbf{U}^* = \mathbf{0}$ as their solutions:

$$\text{Case1}: \quad \begin{array}{c} \text{BC}: \mathbf{U}^* = \mathbf{0},\ \nabla\Theta^* = \mathbf{0} \text{ at the boundaries} \\ \text{IC}: \mathbf{U}^* = \dot{\mathbf{U}}^* = \mathbf{0},\ \Theta^* = \dot{\Theta}^* = 0 \text{ at } t = 0 \end{array} \tag{19}$$

$$\text{Case2}: \quad \begin{array}{c} \text{BC}: \nabla\mathbf{U}^* = \mathbf{0},\ \Theta^* = 0 \text{ at the boundaries} \\ \text{IC}: \mathbf{U}^* = \dot{\mathbf{U}}^* = \mathbf{0},\ \Theta^* = \dot{\Theta}^* = 0 \text{ at } t = 0 \end{array} \tag{20}$$

For these cases, the computation of the body forces according to Eq. (18) will be exemplary demonstrated in the next section. In both cases, the mechanical and thermal initial conditions of the force and the thermal problem are taken as equal. At the boundaries, either the displacements (Case 1) or the stresses (Cases 2) are equal in the force and the thermal problem. Furthermore, either the temperature (Case 2) or the heat flux (Case 1) must be equal at the boundary. The two cases stated in Eqs. (19) and (20) thus cover a fairly wide range of problems with possibly inhomogeneous boundary and initial conditions. In order to compute a suitable solution of Eq. (18) that is consistent with the above derivation, boundary conditions and initial conditions are needed. As will be seen in the subsequent analytical examples, these conditions are provided by the thermal conditions in Eqs. (19) and (20). Of course, one could have posed the analogy in alternative setting. For example, one may ask for a distribution of sources of heat in order that the displacements are equal in the force and the displacement problem, where this time the body forces are given. The solution for this scenario follows from Eq. (18), which now can be integrated directly for r. Moreover, one might ask for body forces, in order that the temperature in both problems becomes equal. The displacements then will be different in general. The strategy for solving this problem can follow closely the above considerations. The problem of an equal temperature however appears to be of little practical relevance, since the temperature associated with the force problem can be expected to be small. Instead of writing down the corresponding relations, we therefore turn to an exemplary justification of the body force analogy for equal displacements.

4 Analytical Examples

We consider a one-dimensional body which occupies the region $(0 \leq X \leq L)$ and the boundaries are $(X = 0)$ and $(X = L)$. In this case, it becomes $\lambda = 0$ so that $m = 2\mu\alpha$. First, we consider Case 1, Eq. (19). In order to be consistent with Eq. (13), the following boundary conditions and initial conditions for $\bar{b}$ should obviously be used:

$$\text{BC}: \overline{b} = 0 \text{ at } X = 0 \text{ and } X = L, \quad \text{IC}: \overline{b} = 0, \ \dot{\overline{b}} = 0 \text{ at } t = 0 \tag{21}$$

As an example, the source of heat is taken as:

$$r = r_0 \cos \frac{\pi X}{L} \exp(-at) \tag{22}$$

where r_0 and a are arbitrary positive constants. We put:

$$\overline{b} = \overline{b_0} \sin \frac{\pi X}{L} \exp(-at) \tag{23}$$

which is satisfied the boundary conditions. It seems that the initial conditions are not be satisfied. In this case, Heaviside unit step function that is discontinuous at time $t = 0$ is omitted in order to avoid the difficulty in differentiation at $t = 0$. By substitution of Eqs. (22) and (23) into differential equation (18) we obtain the body force $\overline{b}$ as

$$\overline{b} = -\frac{2\mu\alpha}{k}\frac{\pi}{L}\frac{1 - (\delta_{1k}t_0 + \delta_{2k}t_1)a}{(\pi/L)^2 - (c/k)a(1 - t_0 a)} r_0 \sin \frac{\pi X}{L} \exp(-at) \tag{24}$$

When Eqs. (22) and (24) are introduced into Eqs. (11) and (12), the following equations are obtained

$$\begin{aligned} &U^*{}_{,XX} - \frac{\rho}{2\mu}\ddot{U}^* + \alpha(\Theta^*{}_{,X} + \delta_{2k}t_1\dot{\Theta}_{,X}) \\ &= -\frac{\alpha}{k}\frac{\pi}{L}\frac{1 - (\delta_{1k}t_0 + \delta_{2k}t_1)a}{(\pi/L)^2 - (c/k)a(1 - t_0 a)} r_0 \sin \frac{\pi X}{L} \exp(-at) \end{aligned} \tag{25}$$

$$\begin{aligned} &\frac{2\mu\alpha\theta_0}{k}(\dot{U}^*{}_{,X} + \delta_{1k}t_0\ddot{U}^*{}_{,X}) + \Theta^*{}_{,XX} - \frac{c}{k}(\dot{\Theta} + t_0\ddot{\Theta}) \\ &= -\frac{1}{k}(1 - \delta_{1k}t_0 a) r_0 \cos \frac{\pi X}{L} \exp(-at) \end{aligned} \tag{26}$$

We take the solutions of differential equations (25) and (26) to be:

$$U^* = A \sin \frac{\pi X}{L} \exp(-at) \tag{27}$$

$$\Theta^* = B \cos \frac{\pi X}{L} \exp(-at) \tag{28}$$

to satisfy the boundary conditions and initial conditions. Then we obtain the displacement and temperature as

$$U^* = 0 \tag{29}$$

$$\Theta^* = \frac{1}{k} \frac{1 - \delta_{1k} t_0 a}{(\pi/L)^2 - (c/k)a(1 - t_0 a)} r_0 \cos \frac{\pi X}{L} \exp(-at) \tag{30}$$

From Eqs. (24) and (30), the assumed relation (13) is proved and expected Eq. (29) is obtained, too. Furthermore, in order to have an independent proof, we insert Eq. (24) into Eqs. (8) and (9) and seek the solution, which then must be equal to the solution of Eqs. (6) and (7) with Eq. (22). Inserting Eq. (24) into Eqs. (8) and (9) and subsequent treatment similar to aforementioned procedure provides the following particular solutions:

$$\overline{U} = -\frac{\alpha}{k} \frac{\pi}{L} \frac{1 - (\delta_{1k} t_0 + \delta_{2k} t_1)a}{\Delta} r_0 \sin \frac{\pi X}{L} \exp(-at) \tag{31}$$

$$\overline{\theta} = \frac{2\mu\alpha^2\theta_0}{k^2} a \left(\frac{\pi}{L}\right)^2 \frac{(1 - \delta_{1k} t_0 a)[1 - (\delta_{1k} t_0 + \delta_{2k} t_1)a]}{(\pi/L)^2 - (c/k)a(1 - t_0 a)} \frac{1}{\Delta} r_0 \cos \frac{\pi X}{L} \exp(-at) \tag{32}$$

where

$$\Delta = \left[\left(\frac{\pi}{L}\right)^2 - \frac{c}{k} a(1 - t_0 a)\right]\left[\left(\frac{\pi}{L}\right)^2 + \frac{\rho}{2\mu} a^2\right] - \frac{2\mu\alpha^2\theta_0}{k}\left(\frac{\pi}{L}\right)^2 a[1 - (\delta_{1k} t_0 + \delta_{2k} t_1)a] \tag{33}$$

Apply the same procedure to Eqs. (6) and (7) with Eq. (22) produces

$$U = -\frac{\alpha}{k} \frac{\pi}{L} \frac{1 - (\delta_{1k} t_0 + \delta_{2k} t_1)a}{\Delta} r_0 \sin \frac{\pi X}{L} \exp(-at) \tag{34}$$

$$\theta = -(1 - \delta_{2k} t_1 a) \frac{\alpha^2}{k} \left(\frac{\pi}{L}\right)^2 \left[\left(\frac{\pi}{L}\right)^2 + \frac{\rho}{2\mu} a^2\right] \times \frac{1 - (\delta_{1k} t_0 + \delta_{2k} t_1)a}{\Delta} r_0 \cos \frac{\pi X}{L} \exp(-at) \tag{35}$$

From these equations, we obtain, for the particular solutions,

$$U^* = U - \overline{U} = 0 \tag{36}$$

$$\Theta^* = \theta - \overline{\theta} = \frac{1}{k}\frac{1-\delta_{1k}t_0 a}{(\pi/L)^2 - (c/k)a(1-t_0 a)} r_0 \cos\frac{\pi X}{L}\exp(-at) \tag{37}$$

Hence, if the initial conditions are the same for U and $\overline{U}$, then the total solutions for displacements are equal. Therefore, the justifications for Eqs. (29) and (30) are reconfirmed. Next, we consider Case 2, Eq. (20). From Eq. (16) the boundary condition becomes

$$\overline{b}_{,X} = -\frac{m}{k}[r + (\delta_{1k}t_0 + \delta_{2k}t_1)\dot{r}] \tag{38}$$

and the initial conditions are

$$\overline{b} = 0, \quad \dot{\overline{b}} = 0 \tag{39}$$

We take the example that

$$r = r_0 \sin\frac{\pi X}{L}\exp(-at) \tag{40}$$

Then the boundary conditions reduce to $\overline{b}_{,X} = 0$ at the boundaries, so that the solution of Eq. (18) is

$$\overline{b} = \frac{2\mu\alpha}{k}\frac{\pi}{L}\frac{1-(\delta_{1k}t_0 + \delta_{2k}t_1)a}{(\pi/L)^2 - (c/k)a(1-t_0 a)} r_0 \cos\frac{\pi X}{L}\exp(-at) \tag{41}$$

Equation (11) with Eq. (34) and Eq. (12) with Eq. (33) are represented as follows.

$$\begin{aligned} &U^*_{,XX} - \frac{\rho}{\mu}\ddot{U}^* + \frac{2\mu\alpha}{\mu}\Theta^*_{,X} \\ &= \frac{2\alpha}{k}\frac{\pi}{L}\frac{1-(\delta_{1k}t_0 + \delta_{2k}t_1)a}{(\pi/L)^2 - (c/k)a(1-t_0 a)} r_0 \cos\frac{\pi X}{L}\exp(-at) \end{aligned} \tag{42}$$

$$\begin{aligned} &\frac{2\mu\alpha\theta_0}{k}(\dot{U}^*_{,X} + t_0\ddot{U}^*_{,X}) + \Theta^*_{,XX} - \frac{c}{k}(\dot{\Theta} + t_0\ddot{\Theta}) \\ &= -\frac{1}{k}[1-(\delta_{1k}t_0 + \delta_{2k}t_1)a] r_0 \sin\frac{\pi X}{L}\exp(-at) \end{aligned} \tag{43}$$

Same procedure to Case 1 produces

$$U^* = 0 \tag{44}$$

$$\Theta^* = \frac{1}{k} \frac{1 - (\delta_{1k} t_0 + \delta_{2k} t_1) a}{(\pi/L)^2 - (c/k) a (1 - t_0 a)} r_0 \cos \frac{\pi X}{L} \exp(-at) \tag{45}$$

From Eqs. (37) and (38), the assumed relation (13) is proved and expected Eq. (37) is obtained, too. The independent proof is omitted here.

5 Conclusions

We have examined the body force analogy for the theory of generalized thermoelasticity including Lord-Shulman's and Green-Lindsay's theories. We considered two thermoelastic fields, where the boundary conditions and initial conditions for the displacement and temperature of these fields are assumed to be equal. The thermoelastic fields, for which the body force or sources of heat are absent, are denoted as the thermal problem and force problem, respectively. In the framework of a theory of generalized thermoelasticity, we have developed a procedure for deriving the body force, which is required to equalize the displacement of the thermal problem. In our derivations, we have required the fields under considerations to be sufficiently smooth, such that the necessary differential operations are justified. The obtained body force analogy nevertheless should be of considerable interest, not only from a theoretical but also from an industrial point of view. Two analytical examples were shown and the efficiency of present procedure could be confirmed for these examples.

References

1. Furukawa, T., Irschik, H.: Body-force analogy for one-dimensional coupled dynamic problems of thermoelasticity. J. Therm. Stress. **28**, 455–464 (2005)
2. Furukawa, T., Irschik, H.: Extension of body force analogy to generalized thermoelasticity. In: Proceedings of 4th European Conference on Structural Control, St. Petersburg, vol. 1, pp. 250–257 (2008)
3. Green, A.E., Lindsay, K.A.: Thermoelasticity. J. Elast. **2**, 1–7 (1972)
4. Irschik, H., Pichler, U.: Dynamic shape control of solids and structures by thermal expansion strains. J. Therm. Stress. **24**, 565–576 (2001)
5. Irschik, H., Pichler, U.: An extension of Neumann's method for shape control of force-induced elastic vibrations by eigenstrains. Int. J. Solids Struct. **41**, 871–884 (2004)
6. Lord, H., Shulman, Y.: A generalized dynamical theory of thermoelasticity. J. Mech. Phys. Solids **15**, 299–309 (1967)
7. Noda, N., Hetnarski, R.B., Tanigawa, Y.: Thermal Stresses, 2nd edn. Taylor and Francis, New York (2003)

Active Vibration and Noise Control of a Car Engine: Modeling and Experimental Validation

Ulrich Gabbert and Stefan Ringwelski

Abstract The paper presents an overall design approach for smart light-weight structures made of metal sheets or fiber reinforced plastics, equipped with thin piezoelectric wafers as actuators and sensors to control vibration and noise. The design process is based on an overall finite element model, which includes the passive structure, the piezoelectric wafers attached to the structure or embedded between the layers of a composite and the controller as well. In active noise control the vibrating structure interacts with the surrounding fluid, which is also included into the overall model. In order to evaluate the quality of the approach, test simulations are carried out and the results are compared with experimental data. As a test case, a smart car engine with surface-mounted piezoelectric actuators and sensors for active noise reduction is considered. A comparison between the measured values and those predicted by the coupled finite element model shows a good agreement.

1 Introduction

In the past years, significant progress has been made in the field of vibration and noise control in automotive engineering. The control of noise and vibration is essential in the design process of an automobile, since it contributes to comfort, efficiency and safety. There are two different approaches to achieve noise and vibration attenuation. In the widely used first approach the vibration and the sound emission of structures are reduced passively by modifying the structural geometry or by applying additional damping materials. These methods are best suited to a frequency range above 1,000 Hz. The second approach is the application of active control techniques to reduce unwanted structural vibration and noise. These

U. Gabbert (✉) · S. Ringwelski
Institute of Mechanics, Otto-von-Guericke University of Magdeburg, Magdeburg, Germany
e-mail: ulrich.gabbert@ovgu.de; stefan.ringwelski@ovgu.de

Alexander K. Belyaev et al. (eds.), *Mechanics and Model-Based Control of Advanced Engineering Systems*, DOI 10.1007/978-3-7091-1571-8_14,

© Springer-Verlag Wien 2014

techniques are usually employed in a frequency range between 50 and 1,000 Hz. In active noise reduction usually smart materials are attached to structures as actuators and sensors connected by a control unit which enables the system to reduce structural vibrations with the objective to reduce simultaneously also the sound radiation caused by structural vibrations.

Considering passenger cars, the power train represents one of the main noise sources, especially during idling, slow driving speeds and full load acceleration. One major contributor to the overall power train noise emission is the engine oil pan. Therefore, the paper aims to design a smart car engine with piezoelectric actuators and sensors attached on the surface of the oil pan for active noise reduction.

Over the past years, some researchers have already studied active noise reduction approaches at car oil pans [16] and a truck oil pans [5] using distributed piezoelectric actuators. However in these studies, the oil pan was treated separately and no attempt was made to consider the interactions between the crankcase and the oil pan. Naake et al. [13] employed piezoceramic patch actuators to minimize the vibrations of a car windshield. The goal of this control approach was to achieve a reduced acoustic pressure level inside the cabin. For the same reasons a similar approach was adopted by Weyer and Monner [20] and Nestorović et al. [14].

The development and industrial application of smart systems for active noise and vibration control require efficient and reliable simulation tools. Virtual models are of particular interest in the design process, since they enable the testing of several control strategies and they are required to determine optimal sensor and actuator locations. There are several numerical approaches [6, 7, 10, 11, 21] available for predicting the behavior of active systems, which include the modeling of the mechanical structure, the piezoelectric actuators and sensors, the surrounding fluids as well as the applied control algorithm.

In the present paper an overall finite element approach is proposed. For modeling thin walled structures shell elements are used, the piezoelectric patches are modelled with special electromechanical coupled shell elements, and the acoustic fluid is modelled with 3D fluid elements, which include the coupling conditions with the structural shell elements. The far field is described with infinite elements developed on the basis of the doubly asymptotic approximation [4]. This approach results in a symmetric form of the coupled electro-mechanical-acoustic system of the equations of motion, including the electric potential, the displacements and the velocity potential as nodal degrees of freedom.

For controller design purposes a reduced model is derived based on selected structural and acoustic eigenmodes.

In order to check the validity of the approach, numerical and experimental studies are carried out using a car engine with surface-mounted piezoelectric actuators. The simulated and the measured data are in a good agreement showing that the developed overall simulation approach can be used for industrial design purposes.

Finally, the quality of the designed active oil pan is checked by experimental testing on a fired engine using a dSPACE controller board which determines the necessary control outputs for the actuators. In the experiment, engine run ups are measured on an engine test bench for the uncontrolled and controlled case.

2 Finite Element Modeling

2.1 Finite Element Model of Piezoelectric Shell Structures

In this section, a simple piezoelectric composite Mindlin-type shell element for analyzing laminated plates with integrated piezoelectric actuators and sensors is briefly discussed. More details regarding the development and the implementation of this finite element can be found in [17] and [3].

In the shell element, it is assumed that the thickness of the layers is the same at each node and the deformations are small. Additionally, it is presumed that the modeled composite laminate plate consists of perfectly bonded layers and the bonds are infinitesimally thin as well as nonshear-deformable. The shell element has six degrees of freedom $u_1, u_2, u_3, \theta_{x2}, \theta_{x1}, \theta_{x3}$ at each node for describing the elastic behavior and additionally one electric potential degree of freedom per layer to model the piezoelectric effect. The strain displacement relations for the used plane shell element are based on the Mindlin first order shear deformation theory.

The poling direction of the piezoelectric layers is assumed to be coincident with the thickness direction, which means that the electric field acts only perpendicular to the layers. Moreover, the difference in the electric potential φ is supposed to be constant in each layer of the shell element. The electric field, which varies linearly through the thickness of a piezoelectric layer, causes an in-plane expansion or contraction. Thus, for modeling the electric field only one electric degree of freedom per layer has to be specified within the element. In Marinković et al. [12] it is shown, that these assumptions are accurate enough in thin structure applications.

The coupled electromechanical behavior of piezoelectric materials in a low voltage regime can be modeled with sufficient accuracy by means of the linearized constitutive equations. In matrix form, the constitutive relations for a piezoelectric layer are defined as [15]

$$\boldsymbol{\sigma} = \mathbf{Q}\boldsymbol{\varepsilon} - \mathbf{e}E\ , \ D = \mathbf{e}^T\boldsymbol{\varepsilon} + \boldsymbol{\kappa}E. \tag{1}$$

Here σ denotes the stress vector, D is the dielectric displacement in thickness direction, $\mathbf{Q}$ and $\mathbf{e}$ are the plane-stress reduced elastic stiffness and the piezoelectric matrices, respectively. The coefficient $\boldsymbol{\kappa}$ represents the plane-stress reduced dielectric permittivity of the piezoelectric layer and E is the electric field of the shell element. The piezoelectric constitutive relations given above are used within the weak form of the mechanical equilibrium equations [19] to derive the electromechanical FE equations of a piezoelectric layer by applying a standard Galerkin procedure. After adding the local equations of all layers and elements to a global model, the resulting system of coupled algebraic equations can be expressed as

$$\begin{bmatrix} \mathbf{M}_u & \mathbf{0} \\ \mathbf{0} & \mathbf{0} \end{bmatrix} \begin{bmatrix} \ddot{\mathbf{u}} \\ \ddot{\boldsymbol{\varphi}} \end{bmatrix} + \begin{bmatrix} \mathbf{C}_u & \mathbf{0} \\ \mathbf{0} & \mathbf{0} \end{bmatrix} \begin{bmatrix} \dot{\mathbf{u}} \\ \dot{\boldsymbol{\varphi}} \end{bmatrix} + \begin{bmatrix} \mathbf{K}_u & \mathbf{K}_{u\varphi} \\ \mathbf{K}_{u\varphi}^T & -\mathbf{K}_{\varphi} \end{bmatrix} \begin{bmatrix} \mathbf{u} \\ \boldsymbol{\varphi} \end{bmatrix} = \begin{bmatrix} \mathbf{f}_u \\ \mathbf{f}_{\varphi} \end{bmatrix}, \tag{2}$$

where $\mathbf{u}$ is the vector with the nodal structural displacements, and $\boldsymbol{\varphi}$ is the vector with the nodal values of the electric potentials. The matrices $\mathbf{M}_u$, $\mathbf{K}_u$ and $\mathbf{K}_\varphi$ are the structural mass, the structural stiffness and the dielectric matrix, respectively. The piezoelectric coupling arises in the piezoelectric coupling matrix $\mathbf{K}_{u\varphi}$. For convenience, a Rayleigh damping is introduced into the system of Eqs. (2), assuming that the matrix $\mathbf{C}_u$ is a linear combination of the matrices $\mathbf{M}_u$ and $\mathbf{K}_u$. The external loads are stored in the mechanical load vector $\mathbf{f}_u$ and in the electric load vector $\mathbf{f}_\varphi$. In the following, it is assumed that the vector $\mathbf{u}$ contains the entire nodal displacements of the mechanical structure as well as the piezoelectric actuators and sensors.

2.2 *Finite Element Modeling of the Acoustic Fluid*

In this section, a FE formulation is presented to model the finite fluid domains around the smart shell structure. The homogeneous and inviscid acoustic fluid is modeled by using the linear acoustic wave equation [8]

$$\frac{1}{c^2}\ddot{\Phi} - \Delta\Phi = 0. \tag{3}$$

In the present study, a hexahedron element is chosen to discretize the fluid domains. The velocity potential is considered as the nodal degree of freedom in the finite element instead of the acoustic pressure p, since Everstine [2] has recommended introducing the velocity potential Φ as a new degree of freedom in order to get symmetric matrices. In Eq. (3) Δ is the Laplacian operator and c is the speed of sound.

To take into account an infinite outer fluid region, the doubly asymptotic approximation [4] is used. The behavior of the outer fluid is considered only in the low and the high frequency range. At low frequencies the fluid of the outer region is assumed to be incompressible, and in the high frequency range plane waves are considered.

Following the standard finite element procedure, the matrix equation of the discretized fluid domain becomes

$$\mathbf{M}_a\ddot{\boldsymbol{\Phi}} + (\mathbf{C}_a + \mathbf{C}_0)\,\dot{\boldsymbol{\Phi}} + (\mathbf{K}_a + \mathbf{K}_0)\,\boldsymbol{\Phi} = \mathbf{f}_a, \tag{4}$$

with the acoustic mass matrix $\mathbf{M}_a$, the acoustic damping matrix $\mathbf{C}_a$, the acoustic stiffness matrix $\mathbf{K}_a$, the acoustic load vector $\mathbf{f}_a$, and the matrices $\mathbf{C}_0$ and $\mathbf{K}_0$, which are additional matrices taking into account the coupling with special infinite or semi-infinite fluid elements for describing the fare field. For more details see Lefèvre [9]. Similar coupling terms also occur if the fare field is described with boundary elements, which can also be coupled with 3D fluid finite elements applied for approximating the near field [17].

2.3 The Vibro-Acoustic Coupling

The dynamic behavior of thin lightweight structures with fluid loading is strongly affected by the interaction between the subsystems. The purpose of the following section is to present a overall FE approach to model the coupling that occurs in smart systems with a fluid-structure interface.

The acoustic pressure represents an additional surface distributed load acting normal to the surface of the piezoelectric structure, such resulting in an additional load vector $\mathbf{f}_{uc}$, which appears on the right hand side of Eq. (3). This additional load vector acting at the fluid-structure interface can be expressed as

$$\mathbf{f}_{uc} = \mathbf{C}_{uc}\dot{\boldsymbol{\Phi}}, \tag{5}$$

where $\mathbf{C}_{uc}$ is the coupling matrix at the fluid-structure interface. In an analogous manner, the vibrating structure interacts with the fluid. This influence is described by an additional load vector $\mathbf{f}_{ac}$, which appears on the right hand side of Eq. (4). This new acoustic load vector can be expressed by an expression similar to Eq. (5) as

$$\mathbf{f}_{ac} = -\frac{1}{\rho_0}\mathbf{C}_{uc}\dot{\mathbf{u}}, \tag{6}$$

with $\mathbf{C}_{uc}$ as the coupling matrix regarding structural vibrations. The variable ρ_0 stands for the density of the acoustic medium.

Equations (2) and (4) are coupled by introducing the Eqs. (5) and (6), which results in symmetric matrices by multiplying all lines related to the fluid degrees of freedom with $(-\rho_0)$. If the element matrices are assembled into the global system matrices, the semi discrete form of the equation of motion of the coupled electro-mechanical-acoustic field problem can be written as

$$\begin{bmatrix} \mathbf{M}_u & \mathbf{0} & \mathbf{0} \\ \mathbf{0} & \mathbf{0} & \mathbf{0} \\ \mathbf{0} & \mathbf{0} & -\rho_0\mathbf{M}_a \end{bmatrix} \begin{bmatrix} \ddot{\mathbf{u}} \\ \ddot{\boldsymbol{\varphi}} \\ \ddot{\boldsymbol{\Phi}} \end{bmatrix} + \begin{bmatrix} \mathbf{C}_u & \mathbf{0} & -\mathbf{C}_{uc} \\ \mathbf{0} & \mathbf{0} & \mathbf{0} \\ -\mathbf{C}_{uc}^T & \mathbf{0} & -\rho_0\left(\mathbf{C}_a + \mathbf{C}_0\right) \end{bmatrix} \begin{bmatrix} \dot{\mathbf{u}} \\ \dot{\boldsymbol{\varphi}} \\ \dot{\boldsymbol{\Phi}} \end{bmatrix}$$
$$+ \begin{bmatrix} \mathbf{K}_u & \mathbf{K}_{u\varphi} & \mathbf{0} \\ \mathbf{K}_{u\varphi}^T & -\mathbf{K}_\varphi & \mathbf{0} \\ \mathbf{0} & \mathbf{0} & -\rho_0\left(\mathbf{K}_a + \mathbf{K}_0\right) \end{bmatrix} \begin{bmatrix} \mathbf{u} \\ \boldsymbol{\varphi} \\ \boldsymbol{\Phi} \end{bmatrix} = \begin{bmatrix} \mathbf{f}_u \\ \mathbf{f}_\varphi \\ -\rho_0\mathbf{f}_a \end{bmatrix}. \tag{7}$$

3 Controller Design

The numerical simulation of smart structures within the finite element frame requires an overall finite element model including the passive structure with the surrounding fluid, the active sensor and actuator elements, as well as an appropriate model of the controller. For the controller design numerous techniques are at disposal, and the application of an appropriate control law is closely related to the requirements of the specific problem. For a model based controller design the large finite element models have to be reduced. One of the standard techniques is the modal reduction based on preselected eigenmodes of the system. To apply this technique, the coupled system of motion (7) is rewritten in the following compact form

$$\tilde{\mathbf{M}}\ddot{\mathbf{r}} + \tilde{\mathbf{C}}\dot{\mathbf{r}} + \tilde{\mathbf{K}}\mathbf{r} = \tilde{\mathbf{f}}, \tag{8}$$

Introducing the state space vector

$$\mathbf{z} = \begin{bmatrix}\mathbf{r}\ \dot{\mathbf{r}}\end{bmatrix}^T = \begin{bmatrix}\mathbf{u}\ \boldsymbol{\varphi}\ \boldsymbol{\Phi}\ \dot{\mathbf{u}}\ \dot{\boldsymbol{\varphi}}\ \dot{\boldsymbol{\Phi}}\end{bmatrix}^T, \tag{9}$$

from Eq. (9) it follows

$$\begin{bmatrix}\tilde{\mathbf{C}} & \tilde{\mathbf{M}}\\ \tilde{\mathbf{M}} & \mathbf{0}\end{bmatrix}\dot{\mathbf{z}} + \begin{bmatrix}\tilde{\mathbf{K}} & \mathbf{0}\\ \mathbf{0} & -\tilde{\mathbf{M}}\end{bmatrix}\mathbf{z} = \tilde{\mathbf{B}}\dot{\mathbf{z}} + \tilde{\mathbf{A}}\mathbf{z} = \begin{bmatrix}\tilde{\mathbf{f}}\\ \mathbf{0}\end{bmatrix}. \tag{10}$$

From Eq. (10) the linear eigenvalue problem can be derived as

$$\left(\tilde{\mathbf{A}} - \lambda_j\tilde{\mathbf{B}}\right)\hat{\mathbf{z}}_j = \mathbf{0}. \tag{11}$$

The solution of Eq. (11) results in the modal matrix $\mathbf{Q}$ with $2k$ pairs of conjugate complex eigenvectors

$$\mathbf{Q} = \begin{bmatrix}\hat{\mathbf{z}}_1\ \hat{\mathbf{z}}_2\ \ldots\ \hat{\mathbf{z}}_{2k}\end{bmatrix}. \tag{12}$$

If the modal matrix $\mathbf{Q}$ is ortho-normalized with $\mathbf{Q}^T\tilde{\mathbf{B}}\mathbf{Q} = \mathbf{I} = diag(1)$ and $\mathbf{Q}^T\tilde{\mathbf{A}}\mathbf{Q} = \boldsymbol{\Lambda} = diag(\Lambda)$, and new coordinates $\mathbf{z} = \mathbf{Q}\mathbf{q}$ are introduced in Eq. (12), the reduced state space form is obtained as

$$\dot{\mathbf{q}} + \boldsymbol{\Lambda}\mathbf{q} = \mathbf{Q}^T\begin{bmatrix}\tilde{\mathbf{f}}\\ \mathbf{0}\end{bmatrix}. \tag{13}$$

If the right hand side of Eq. (13) is subdivided into the control forces and the external forces and the measurement equation (15) is added, the following set of equations, which can be used to design an appropriate controller, is obtained as

$$\dot{\mathbf{q}} = -\boldsymbol{\Lambda}\mathbf{q} + \mathbf{Q}^T\begin{bmatrix}\bar{\mathbf{B}}\\ \mathbf{0}\end{bmatrix}\mathbf{u}(t) + \mathbf{Q}^T\begin{bmatrix}\bar{\mathbf{E}}\\ \mathbf{0}\end{bmatrix}\mathbf{f}(t) = \mathbf{A}\mathbf{q} + \mathbf{B}\mathbf{u}(t) + \mathbf{E}\mathbf{f}(t), \tag{14}$$

$$\mathbf{y}(t) = \mathbf{C}\mathbf{q} + \mathbf{D}\mathbf{u}(t) + \mathbf{F}\mathbf{f}(t). \tag{15}$$

For controller design purposes the state matrices **A**, **B**, **E**, **C**, **D** and **F** are transferred from the finite element approach to Matlab/Simulink via an internal data interface. Based on the model matrices different controller, such as the velocity feedback control, the optimal LQ control and the modal reference adaptive control (MRAC) have been developed and successfully tested.

4 Smart Car Engine for Active Noise Reduction

The purpose of this section is to demonstrate the applicability of the proposed modeling approach. For this reason, a smart car engine that consists of a crankcase an oil pan with surface-mounted piezoelectric actuators and sensors for active noise reduction is designed.

4.1 Dominant Mode Shapes

In order to design a smart stripped car engine for active noise and vibration control, it is essential to identify the most dominant mode shapes. This step is carried out by means of harmonic FE simulations using a Fourier transformed version of Eq. (2).

A point force excitation at the oil pan flange is chosen to excite all eigenmodes in a frequency range up to 1,200 Hz. Figure 1 shows the resulting frequency response function (FRF) between the structural displacement at the center of the oil pan bottom and the excitation force at the flange. In addition, the mode shapes that are associated with the respective resonance frequencies are illustrated. In Fig. 1, it can be seen that the first and the third eigenmode are pure bending modes of the oil pan bottom. The second and the fourth mode are global bending modes of the whole stripped car engine. Under real operating conditions these bottom modes are the main contributor to the overall sound emission. Due to this fact, the present approach aims to control only these modes.

4.2 Definition of the Actuator Positions and Modeling

An often used method for the actuator placement is based on the assumption that an actuator is placed well when it is able to influence significantly the shape of the structural modes. This means that an in-plane actuator should be placed at positions on the surface of the structure, where the strains and the corresponding electric potentials are the highest [1, 18]. In case of the stripped car engine, the first and the third eigenmode are considered. The modal strains of these modes can be derived from the structural part $\hat{\mathbf{u}}_j$ of the complex eigenvectors $\hat{\mathbf{z}}_j$ with $j = 1$ and 3.

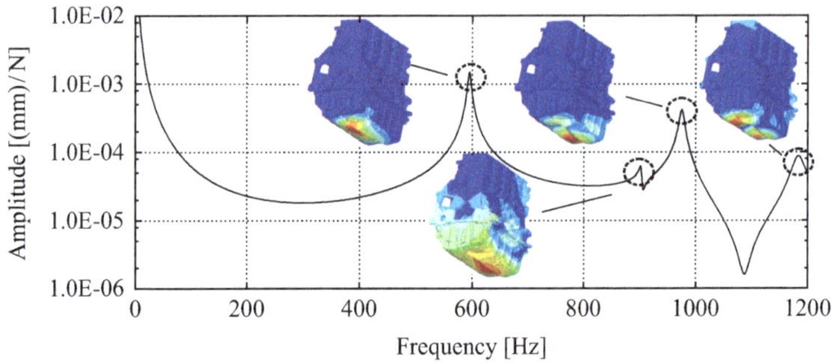

Fig. 1 Computed FRF of the stripped car engine

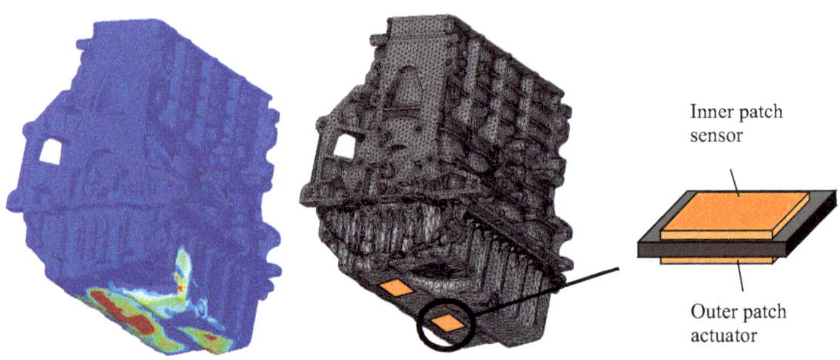

Fig. 2 Contour plot of the superposed field and FE mesh with selected actuator and sensor position

The structural displacements $\hat{\mathbf{u}}_j$ are associated with the modal strains $\hat{\boldsymbol{\varepsilon}}_j$ by the relationship

$$\hat{\boldsymbol{\varepsilon}}_j = \mathbf{B}_u \hat{\mathbf{u}}_j, \tag{16}$$

where **B** is the matrix that calculates the strains of a finite element surface area using nodal displacements. The modal strains $\hat{\boldsymbol{\varepsilon}}_j$ are related with the electric potentials $\hat{\varphi}_j$ by the constitutive equations of piezoelectricity [17]. By means of a multiplicative superposition of the electric potentials $\hat{\varphi}_j$ one obtains the super-posed field

$$\hat{\varphi}_{max} = \prod_{j=1,3} \hat{\varphi}_j. \tag{17}$$

In contrast to an additive superposition, the multiplicative superposition makes sure that the actuators are not placed on node lines. A contour plot of the super-posed field $\hat{\varphi}_{max}$ allows the definition of optimal actuator positions. Two actuator positions have been selected according to the contour plot visible in Fig. 2 (picture at the left hand side), where the orange areas mark the preferred regions for placing actuators. The orange rectangular areas of the FE mesh (picture in the center of

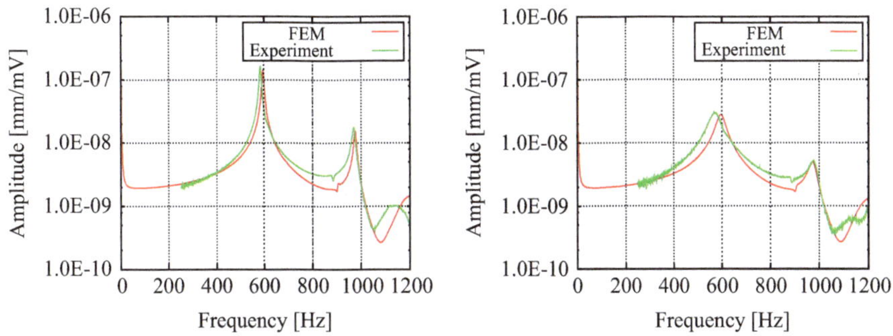

Fig. 3 Uncontrolled (*left*) and controlled (*right*) FRFs of the stripped car engine

Fig. 2) mark the two selected actuator positions. Additionally, two sensors are placed on the opposite side of the oil pan. The collocated design of the piezoelectric actuator/sensor pairs guarantees control stability. It should be noted that more than two collocated actuator/sensor pairs are possible.

4.3 Numerical and Experimental Studies

In the following section numerical simulations are carried out and the results are compared with experimental data to test the performance of the system designed. In the analysis the piezoceramic actuators and sensors are modeled using multilayer shell elements. A velocity feedback control is chosen to compute input signals for the actuators that are bonded on the outer surface of the oil pan. Standard tetrahedral elements are used to model the irregular-shaped geometry of the crankcase and the oil pan. The surrounding fluid volume is approximated with a mixture of tetrahedral and hexahedral elements. These elements are coupled by the above presented fluid-structure coupling matrices with the structural finite elements used for modeling the engine. For the far field approximation the above mentioned infinite elements are applied. In all these elements, quadratic shape functions are employed. Figure 2 illustrates the structural FE mesh of the stripped car engine.

For comparison purposes uncontrolled and controlled FRFs are considered. Similar to the computations in Sect. 4.1 a harmonic point force excitation at the oil pan flange has been used to excite the system, and the response is regarded in terms of the structural displacement at the center of the oil pan bottom. In Fig. 3, it can be observed that the measured data and the numerical predictions agree very well. Additionally, the results in Fig. 3 show that a significant damping at the dominating resonance frequencies is achieved, due to the implementation of the velocity feedback control. The amplitudes are reduced by more than 14 dB at 595 Hz and by about 10 dB at 975 Hz.

Figure 4 shows the deformed shape of the car engine for the uncontrolled and the controlled case at the first resonance frequency (595 Hz). It can be seen that

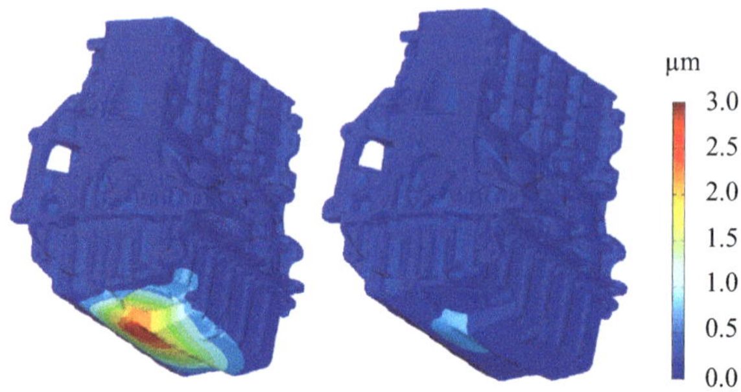

Fig. 4 Uncontrolled (*left*) and controlled (*right*) deformed shape of the stripped car engine at the first eigenfrequency

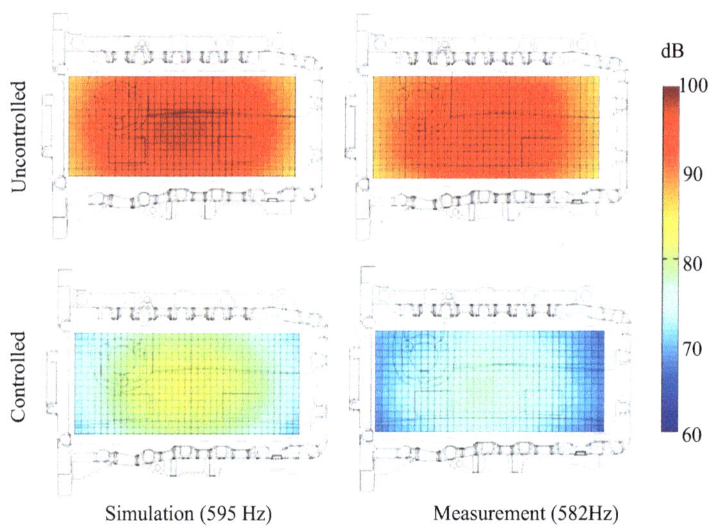

Fig. 5 Sound pressure distribution of the controlled and uncontrolled stripped engine

the displacements are reduced by approximately 16 dB. In order to analyze the near-field airborne noise from the outer surface of the oil pan, the corresponding pressure fields are computed. On the left-hand side of Fig. 5 the computed sound pressure distribution of the uncontrolled and controlled stripped engine are plotted. For visualization a plane approximately 50 mm apart from the bottom surface was chosen. To verify the simulated data, measurements have been carried out in a free-field room with the help of a uniformly distributed microphone-array. The acoustic field of the structure was scanned at 32 microphone positions with 50 mm measuring grid spacing between. The corresponding measurements are shown on the right-hand side of Fig. 5. From Fig. 5 can be seen that the simulation results correlate

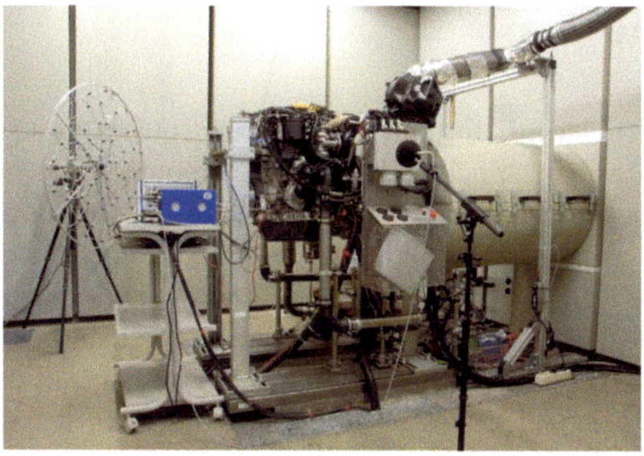

Fig. 6 Four-cylinder diesel engine in an anechoic room

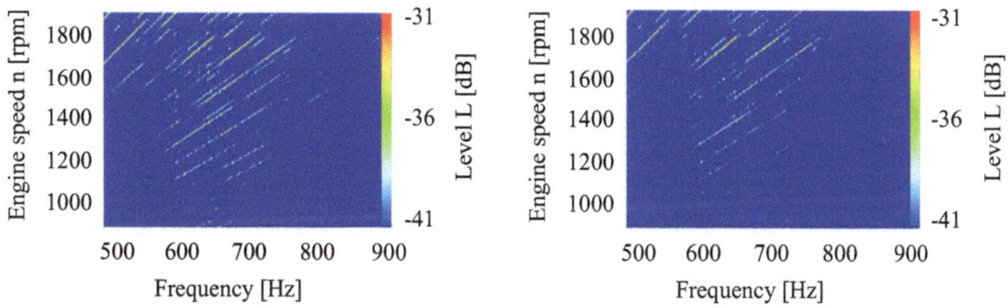

Fig. 7 Campbell diagrams of the uncontrolled (*left*) and controlled (*right*) diesel engine

well with the experimental results. Furthermore it can be noticed that due to the controller influence the sound pressure level is reduced by approximately 16 dB, which indicates the noise reduction potential of the designed system.

4.4 Engine Measurements on a Test Bench

In order to evaluate the quality of the smart system designed, several experimental tests on an acoustic engine test bench have been carried out under real operating conditions. In Fig. 6 the altered four-cylinder common rail diesel engine is shown.

In order to get an overview of the uncontrolled and controlled behavior at higher engine speeds, engine run-ups (900–3,000 rpm) were made and evaluated by means of appropriate Campbell diagrams. A displacement patch sensor was used to generate the diagrams, since this sensor represents the entire vibration characteristics of the oil pan. The measurements, shown in Fig. 7, reveal that due to the controller influence the displacement level amplitudes in the resonance frequency regions are reduced by approximately 4 dB, which indicates the noise reduction potential of the designed system.

5 Conclusions

In this paper, a coupled FE formulation has been presented to simulate a fluid-loaded smart lightweight structure with surface-mounted piezoelectric actuators and sensors. Besides the passive structure, the finite element model includes active piezoelectric elements, the acoustic fluid, the vibro-acoustic coupling and the controller influence as well. Piezoelectric layered shell type finite elements have been extended to include the vibro-acoustic coupling with 3D acoustic finite elements and infinite elements for the far field. Because of the large number of degrees of freedom of the FE model a modal truncation technique based on a complex eigenvalue analysis is performed. The reduced model is transformed into the state space form. The developed approach is applied to design the smart car engine. For the design process structural FE simulations of the car engine were carried out to identify the most dominant mode shapes within a frequency range of 0–1,200 Hz. Based on these results optimal actuator positions were calculated. Additionally, the exterior noise radiation of a stripped car engine was numerically analyzed by applying a mixture of finite and infinite elements. A velocity feedback control algorithm in a real collocated design was used to obtain a high active damping effect. To demonstrate the applicability and validity of the developments, test simulations are carried out, which were compared with measurements performed on a laboratory set-up. This comparison between the experimental and numerical results shows a good agreement. In order to demonstrate that the designed system works also under real operating conditions tests on a fired engine have been carried out. With a velocity feedback control, attenuations up to 4 dB in the vibration level were achieved at the resonance frequency regions of the most dominant modes.

References

1. Bin, L., Yugang, L., Xuegang, Y., Shanglian, H.: Maximal modal force rule for optimal placement of point piezoelectric actuators for plates. J. Intell. Mater. Syst. Struct. **11**(7), 512–515 (2000)
2. Everstine, G.C.: Finite element formulations of structural acoustics problems. Comput. Struct. **65**(3), 307–321 (1997)
3. Gabbert, U., Berger, H., Köppe, H., Cao, X.: On modelling and analysis of piezoelectric smart structures by the finite element method. Appl. Mech. Eng. **5**(1), 127–142 (2000)
4. Geers, T.L.: Doubly asymptotic approximations for transient motions of submerged structures. J. Acoust. Soc. Am. **64**(5), 1500–1508 (1978)
5. Heintze, O., Misol, M., Algermissen, S., Hartung, C.F.: Active structural acoustic control for a serial production truck oil pan: experimental realization. In: Proceedings of the Adaptronic Congress, Berlin, pp. 147–153 (2008)
6. Kaljević, I., Saravanos, D.A.: Steady-state response of acoustic cavities bounded by piezoelectric composite shell structures. J. Sound Vib. **204**(3), 459–476 (1997)
7. Khan, M.S., Cai, C., Hung, K.C., Varadan, V.K.: Active control of sound around a fluid-loaded plate using multiple piezoelectric elements. Smart Mater. Struct. **11**, 346–354 (2002)

8. Kollmann, F.G.: Maschinenakustik. Grundlagen, Meßtechnik, Berechnung, Beeinflussung. Springer, Berlin (2000)
9. Lefèvre, J.: Finite element simulation of adaptive light-weight structures for vibration and noise reduction (in German). VDI Verlag, Düsseldorf (2007)
10. Lefèvre, J., Gabbert, U.: Finite element modelling of vibro-acoustic systems for active noise reduction. Technol. Mech. **25**(3–4), 241–247 (2005)
11. Lerch, R., Landes, H., Friedrich, W., Hebel, R., Hoss, A., Kaarmann, H.: Modelling of acoustic antennas with a combined finite-element-boundary-element method. In: Proceedings of the IEEE Ultrasonics Symposium, Tucson, pp. 643–654 (1992)
12. Marinković, D., Köppe, H., Gabbert, U.: Numerically efficient finite element formulation for modeling active composite laminates. Mech. Adv. Mater. Struct. **13**, 379–392 (2006)
13. Naake, A., Schmidt, K., Meschke, J., Weyer, T., Knorr, A., Weiser, J., Rehfeld, M., Rödig, T.: Vehicle windshield with active noise reduction. In: Proceedings of the Adaptronic Congress, Göttingen, vol. 22, pp. 1–5 (2007)
14. Nestorović, T., Seeger, F., Köppe, H., Gabbert, U.: Controller design for the active vibration suppression of a car roof. In: Proceedings of the International Congress Motor Vehicles & Motors, Kragujevac (2006)
15. Pinto Correia, I.F., Mota Soares, C.M., Mota Soares, C.A., Normam, J.H.: Active control of axisymmetric shells with piezoelectric layers: a mixed laminated theory with a high order displacement field. Comput. Struct. **80**(1/2), 2265–2275 (2002)
16. Redaelli, M., Manzoni, S., Cigada, A., Wimmel, R., Siebald, H., Fehren, H., Schiedewitz, M., Wolff, K., Lahey, H.-P., Nussmann, C., Nehl, J., Naake, A.: Different techniques for active and passive noise cancellation at powertrain oil pan. In: Proceedings of the Adaptronic Congress, Göttingen, vol. 21, pp. 1–8 (2007)
17. Ringwelski, S., Gabbert, U.: Modeling of a fluid-loaded smart shell structure for active noise and vibration control using a coupled finite element-boundary element approach. Smart Mater. Struct. **19**(10), 105009 (2010)
18. Seeger, F.: Simulation und Optimierung adaptiver Schalenstrukturen. VDI Verlag, Düsseldorf (2004)
19. Verhoosel, C.V., Gutierrez, M.A.: Modelling inter- and transgranular fracture in piezoelectric polycrystals. Eng. Fract. Mech. **76**(6), 742–760 (2009)
20. Weyer, T., Monner, H.P.: PKW-Innenlärmreduzierung durch aktive Beruhigung der durch die Motorharmonischen erregten Dachblech-Schwingungen. Motor und Aggregateakustik (2003)
21. Zhang, Z., Chen, Y., Yin, X., Hua, H.: Active vibration isolation and underwater sound radiation control. J. Sound Vib. **318**, 725–736 (2008)

Magnetic Techniques for Estimating Elastic and Plastic Strains in Steels Under Cyclic Loading

E.S. Gorkunov, R.A. Savrai, and A.V. Makarov

Abstract The effect of high-cycle fatigue loading (elastic deformation) of high-carbon steel (1.03 wt% C) on the behaviour of the tangential component of the magnetic induction vector of a specimen in the residual magnetization state has been studied. It has been found that the magnetic measurement technique allows both structural changes and cracks resulting from the fatigue degradation of high-carbon pearlitic steel to be recorded. The effect of cyclic loading in the low-cycle fatigue region (plastic deformation) on the variations in the coercive force and residual magnetic induction of annealed medium-carbon steel (0.45 wt% C) for the major and minor magnetic hysteresis loops has also been studied. The sensitivity of the magnetic characteristics to both large and small plastic strains accumulated during cyclic loading has been established.

1 Introduction

In most machines and structures components operate under conditions of cyclically varying loads with regularly or irregularly alternating cycles and different levels of stresses in the cycles. The level of stresses under different conditions may correspondingly vary over a wide range and cause both elastic and plastic strains in the components. This significantly complicates the study of fatigue resistance, the prediction of durability and the determination of residual life, and it requires a vast amount of experimental data and full-scale tests. It is therefore necessary to apply methods that, along with collecting experimental data, would enable one to understand the physical background of fatigue as a phenomenon. In this connection,

E.S. Gorkunov (✉) · R.A. Savrai · A.V. Makarov
Institute of Engineering Science, RAS (Ural Branch), 34 Komsomolskaya str., 620049 Ekaterinburg, Russia
e-mail: ges@imach.uran.ru

Alexander K. Belyaev et al. (eds.), *Mechanics and Model-Based Control of Advanced Engineering Systems*, DOI 10.1007/978-3-7091-1571-8_15,
© Springer-Verlag Wien 2014

the application of nondestructive test methods, particularly magnetic ones, is very promising.

In the course of fatigue loading, repeated elastic deformation induces structural changes and accumulated damage thus affecting the magnetic properties of a specimen. To detect fatigue damage of the kind, it is promising to apply magnetic inspection of articles in the residual magnetization state with the use of highly sensitive transducers enabling one to make local measurements of stray magnetic fields, including those arising at defects [1, 2].

Among the magnetic characteristics used for evaluating the physical-mechanical properties of a material under plastic deformation, one can select coercive force and residual magnetic induction, which are affected by changes in dislocation density and the appearance of discontinuities (plastic loosening, microcracks and pores) [3, 4]. In this connection, it seems reasonable to divide changes in the magnetic characteristics and microstructure of a material under plastic deformation into three stages, namely, (1) at small strains; (2) at medium strains; (3) after large strains [4].

Thus, the aim of this work is to study the effect of cyclic loading of (1) high-carbon pearlitic steel (1.03 wt% C) under high-cycle fatigue (elastic deformation) on the behaviour of the tangential component of the magnetic induction vector for a specimen in the residual magnetization state; (2) annealed medium-carbon steel (0.45 wt% C) under low-cycle fatigue (plastic deformation) on the behaviour of coercive force and residual magnetic induction for the major and minor magnetic hysteresis loops.

2 Experimental Procedure and Material

Commercially cast high-carbon (1.03 wt% C) and medium-carbon (0.45 wt% C) steels were studied. The structure of fine-lamellar pearlite in high-carbon steel was obtained by 5 min isothermal holding of specimens (preheated for 15 min to 1,050 °C) in a salt bath at 500 °C followed by cooling in water. The specimens were then annealed in a salt bath at 650 °C for 10 min. Medium-carbon steel was annealed at 800 °C for 8 h and then cooled slowly with an oven to room temperature.

The specimens made from high-carbon pearlitic steel were cyclically loaded with controlled stress values $\Delta\sigma = 2\sigma_a = 0.65\sigma_{0.2}$ (where $\sigma_{0.2}$ is offset yield strength under static tension), the cycle asymmetry coefficient $R_\sigma = 0$ (intermittent zero-to-tension stress cycle) and a loading frequency of 10 Hz, the cycle stress variations obeying the sine law. The specimen was loaded step by step, with the numbers of loading cycles $N = 20{,}000, 60{,}000, 100{,}000, 160{,}000$. The specimens made from annealed medium-carbon steel were cyclically loaded with a controlled value of total strain $\epsilon_{tot} = 2\epsilon_a = \epsilon_{el} + \epsilon_{pl} = 0.0076$ (where ϵ_a is total strain amplitude, ϵ_{el} is the elastic part of the strain, ϵ_{pl} is the plastic part of the strain), a pulsating deformation cycle, the triangular change of the strain amplitude and a loading frequency of 0.5 Hz. The testing was conducted so that the cycle asymmetry coefficients ($R_\epsilon = \epsilon_{min}/\epsilon_{max}$, $R_\sigma = S_{min}/S_{max}$) met the condition $R_\epsilon = R_\sigma = 0$. The specimens were cyclically loaded with $N = 5, 10, 50, 200$ and 400 cycles.

Magnetic measurements of the tangential component of the magnetic induction vector B_t for high-carbon pearlitic steel were made with a flux-gate transducer by scanning the surface of a residually magnetized specimen along its axis at a velocity of $2\,mm\,s^{-1}$. The coercive force and residual magnetic induction of annealed medium-carbon steel were measured for the major magnetic hysteresis loop ($H_{max} = 60$ kA/m) and minor ones corresponding to the maximum magnetic induction of a hysteresis cycle $b_{max} = 1$, 0.4, 0.1 and 0.05 T respectively.

3 Results and Discussion

Let us consider the effect of high-cycle fatigue loading (elastic deformation) of high-carbon pearlitic steel on the variation of the tangential component of the magnetic induction vector B_t. It follows from Fig. 1 that the fatigue tests are accompanied by the appearance of nonuniformity in the distribution of the tangential component of the magnetic induction vector along the specimen length in the residual magnetization state; this is noted even at $N = 60{,}000$. After fatigue loading with $N = 100{,}000$, a noticeable increase in B_t was observed. This may be due to structural changes occurring in the pearlitic steel during fatigue loading, in particular, to the spheroidization of cementite lamellae. The spheroidization of fine cementite lamellae during cyclic tension under high-cycle fatigue conditions is due to the cooperative effect of elastic tensile stresses, microplastic strain and local heating accelerating the diffusion of iron and carbon atoms [5].

With the number of cycles $N = 160{,}000$, a clearly pronounced peak is observed in the distribution of the tangential component of the magnetic induction vector B_t (see Fig. 1). It is due to the appearance of a main fatigue crack initiated at the stress concentrator on the specimen surface (Fig. 2). Note that the width of the fatigue crack opening is less than 1 μm, and this testifies to the feasibility of the nondestructive inspection of the fatigue degradation of steels under high-cycle fatigue loading (elastic deformation).

Consider now the effect of low-cycle fatigue loading (plastic deformation) of annealed medium-carbon steel on its magnetic behaviour.

The initial annealed state is characterized by minimum values of the coercive force, see Table 1.

Under cyclic loading, the increasing structural imperfection density is accompanied by increasing values of the critical fields of interaction between domain walls and defects [6]. This results in the growth of coercive force values for both major and minor magnetic hysteresis loops (Fig. 3a, b). Note a considerable difference between the behaviour of the coercive force in strong fields and that in weak ones. For weak fields (0.1 T and lower, when the maximum hysteresis loop field $h_{max} < H_c$), the coercive force for minor magnetic hysteresis loops continuously rises during the whole deformation process, with a sharp increase observed at the early stage of

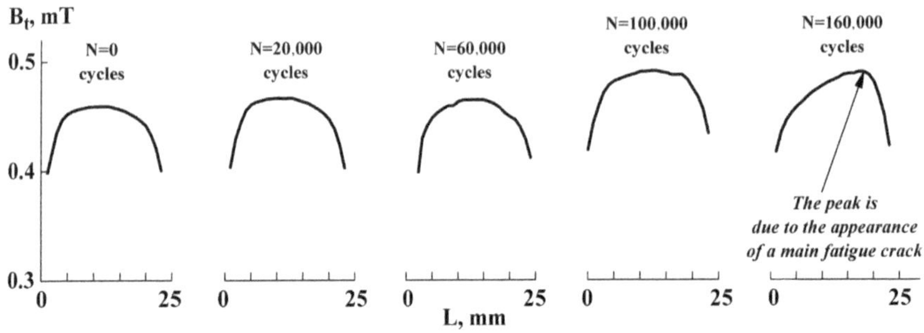

Fig. 1 Distribution of the tangential component of the magnetic induction vector B_t along the gage length of a specimen in the residual magnetization state before loading (N = 0) and after loading with a given numbers of cycles. The test results are given for the specimen side where a main fatigue crack was found

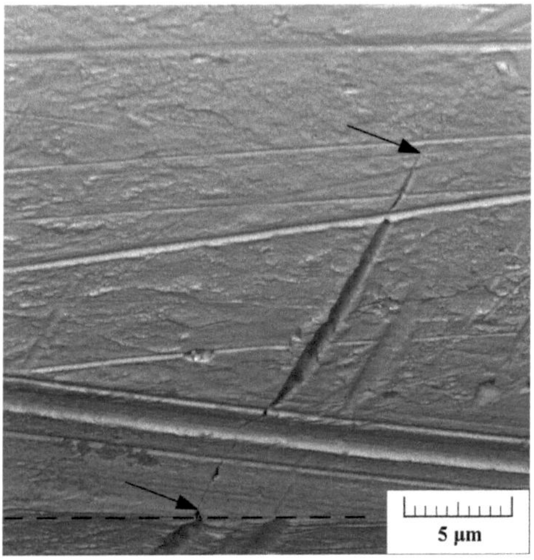

Fig. 2 An electron micrograph (SEM) of the specimen surface (after loading with the number of cycles $N = 160{,}000$). The *arrows* show a fatigue crack. The *dashed line* corresponds to the specimen edge parallel to the loading axis

Table 1 The values of coercive force (h_c, H_c) and residual magnetic induction (b_r, B_r) for medium-carbon steel in the initial annealed state (the number of cycles $N = 0$, accumulated plastic strain $\epsilon_\Sigma = 0$)

$b_{max} = 0.05\,T$		$b_{max} = 0.1\,T$		$b_{max} = 0.4\,T$		$b_{max} = 1\,T$		$H = 60$ kA/m (main loop)	
h_c, A/m	b_r, T	h_c, A/m	b_r, T	h_c, A/m	b_r, T	h_c, A/m	b_r, T	H_c, A/m	B_r, T
15.55	0.017	31.8	0.043	99.7	0.3	146.2	0.766	194.5	1.159

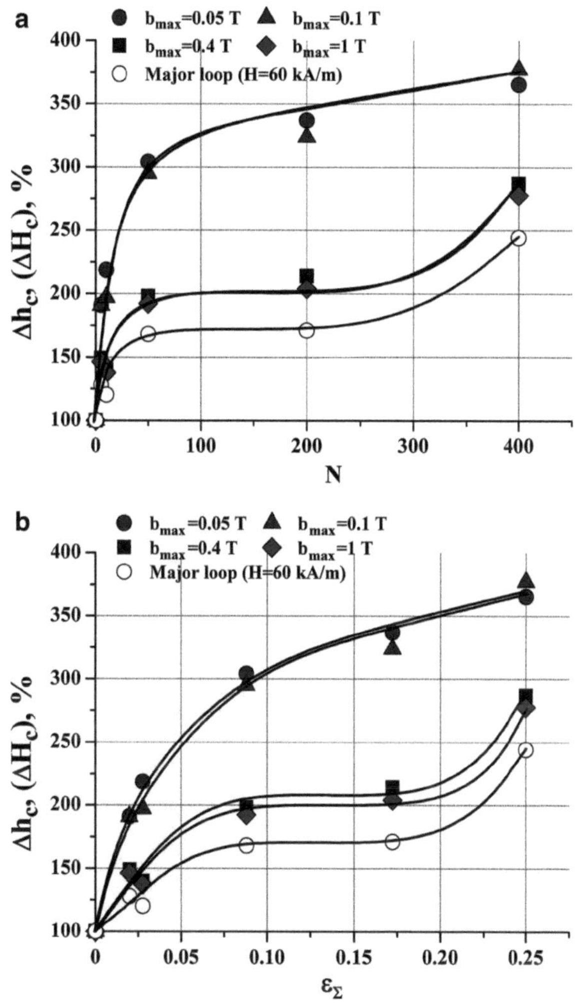

Fig. 3 Relative change in the coercive force as dependent on the number of loading cycles (**a**) and accumulated plastic strain (**b**) for medium-carbon steel specimens. Coercive force values in the initial annealed state are taken to be 100 %, see Table 1

deformation and a subsequent smoother growth at larger strains. The reason is that, in weak magnetic fields, domain walls interact mainly with single dislocations or structural imperfections having small critical fields $H_{cr} \approx h_{max}$.

Under magnetization reversal in stronger fields, one can see the stabilization of coercive force values as the accumulated plastic strain varies between 0.07–0.1 and 0.15–0.17. The stabilization may be caused by the formation of a cellular dislocation structure. In this case, domain walls interact with both individual dislocations and cell walls [6]. The further deformation is again accompanied by increasing coercive force values, and this is attributable to the appearance of microscopic pores, which grow in size and number up to fracture. Being a source of stray fields, the microscopic pores hinder magnetization reversal according to the inclusion theory [7].

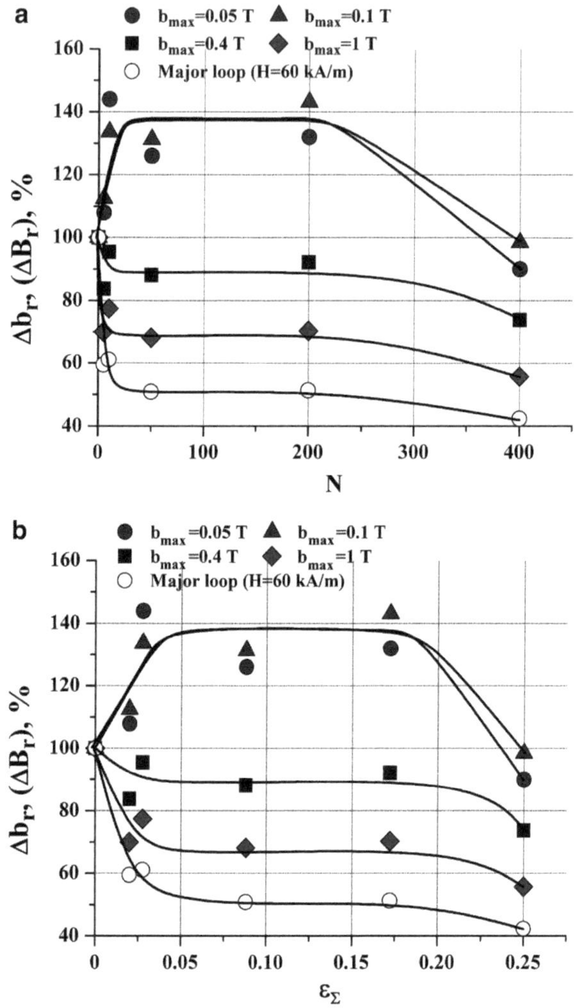

Fig. 4 Relative change in residual magnetic induction as dependent on the number of loading cycles (**a**) and accumulated plastic strain (**b**) for medium-carbon steel specimens. Residual induction values in the initial annealed state are taken to be 100 %, see Table 1

In the early stage of deformation, at a total plastic strain of 0.07–0.1 and lower, the values of residual magnetic induction decrease for the major loop and the minor loops corresponding to the maximum magnetic induction $b_{max} = 1$ and 0.4 T, and (at a total plastic strain of 0.02–0.03 and lower) they increase for the minor loops corresponding to the maximum magnetic induction $b_{max} = 0.1$ and 0.05 T (Fig. 4a, b).

Thus, as is the case with the coercive force, there is a difference between the variation of residual magnetic induction in week fields and that in strong ones. On the one hand, segments with large local microstresses (in particular, dislocations) are the sites where magnetization reversal centers are easily generated [7]. This contributes to a decrease in residual magnetic induction for strong fields. On the other hand, dislocation redistribution induces regions with a less imperfect structure. Magnetization in weak fields is accompanied by small domain wall displacements, which do not seem to be larger than these regions. Therefore the values of residual magnetic induction for weak fields grow up to the complete formation of a cellular dislocation structure in the whole bulk of the material. The values of residual magnetic induction stabilize thereafter up to a strain of 0.15–0.17. The further deformation is accompanied by decreasing residual magnetic induction regardless of the field magnitude, this being attributable to the appearance of microscopic pores. On macrodefects, (cracks and pores) there appear stray fields [7] directed oppositely to the magnetizing field, and this finally results in lower values of residual induction.

4 Conclusion

It has been found that the high-cycle fatigue loading (elastic deformation) of high-carbon steel specimens (1.03 wt% C) up to fatigue cracking causes an increase in the tangential component of the magnetic induction vector of a specimen in the residual magnetization state and the nonuniform distribution of the component along the specimen length. This is due to structural changes (spheroidization of cementite lamellae) occurring in high-carbon pearlitic steel under high-cycle fatigue loading. The formation of a main fatigue crack is accompanied by the appearance of a clearly pronounced peak in the distribution of the tangential component of the magnetic induction vector. The possibility of inspecting the fatigue degradation of steels under cyclic loading in the region of high-cycle fatigue has been shown.

The dependences describing the behaviour of the coercive force and residual magnetic induction for the major and minor magnetic hysteresis loops as a function of the number of loading cycles (the value of accumulated plastic strain) have been obtained for cyclically loaded specimens made from annealed medium-carbon steel (0.45 wt% C) in the low-cycle fatigue region. The dependences testify to the sensitivity of the magnetic characteristics studied to both large and small plastic strains. The possibility of using the values of magnetic parameters to inspect plastic strain accumulated during cyclic loading has been demonstrated.

Acknowledgements The work was supported by project No. 12-P-1-1027 according to RAS Presidium program No. 25.

References

1. Gorkunov, E.S., Novikov, V.F., Nichipuruk, A.P., Nassonov, V.V., Kadrov, A.V., Tatlybaeva, I.N.: Residual magnetization stability of heat-treated steel articles to the action of elastic deformations. Defektoskopiya (2), 68–76 (1991)
2. Gloria, N.B.S., Areiza, M.C.L., Miranda, I.V.J., Rebello, J.M.A.: Development of a magnetic sensor for detection and sizing of internal pipeline corrosion defects. NDT&E Int. **42**(8), 669–677 (2009)

3. Shah, M.B., Bose, M.S.C.: Magnetic NDT technique to evaluate fatigue damage. Phys. Stat. Solidi (a). **86**(1), 275–281 (1984)
4. Babich, V.K., Pirogov, V.A.: On the features of coercive force change under deformation of annealed carbon steels. Phys. Met. Metall. **28**(3), 447–453 (1969)
5. Makarov, A.V., Savrai, R.A., Schastlivtsev, V.M., Tabatchikova, T.I., Yakovleva, I.L., Egorova, L.Y.: Structural features of the behavior of a high-carbon pearlitic steel upon cyclic loading. Phys. Met. Metall. **111**(1), 95–109 (2011)
6. Vicena, F.: On the influence of dislocations on the coercive force of ferromagnetics. Czechosl. J. Phys. **5**(4), 480–499 (1955)
7. Mikheev, M.N., Gorkunov, E.S.: Magnetic Methods of Structure Analysis and Nondestructive Testing. Nauka, Moscow (1993)

Induction Machine Torque Control with Self-Tuning Capabilities

Bojan Grcar, Anton Hofer, Gorazd Stumberger, and Peter Cafuta

Abstract In any torque control concept for induction machines the capability to produce a desired torque with the minimum stator current (maximum torque per ampere ratio) becomes a demanding requirement. In this contribution it is shown how a recently published torque control strategy which provides the maximum torque per ampere ratio feature can be augmented by some self-tuning capabilities. Since the behaviour of the proposed controller is mainly influenced by the rotor time constant which is subjected to variations due to thermal influences on the rotor resistance a method for an on-line tuning of this parameter is developed. It is based on the extremum seeking approach, which is well known from adaptive control. The performance of the control concept is demonstrated by experimental results.

1 Introduction

The squirrel cage induction machine can be regarded as the working horse in industrial applications although it constitutes a demanding control problem. Many control concepts have been established for this type of electrical machinery, the most popular being the field oriented approach [2] and the direct torque control [8]. In recent years energy efficiency has become increasingly important. Thus besides the required dynamic performance the so-called maximum torque per ampere ratio, i.e. the ability to produce a desired torque with the minimum stator current amplitude, becomes a stringent design goal. Some strategies have already been proposed

B. Grcar · G. Stumberger · P. Cafuta
Faculty of Electrical Engineering and Computer Science, University of Maribor, 2000 Maribor, Slovenia

A. Hofer (✉)
Institute of Automation and Control, Graz University of Technology, 8010 Graz, Austria
e-mail: anton.hofer@tugraz.at

Alexander K. Belyaev et al. (eds.), *Mechanics and Model-Based Control of Advanced Engineering Systems*, DOI 10.1007/978-3-7091-1571-8_16,
© Springer-Verlag Wien 2014

which augment the classical control concepts by some superordinate optimization schemes in order to cope with this energy efficiency problem. Nevertheless a control law would be preferable which a priori satisfies the maximum torque per ampere ratio requirement. Such a torque control concept has been proposed in [5] and extended in [6]. It is a characteristic feature of this approach that the torque error is directly mapped into the stator current vector without any rotor flux control loop. The amplitude and the rotation speed of the stator current vector are modulated appropriately such that the maximum torque per ampere feature is valid not only in steady state operating conditions but also during transients. In order to overcome a singularity problem at some set points the control is switched to a simple time optimal flux control in these cases.

The proposed controller depends only on a small number of parameters, the most important one being the rotor time constant. Mainly due to variations of the rotor resistance this parameter must be considered as uncertain. Therefore it is the purpose of this contribution to provide an on-line tuning facility for this critical item. It will be shown that the so-called extremum seeking method [1] can be applied to this parameter adaption problem.

2 Torque Controller with Maximum Torque per Ampere Ratio

The development of the torque control concept is based on the standard two axis model of the induction machine (see e.g. [7]) assuming linear magnetics. Here $i_{sd}, i_{sq}, u_{sd}, u_{sq}$ and ψ_{rd}, ψ_{rq} denote the components of the stator current vector, the stator voltage vector and the rotor flux vector in an arbitrary $d-q$ reference frame rotating with the angular velocity ω_a while T_{el} denotes the torque produced by the machine

$$
\begin{aligned}
\frac{d}{dt}\begin{bmatrix} i_{sd} \\ i_{sq} \end{bmatrix} &= \begin{bmatrix} -\frac{1}{\tau_\sigma} & \omega_a \\ -\omega_a & -\frac{1}{\tau_\sigma} \end{bmatrix}\begin{bmatrix} i_{sd} \\ i_{sq} \end{bmatrix} + \frac{M}{L_r L_\sigma}\begin{bmatrix} \frac{1}{\tau_r} & \omega_m \\ -\omega_m & \frac{1}{\tau_r} \end{bmatrix}\begin{bmatrix} \psi_{rd} \\ \psi_{rq} \end{bmatrix} + \frac{1}{L_\sigma}\begin{bmatrix} u_{sd} \\ u_{sq} \end{bmatrix} \\
\frac{d}{dt}\begin{bmatrix} \psi_{rd} \\ \psi_{rq} \end{bmatrix} &= \begin{bmatrix} -\frac{1}{\tau_r} & \omega_a - \omega_m \\ -\omega_a + \omega_m & -\frac{1}{\tau_r} \end{bmatrix}\begin{bmatrix} \psi_{rd} \\ \psi_{rq} \end{bmatrix} + \frac{M}{\tau_r}\begin{bmatrix} i_{sd} \\ i_{sq} \end{bmatrix} \qquad (1) \\
T_{el} &= \frac{M}{L_r}\left(\psi_{rd} i_{sq} - \psi_{rq} i_{sd}\right).
\end{aligned}
$$

An induction machine with two poles only is considered for simplicity. The parameters of this model are the rotor resistance R_r, the rotor inductance L_r, the stator resistance R_s, the stator inductance L_s, the mutual inductance M and the angular velocity of the rotor ω_m. Further parameters are derived by the following relations: $L_\sigma = L_s - M^2/L_r$, $R_\sigma = R_r + M^2 R_r/L_r^2$, the leakage time constant $\tau_\sigma = L_\sigma/R_\sigma$ and the rotor time constant $\tau_r = L_r/R_r$. Now the rotor reference frame is selected, i.e $\omega_a = \omega_m$ and a *perfect* current control loop is assumed, i.e. the stator current vector (perfectly tracking its prescribed reference signal) can be considered as the control input. Since usually the machine torque T_{el} has to be estimated, this fact is taken into account by a linear first order dynamic with time constant $\tau_f \ll \tau_r$ for the estimated machine torque $\hat{T}_{el}$. Under these assumptions a simplified model for the induction machine can be derived as

$$\frac{d}{dt}\begin{bmatrix} \psi_\gamma \\ \psi_\delta \end{bmatrix} = -\frac{1}{\tau_r}\begin{bmatrix} \psi_\gamma \\ \psi_\delta \end{bmatrix} + \frac{M}{\tau_r}\begin{bmatrix} i_\gamma \\ i_\delta \end{bmatrix} \tag{2}$$

$$\frac{d\hat{T}_{el}}{dt} = -\frac{1}{\tau_f}\hat{T}_{el} + \frac{M}{L_r\tau_f}\left(\psi_\gamma i_\delta - \psi_\delta i_\gamma\right)$$

where $i_\gamma, i_\delta, \psi_\gamma, \psi_\delta$ denote the components of the stator current vector and the rotor flux vector in the rotor reference frame. This type of mathematical model is termed nonholonomic integrator with drift (Brockett or Heisenberg system, see e.g. [3]) which is well known to establish a challenging control problem. The nonholonomic nature of the model is given by the constraint $T_{el} = \left(\psi_\gamma\dot{\psi}_\delta - \psi_\delta\dot{\psi}_\gamma\right)\tau_r/L_r =$ const. which has to be satisfied if a desired *constant* torque has to be produced. In order to meet this constraint the rotor flux vector ψ_r and also the stator current vector i_s must trace out *closed orbits*. The simplest solution to this problem is to choose harmonic functions for the control inputs $i_\gamma = A\cos\omega t$ and $i_\delta = A\sin\omega t$. Now the crucial question arises, how the value of ω has to be selected such that the desired constant torque T_{el}^* is obtained with *minimum* amplitude A (maximum torque per ampere ratio) in steady state. Simple computations given in [6] reveal that the solution can be derived as

$$\omega^* = \operatorname{sgn}\left(T_{el}^*\right)\frac{1}{\tau_r},\ \ A^* = \frac{1}{M}\sqrt{2L_r\left|T_{el}^*\right|},\ \ \|\psi_r^*\| = \sqrt{L_r\left|T_{el}^*\right|},\ \ \varphi^* = \operatorname{sgn}(T_{el}^*)\frac{\pi}{4} \tag{3}$$

where φ^* denotes the optimal angle between the stator current vector and the rotor flux vector.

In [5] a nonlinear state controller has been proposed which intrinsically provides the maximum torque per ampere ratio feature. Using the following abbreviations $\Delta T := T_{el}^* - \hat{T}_{el}$ (torque error), $s_T := \operatorname{sgn}\left(T_{el}^*\right)$ and the design parameter $k_P > 0$ the control law can be written as:

$$\begin{bmatrix} i_\gamma \\ i_\delta \end{bmatrix} = \frac{\tau_r}{M}\begin{bmatrix} \dfrac{1}{\tau_r} + s_T k_P \Delta T & -(s_T\dfrac{1}{\tau_r} + k_P\Delta T) \\ s_T\dfrac{1}{\tau_r} + k_P\Delta T & \dfrac{1}{\tau_r} + s_T k_P \Delta T \end{bmatrix}\begin{bmatrix} \psi_\gamma \\ \psi_\delta \end{bmatrix} \tag{4}$$

Inserting this controller into the mathematical model (2) of the plant leads to the description of the closed loop system:

$$\frac{d}{dt}\begin{bmatrix}\psi_\gamma \\ \psi_\delta \\ \hat{T}_{el}\end{bmatrix} = \begin{bmatrix} s_T k_P \Delta T \psi_\gamma - (s_T \dfrac{1}{\tau_r} + k_P \Delta T)\psi_\delta \\ (s_T \dfrac{1}{\tau_r} + k_P \Delta T)\psi_\gamma + s_T k_P \Delta T \psi_\delta \\ -\frac{1}{\tau_f}\hat{T}_{el} + \frac{\tau_r}{L_r \tau_f}(s_T \dfrac{1}{\tau_r} + k_P \Delta T)(\psi_\gamma^2 + \psi_\delta^2) \end{bmatrix} \tag{5}$$

It can be shown by a nonlinear state transformation and a suitable Lyapunov function that the closed loop system has the following properties: For any $\psi_r(0) \neq 0$ and any $\hat{T}_{el}(0)$ and any constant $T_{el}^* \neq 0$ the solutions of the closed loop system converge to

$$\overline{\psi}_\gamma = \|\psi_r^*\| \cos(\omega^* t - \varphi^*) \qquad \overline{i}_\gamma = A^* \cos \omega^* t$$

$$\overline{\psi}_\delta = \|\psi_r^*\| \sin(\omega^* t - \varphi^*) \qquad \overline{i}_\delta = A^* \sin \omega^* t$$

Furthermore $\overline{\hat{T}}_{el} = T_{el}^*$ and the angle between i_s and ψ_r is *always* $s_T \pi/4$ i.e. the maximum torque per ampere ratio is valid even during *transients* of the control system.

Obviously the proposed control law has a problem of singularity at $\psi_r(0) = 0$ or in the case $T_{el}^* = 0$. In order to cope with this difficulty the control concept has been extended by a time optimal flux controller in [6]. Considering the fact that the stator current vector has to be kept bounded during operation, i.e. $\|i_s\| \leq i_{\max}$, where $i_{\max}$ denotes some positive constant (consequently the desired torque is also limited by $|T_{el}^*| \leq M^2 i_{\max}^2/(2L_r)$), the time optimal flux controller can be determined as follows: Choose $i_\gamma = v \cos\beta$, $i_\delta = v \sin\beta$ and set

$$v = \begin{cases} i_{\max} & \text{if } \|\psi_r\| \neq \|\psi_r^*\| \\ \frac{1}{M}\sqrt{2L_r |T_{el}^*|} & \text{if } \|\psi_r\| = \|\psi_r^*\| \end{cases} \qquad \beta = \begin{cases} \theta & \text{if } \|\psi_r\| < \|\psi_r^*\| \\ \theta + s_T \dfrac{\pi}{4} & \text{if } \|\psi_r\| = \|\psi_r^*\| \\ \theta + \pi & \text{if } \|\psi_r\| > \|\psi_r^*\| \end{cases} \tag{6}$$

where $\theta := \arctan \frac{\psi_\delta}{\psi_\gamma}$ with the addition $\theta = 0$ if $\psi_r = 0$. This controller provides the following behaviour of the closed loop system: The desired flux amplitude $\|\psi_r^*\|$ is reached in minimum time starting from any $\psi_r(0)$. Immediately afterwards the desired torque T_{el}^* is produced with minimum current amplitude $\|i_s\| = A^*$.

In order to combine both controllers a simple selection logic consisting of a flip-flop is introduced as shown in Fig. 1. The inputs R and S of the flip-flop are determined by the following conditions, where $\varepsilon_1, \varepsilon_2$ denote small positive constants:

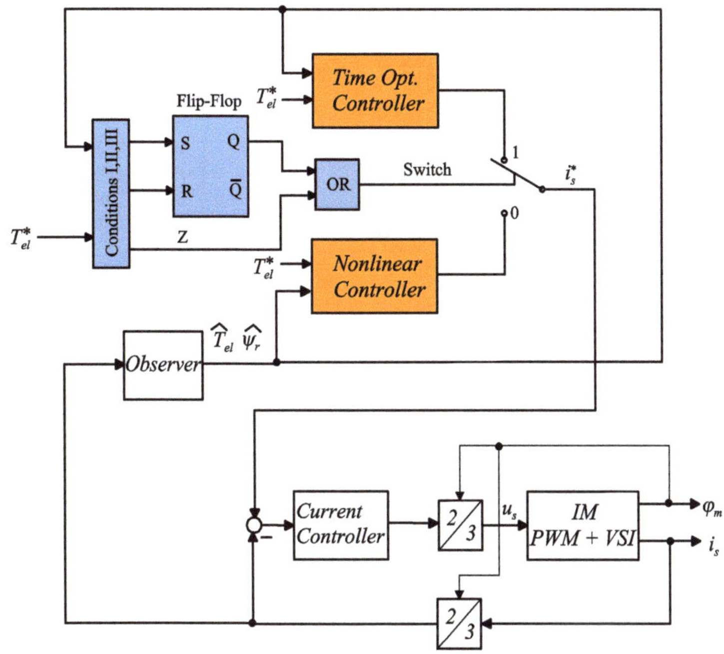

Fig. 1 Torque control system

$$\text{If } \hat{\psi}_\gamma^2 + \hat{\psi}_\delta^2 \leq \varepsilon_1 \text{ then } S = 1 \text{ otherwise } S = 0$$

$$\text{If } \left| T_{el}^* - \hat{T}_{el} \right| \leq \varepsilon_2 \text{ then } R = 1 \text{ otherwise } R = 0$$

$$\text{If } T_{el}^* = 0 \text{ then } Z = 1 \text{ otherwise } Z = 0$$

As full state information is required in (4) and (6) the control concept has to be augmented by a state observer providing estimates of the rotor flux vector (denoted by $\hat{\psi}_r$ in Fig. 1) and the produced machine torque $\hat{T}_{el}$. In its simplest form the observer is based on the plant model (2) together with the measured stator current. More sophisticated concepts are also available for this purpose.

3 On-Line Tuning of the Controller Parameter

The rotor time constant $\tau_r = L_r/R_r$ is the essential parameter in the proposed nonlinear controller (4) as well as in the observer. Mainly due to thermal effects i.e. variations of the rotor resistance R_r this parameter must be considered as uncertain. A mismatch between the estimated value $\hat{\tau}_r$ used in the controller and in the observer and its true value has the consequence that the maximum torque per ampere ratio is

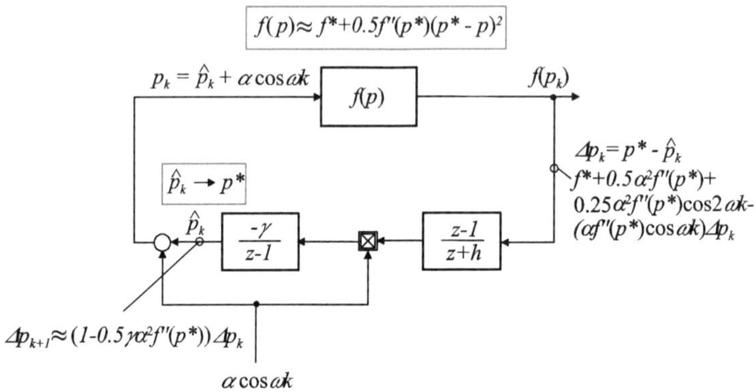

Fig. 2 Extremum seeking scheme

lost, i.e. if $\hat{\tau}_r \neq \tau_r$ we get

$$\overline{T}_{el} = \frac{2\tau_r\hat{\tau}_r}{\tau_r^2 + \hat{\tau}_r^2} T_{el}^*. \tag{7}$$

In the sequel a simple strategy is presented which provides an on-line tuning of $\hat{\tau}_r$. It is based on the observation that from (3) the optimal values $\|\psi_r^*\|$ and $\|i_s^*\| = A^*$ do *not* depend on the parameter τ_r. This leads to the following idea: Under the assumption that the rotor angular velocity ω_m is measured and the load torque is constant, adjust $\hat{\tau}_r$ via on-line optimization such that for a given constant T_{el}^* the steady state rotor speed $\bar{\omega}_m$ is maximized. The on-line optimization of the parameter $\hat{\tau}_r$ can be done by extremum seeking (see e.g. [1]) which is a well known method in adaptive control. A discrete time version of this strategy is shown in Fig. 2. It is assumed that the smooth nonlinear scalar function $f(p)$ of the scalar parameter p has a minimum value f^* at $p = p^*$. Therefore in a suitable neighborhood of p^* the function can be approximated by

$$f(p) \approx f^* + 0.5f''(p^*)(p^* - p)^2. \tag{8}$$

At time instant $t = kT$, where T denotes the sampling period of the optimization algorithm, the parameter $p_k := p(kT)$ is chosen as $p_k = \hat{p}_k + \alpha \cos \omega k$. Here $\hat{p}_k$ denotes the actual estimate of p^* which is perturbed by a 'harmonic' sequence with suitably chosen parameters $\alpha > 0$ and $0 < \omega < \pi$. With $\Delta p_k := p^* - \hat{p}_k$ the value $f(p_k)$ can be written as

$$f(p_k) \approx f^* + 0.5\alpha^2 f''(p^*) + 0.25\alpha^2 f''(p^*)\alpha \cos 2\omega k - \alpha\Delta p_k f''(p^*)\cos \omega k \tag{9}$$

Now the constant part of (9) is assumed to be eliminated by the high pass filter $\frac{z-1}{z+h}$ whose output is modulated by the external perturbation signal $\alpha \cos \omega k$. Finally the update of $\hat{p}_k$ is carried out by the discrete time integrator $\frac{-\gamma}{z-1}$ which also acts as a low pass filter leading to

$$\Delta p_{k+1} \approx \left(1 - 0.5\gamma\alpha^2 f''(p^*)\right) \Delta p_k. \tag{10}$$

A rigorous proof of the convergence of $\hat{p}_k$ to the optimizer p^* in this scheme is a very demanding problem (see e.g. [4]) but from (10) a simple sufficient condition can be derived

$$0.5\gamma\alpha^2 f''(p^*) < 1.$$

It is obvious that in the present application $\hat{\tau}_r$ gets the role of the parameter p and the function to be minimized is given as $f(\hat{\tau}_r) = -|\bar{\omega}_m|$.

4 Experimental Results

The proposed control concept has been applied to an induction machine with the following data: $R_r = 2.9\ [\Omega]$, $R_s = 1.9\ [\Omega]$, $L_r = L_s = 0.23$ [H], $M = 0.22$ [H], $J = 0.05$ [kg m^2], $i_{\max} = 20$ [A], $p = 2$ (number of pole pairs). Thus the nominal value for the rotor time constant is given as $\tau_r = L_r/R_r = 0.08$ [s]. A discrete time version of the controller has been realized on a dSPACE PPC board with the sampling period of $T_d = 250$ [µs], while the extremum seeking for the adjustment of $\hat{\tau}_r$ was running with a sampling period of $T = 3$ [s]. In the first experiment – the results are shown in Fig. 3 – a desired torque $T_{el}^* = 5$ [Nm] has been chosen for $t \geq 0$. It becomes evident that the controller forces the rotor flux linkage vector to reach the desired magnitude $\|\psi_r^*\| = \sqrt{L_r |T_{el}^*|} = 1.07$ [Vs] in minimum time starting from $\psi_r(0) = 0$ and afterwards the desired torque is produced with the minimum current vector magnitude $A^* = \sqrt{2L_r |T_{el}^*|/M} = 6.9$ [A].

In the second experiment an adaption of the uncertain parameter $\hat{\tau}_r$ by the extremum seeking procedure has been performed. The results are summarized in Fig. 4. In the first plot the inverse of τ_r is depicted in order to improve visibility. Initially the value $\hat{\tau}_r^{-1} = 6$ has been set and after a few iterations convergence to the correct value $\tau_r^{-1} \approx 12.6$ can be observed. The parameters of the extremum seeking have been chosen as $\alpha = 2, \gamma = 0.075, \omega = 0.75\pi$ and $h = 0.2$.

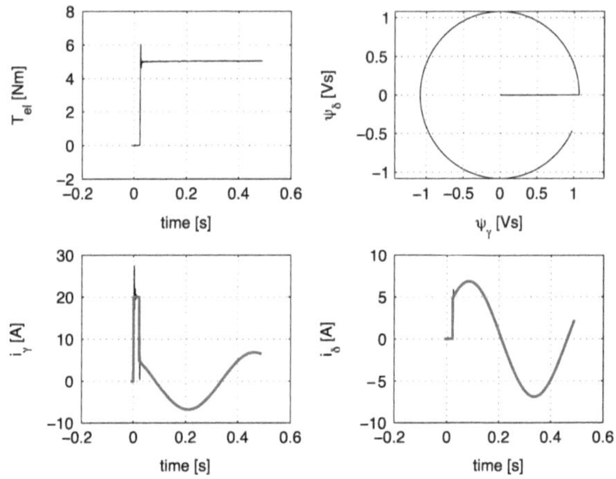

Fig. 3 Starting from singularity $\psi_r(0) = 0$

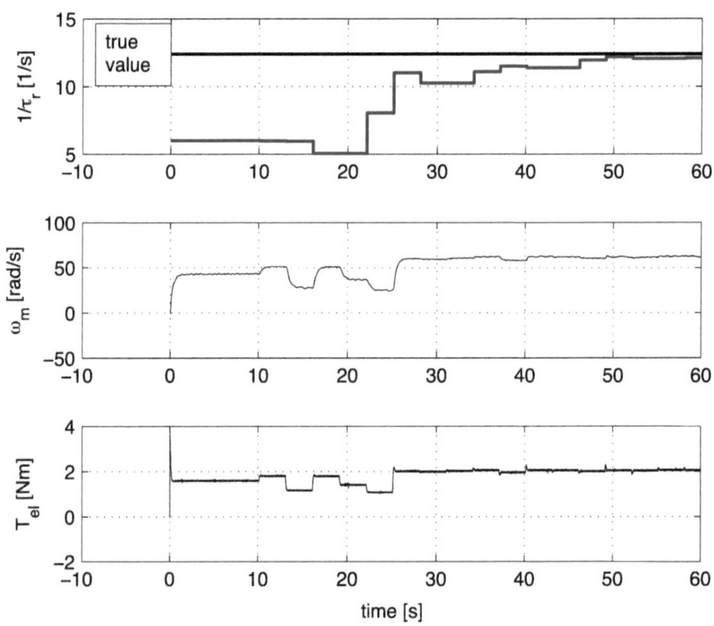

Fig. 4 Adaption of $\hat{\tau}_r$ by extremum seeking

5 Conclusion

An induction machine torque control concept which guarantees the maximum torque per ampere feature has been presented. It is conceptually different form the well established field oriented control and the direct torque control approaches.

The proposed nonlinear controller provides suitable amplitude and frequency modulation of the stator current vector which force the rotor flux linkage vector to trace out periodic orbits. The inherent problem of singularity is avoided by an additional time-optimal controller for the rotor flux linkage vector. In order to cope with parameter variations concerning the rotor time constant an on-line tuning strategy using extremum seeking is proposed. The estimated value of the rotor time constant which is used in the flux observer and in the nonlinear controller is adapted such that for a prescribed torque value the measured rotor speed is maximized. Modifications of the control concept in order to compensate magnetic saturation effects will be the main focus of future work.

References

1. Ariyur, K.B., Krstic, M.: Real-Time Optimization by Extremum-Seeking Control. Wiley, Hoboken (2003)
2. Blaschke, F.: Das Prinzip der Feldorientierung, die Grundlage für die Transvektor-Regelung von Drehfeldmaschinen. Siemens Zeitschrift **45**(10), 757–768 (1971)
3. Brockett, R.: Pattern generation and the control of nonlinear systems. In: Proceedings of the 13th IFAC World Congress, San Francisco, pp. 257–262 (1996)
4. Choi, J.Y., Krstic, M., Ariyur, K.B., Lee, J.S.: Extremum seeking control of discrete-time systems. IEEE Trans. Autom. Control **47**(2), 318–323 (2002)
5. Grcar, B., Cafuta, P., Stumberger, G., Stankovic, A.M., Hofer, A.: Non-holonomy in induction machine torque control. IEEE Trans. Control Syst. Technol. **19**(2), 367–375 (2011)
6. Grcar, B., Hofer, A., Cafuta, P., Stumberger, G.: A contribution to the control of the non-holonomic integrator including drift. Automatica **48**(11), 2888–2893 (2012)
7. Krause, P.C.: Analysis of Electric Machinery. McGraw-Hill, New York (1986)
8. Takahashi, I., Noguchi, N.: A new quick-response and high-efficiency control strategy of an induction motor. IEEE Trans. Ind. Appl. **22**(5), 820–827 (1986)

On Equivalences in the Dynamic Analysis of Layered Structures

Rudolf Heuer

Abstract Flexural vibrations of layered structures composed of moderately thick elastic layers are studied. Alternative formulations of various higher-order theories are introduced that offer complete analogies between the corresponding initial-boundary value problems and those of homogenized single layer structures of effective parameters. Also the effects of an elastic interlayer slip are considered within appropriate equivalences. Moreover, the boundary value problem of a single damping layer showing fractional viscoelastic behavior is treated, where even closed-form solutions can be found for special load cases.

1 Introduction

Although the earlier theories for laminates were based on the Kirchhoff-Love hypothesis, it was soon recognized that, due to the relative small transverse stiffness of composites, thickness-shear deformations should be included to obtain realistic predictions of flexural behavior. Furthermore, for a composite structure whose material and geometric characteristics approach those of a sandwich element, a uniform transverse-shear strain assumption made in most laminate theories becomes unrealistic as it can be ascertained by comparison with a three-dimensional elasticity analysis. Transverse discontinuous mechanical properties cause displacement fields in the thickness direction, which can exhibit a rapid change in slopes corresponding to each layer interface (zig-zag effect). The transverse stresses must fulfill interlaminar continuity at each layer interface.

In particular, a comparative study of different theories for the dynamic response of laminates is given in [1] and [2]. In [3], Reddy presents a review of equivalent

R. Heuer (✉)
Vienna University of Technology, Vienna A-1040/E2063, Austria
e-mail: rudolf.heuer@tuwien.ac.at

Alexander K. Belyaev et al. (eds.), *Mechanics and Model-Based Control of Advanced Engineering Systems*, DOI 10.1007/978-3-7091-1571-8_17,
© Springer-Verlag Wien 2014

single layer and layerwise laminate theories and discusses their mechanical models by means of the FEM and the mesh superposition technique.

If rigid bond between the laminates cannot be achieved, an interlayer slip occurs, that significantly can affect both strength and deformation of the layered structure. The mechanical behavior of layered beams and plates with flexible connection has been mainly discussed for civil engineering structures, compare [4–6]. Thermal and piezoelectric effects in two-layer beams are treated in [7] and [8], respectively. Murakami [9] proposes a general formulation of the boundary value problem, where any interlayer slip law can be adopted in the beam model. A correspondence between the analyses of sandwich beams with or without interlayer slip has been derived by the author in Refs. [10] and [11].

The present paper shows alternative formulations of various theories for layered structures that offer complete analogies between the corresponding initial-boundary value problems and those of homogenized single layer structures of effective parameters. The structures treated are composed of three moderately thick layers and even the effects of geometrically nonlinear large deflection and elastic interlayer slip are considered within appropriate equivalences.

Regarding the modeling of damping layers the paper introduces the initial-boundary value problem of the fractional viscoelastic Euler-Bernoulli beam, where, as a first step, closed-form solutions are introduced for quasi-static loads.

2 First Order Shear Deformation Laminate Theory

2.1 Layered Beams

Considering a layered beam to bend cylindrically and including the effect of transverse shear by means of Timoshenko's kinematic hypothesis the displacement field is expressed as

$$\begin{pmatrix} u(x,z;t) \\ v(x,z;t) \\ w(x,z;t) \end{pmatrix} = \begin{pmatrix} u^{(0)}(x;t) + z\,\psi(x;t) \\ 0 \\ w(x;t) \end{pmatrix}, \tag{1}$$

where the origin of the Cartesian (x,z)-coordinate system is located in the global elastic centroid of the composite cross-section. x represents the axial beam coordinate and $\psi(x;t)$ denotes the cross-sectional rotation. Thus the non-vanishing total strains at any point of the beam become

$$\begin{pmatrix} \varepsilon \\ \gamma \end{pmatrix} \equiv \begin{pmatrix} \varepsilon_x \\ \gamma_{xz} \end{pmatrix} = \begin{pmatrix} \varepsilon^{(0)} + z\,\psi_{,x} \\ w_{,x} + \psi \end{pmatrix}, \quad \varepsilon^{(0)} = u_{,x}^{(0)}. \tag{2}$$

The constitutive relations for a linear thermo-elastic beam can be formulated according to the generalized Hooke's law,

$$\begin{pmatrix} \sigma(x,z;t) \\ \tau(x,z;t) \end{pmatrix} = \begin{pmatrix} E\left[\varepsilon_0 + z\,\psi_{,x} - \alpha\,\theta\right] \\ G\left(w_{,x} + \psi\right) \end{pmatrix}, \tag{3}$$

where $E = E(x,z)$, $G = G(x,z)$ are time-independent Young's modulus and transverse shear modulus, respectively. $\theta = \theta(x,z)$ represents a change of temperature with respect to a stress-free reference configuration, and α stands for the linear thermal expansion coefficient. Without loss of generality, we assume that $E(x,z) = E(z)$, $G(x,z) = G(z)$, and $\alpha(x,z) = \alpha(z)$ in all further derivations. By means of spatial integration, the stress resultants become

$$\begin{pmatrix} N \\ M \\ Q \end{pmatrix} = \begin{bmatrix} D & 0 & 0 \\ 0 & B & 0 \\ 0 & 0 & S \end{bmatrix} \begin{pmatrix} \varepsilon_0 - \bar{\varepsilon}_\theta \\ \psi_{,x} - \bar{\kappa}_\theta \\ w_{,x} + \psi \end{pmatrix}, \quad D = \int_A E\,dA, \quad B = \int_A E\,z^2 dA. \tag{4}$$

The effective shear rigidity, S, follows from the concept of equivalent strain energy.

$$\bar{\varepsilon}_\theta(x;t) = \frac{1}{D}\int_A E\,\alpha\theta\,dA, \quad \bar{\kappa}_\theta(x;t) = \frac{1}{B}\int_A E\,z\alpha\theta\,dA, \tag{5}$$

denote the cross-sectional means of thermal strain and curvature, respectively.

Applying the conservation of momentum and conservation of angular momentum, expressing the stress resultants by means of Eq. (4) and subsequent elimination of the cross-sectional rotation leads to a single fourth-order differential equation of motion for the beam deflection,

$$B w_{,xxxx} + \mu\ddot{w} - \left(\frac{\mu B}{S} + I\right)\ddot{w}_{,xx} + \frac{\mu I}{S}\ddddot{w} = p - \frac{B}{S}p_{,xx} + \frac{I}{S}\ddot{p} - B\,\bar{\kappa}_{\theta,xx}. \tag{6}$$

$p(x;t)$ is an external load distribution, and the inertia terms, containing the mass density $\rho(x)$, are

$$\mu = \int_A \rho\,dA, \quad I = \int_A \rho\,z^2 dA. \tag{7}$$

Equation (6) represents the equation of motion of a homogenized Timoshenko beam with effective parameters according to Eqs. (4), (5), and (7).

2.2 Symmetric Three-Layer Shallow Shells

For thermally loaded shallow shells composed of three isotropic layers with physical properties symmetrically disposed about the middle surface the corresponding equation of motion becomes, compare [12],

$$K\left(1+\frac{n}{S}\right)\Delta\Delta w - n\left[\Delta w - 2\left(H - \frac{K}{S}\Delta H\right)\right] + \mu\ddot{w} - \frac{\mu K}{S}\Delta\ddot{w} =$$
$$= -K(1+\nu)\Delta\bar{\kappa}_\theta, \qquad (8)$$

where the influence of rotatory inertia has been neglected. The corresponding effective parameters are

$$K = \sum_{k=1}^{3}\int_{z_{k-1}}^{z_k}\frac{E_k z^2}{(1-\nu_k^2)}dz,\quad D = \sum_{k=1}^{3}\int_{z_{k-1}}^{z_k}\frac{E_k}{(1-\nu_k^2)}dz,\quad \nu = \frac{1}{K}\sum_{k=1}^{3}K_k\nu_k, \qquad (9)$$

$$\bar{\varepsilon}_\theta = \frac{1}{D(1+\nu)}\sum_{k=1}^{3}\int_{z_{k-1}}^{z_k}\frac{E_k\,\alpha_k\theta_k}{(1-\nu_k)}dz,\quad \bar{\kappa}_\theta = \frac{1}{K(1+\nu)}\sum_{k=1}^{3}\int_{z_{k-1}}^{z_k}\frac{E_k\,\alpha_k\theta_k}{(1-\nu_k)}z\,dz. \qquad (10)$$

In Eq. (8) the influence according to the *Theory of Second Order* is approximately gathered in a mean hydrostatic tensile in-plane force,

$$n = -\frac{D(1+\nu)}{A}\int_A \bar{\varepsilon}_\theta\,dA. \qquad (11)$$

3 Sandwich Beams with or Without Interlayer Slip

Sandwich structures are commonly defined as three-layer type constructions consisting of two thin face layers of high-strength material attached to a moderately thick core layer of low strength and density. Effects of interlayer slip have been discussed for elastic bonding by Hoischen [4] and Goodman and Popov [6], and for more general interlayer slip laws by Murakami [9]. Heuer [10] presents complete analogies between various models of viscoelastic sandwich structures, even with or without interlayer slip, with homogenized single layer structures of effective parameters.

Figure 1 shows the free-body diagram of a three-layer beam. The kinematic assumptions according to the first order shear-deformation theory are applied to each layer. For symmetrically three-layer beams with perfect bonds the following assumptions are made:

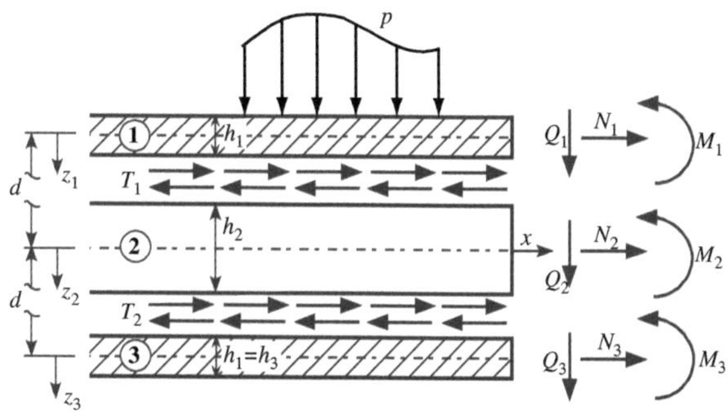

Fig. 1 Geometry and stress resultants of a laterally loaded symmetric three-layer beam

1. The thin faces of high strength material are rigid in shear
2. The individual bending stiffness of the faces are not neglected
3. The bending stiffness of the core is neglected

Alternatively, for sandwich beams with viscoelastic interlayer slip, see [7], the classical assumption of all three layers to be rigid in shear is made, with the shear traction in the physical interfaces of vanishing thickness being proportional to the displacement jumps with a viscoelastic interface stiffness understood. In case of elastic interface stiffness the corresponding equation of motion finally reads

$$w_{,xxxxxx} - \lambda^2 w_{,xxxx} + \frac{\mu}{B_0}\ddot{w}_{,xx} - \lambda^2 \frac{\mu}{B_\infty}\ddot{w} = -\frac{\lambda^2}{B_\infty} p + \frac{1}{B_0} p_{,xx}. \qquad (12)$$

The shear coefficient in (12) is either proportional to the core's shear modulus G_2 in the case of perfectly bonded interfaces, cf. [10],

$$\lambda^2 = \left(\kappa^2 G_2\right) \frac{2b}{h_2} \frac{B_\infty}{D_1 B_0}, \qquad (13)$$

or, for the symmetric three-layer beam with elastic interlayer slip, it becomes proportional to the elastic stiffness k when common to both physical interfaces

$$\lambda^2 = k \frac{B_\infty}{D_1 B_0}. \qquad (14)$$

4 Fractional Viscoelastic Single Layer

4.1 Governing Equations

Considering a single damping layer the following section introduces the initial-boundary value problem of the fractional visco-elastic Euler-Bernoulli beam. Let

$E(t)$ and $D(t)$ the relaxation and the creep function, respectively. $E(t)$ can be interpreted as the stress history for a unit strain $\varepsilon(t) = U(t)$, and $D(t)$ represents the strain history for a unit stress $\sigma(t) = U(t)$, ($U(t)$ being the unit step function). At the beginning of the last century Nutting [13] observed that $E(t)$ is well suited by a power law decay,

$$E(t) = \frac{c_\beta}{\Gamma\left(1-\beta\right)} t^{-\beta}, \qquad 0 < \beta < 1, \tag{15}$$

where $\Gamma(.)$ is the Euler-Gamma function, $c_\beta/\Gamma(1-\beta)$ and β are characteristic coefficients depending on the material at hand. Once $E(t)$ is determined in the form according to Eq. (15) the function $D(t)$ is given as

$$D(t) = \frac{1}{c_\beta \, \Gamma\left(1+\beta\right)} t^{\beta}, \qquad 0 < \beta < 1. \tag{16}$$

Due to Boltzmann superposition principle the stress and strain history may be derived in the form of convolution integrals

$$\sigma(t) = \int\limits_0^t E(t-\bar{t})\dot{\varepsilon}(\bar{t})d\,\bar{t}\,, \quad \varepsilon(t) = \int\limits_0^t D(t-\bar{t})\dot{\sigma}(\bar{t})d\,\bar{t}, \tag{17}$$

that are valid if the system starts at rest, $t = 0$, otherwise $E(t)\varepsilon(0)$ and $D(t)\sigma(0)$ have to be added, respectively. Thus, combining Eqs. (15) and (16) with (17), leads to

$$\sigma(t) = c_\beta \left({}_C D_{0+}^{\beta} \varepsilon\right)(t)\,, \quad \varepsilon(t) = \frac{1}{c_\beta}\left(D_{0+}^{-\beta}\sigma\right)(t), \tag{18}$$

where the symbol $\left({}_C D_{0+}^{\beta} \varepsilon\right)(t)$ is the Caputos fractional derivative defined as, compare [4],

$$\left({}_C D_{0+}^{\beta} \varepsilon\right)(t) = \frac{1}{\Gamma\left(1-\beta\right)} \int\limits_0^t \frac{\dot{\varepsilon}(\bar{t})}{(t-\bar{t})^{\beta}} d\,\bar{t}, \tag{19}$$

while $\left(D_{0+}^{-\beta}\sigma\right)(t)$ is the Riemann-Liouville fractional integral defined as

$$\left(D_{0+}^{-\beta}\sigma\right)(t) = \frac{1}{\Gamma\left(\beta\right)} \int\limits_0^t \frac{\sigma(\bar{t})}{(t-\bar{t})^{1-\beta}} d\,\bar{t}. \tag{20}$$

From Eqs. (19) and (20) we may recognize that for $\beta = 0$ and $\beta = 1$ the purely elastic and viscous fluid behavior is recovered, respectively.

Consider an isotropic homogeneous Bernoulli-Euler beam of length L that exhibits spatially distributed fractional viscoelasticity, the corresponding equation of motion reads, compare [14],

$$\rho(x)\ddot{w}(x,t) + c_\beta \frac{\partial^2}{\partial x^2}\left[I_y(x) \frac{\partial^2}{\partial x^2}\left[\left({}_C D_{0+}^{\beta} w \right)(x,t) \right] \right] = q_z(x,t), \tag{21}$$

where $\rho(x)$ is the mass per unit length and $I_y(x)$ is the moment of inertia of the cross section with respect to the y-axis. The constitutive law for the bending moment is of the form

$$-c_\beta I_y(z) \frac{\partial^2 w(z,t)}{\partial x^2} = \left(D_{0+}^{-\beta} M_y \right)(x,t). \tag{22}$$

4.2 Example Problem

Let us suppose that the loading function varies in such a slow way that the inertial forces may be neglected. In this case the first term at the left hand side of Eq. (21) or Eq. (20) may be cancelled.

The example problem under consideration is a clamped-simply supported beam with an external bending moment $M(t) = M_B U(t)$, at the hinged support $x = L$. For simplicity's sake we suppose that $I_y(x) = I_y = const$. In this case the boundary value problem simplifies to

$$0 \leqslant x \leqslant L: \; c_\beta I_y \frac{\partial^4 w(x,t)}{\partial x^4} = \left(D_{0+}^{-\beta} q_z \right)(x,t),$$

$$x = 0: w = 0, \; \frac{\partial w}{\partial x} = 0; \quad x = L: \; w = 0, \; -c_\beta I_y \frac{\partial^2 w}{\partial x^2} = -\left(D_{0+}^{-\beta} M_B U \right)(t). \tag{23}$$

The closed-form solution for the quasi-static deflection becomes,

$$w(x,t) = \frac{M_B}{4 L I_y} x^2 (L - x) D(t) U(t), \tag{24}$$

and furthermore, the bending moment is determined as

$$M_x(x,t) = -\frac{M_B}{2}\left(1 - \frac{3x}{L}\right) U(t), \tag{25}$$

which coincides with the bending moment distribution evaluated in the purely elastic case, while displacements may be simply obtained from the elastic ones. That means that the first version of "correspondence principle" holds also for the fractional constitutive law.

References

1. Reddy, J.N.: A simple higher-order theory for laminated composite plates. J. Appl. Mech. **51**, 745–752 (1984)
2. Irschik, H.: On vibrations of layered beams and plates. ZAMM **73**, T34–T45 (1993)
3. Reddy, J.N.: An evaluation of equivalent-single-layer and layerwise theories of composite laminates. Comput. Struct. **25**, 21–35 (1993)
4. Hoischen, A.: Verbundträger mit elastischer und unterbrochener Verdübelung. Bauingenieur **29**, 241–244 (1954)
5. Pischl, R.: Ein Beitrag zur Berechnung hölzener Biegeträger. Bauingenieur **43**, 448–451 (1968)
6. Goodman, J.R., Popov, E.P.: Layered beam systems with interlayer slip. J. Struct. Div. ASCE **94**, 2535–2547 (1968)
7. Adam, C., Heuer, R., Raue, A., Ziegler, F.: Thermally induced vibrations of composite beams with interlayer slip. J. Therm. Stress. **23**, 747–772 (2000)
8. Heuer, R., Adam, C.: Piezoelectric vibrations of composite beams with interlayer slip. Acta Mech. **140**, 247–263 (2000)
9. Murakami, H.: A laminated beam theory with interlayer slip. J. Appl. Mech. **51**, 551–559 (1984)
10. Heuer, R.: A correspondence for the analysis of sandwich beams with or without interlayer slip. Mech. Adv. Mater. Struct. **11**, 425–432 (2004)
11. Heuer, R., Adam, C., Ziegler, F.: Sandwich panels with interlayer slip subjected to thermal loads. J. Thermal Stress. **26**, 1185–1192 (2003)
12. Heuer, R.: Large flexural vibrations of thermally stressed layered shallow shells. Nonlinear Dyn. **5**, 25–38 (1994)
13. Nutting, P.G.: A new general law deformation. J. Frankl. Inst. **191**, 678–685 (1921)
14. Di Paola, M., Heuer, R., Pirrotta, A.: Mechanical behavior of fractional visco-elastic beams. In: Eberhardsteiner, J., Böhm, H.J., Rammerstorfer, F.G. (eds.) CD-ROM Proceedings of the 6th European Congress on Computational Methods in Applied Sciences and Engineering (ECCOMAS 2012), Paper No. 2104, 10–14 Sept 2012, Vienna University of Technology, Vienna. ISBN:978-3-9502481-9-7

Turbulent Flow Characteristics Controlled by Polymers

Ruri Hidema, Naoya Yamada, Hiroshi Suzuki, and Hidemitsu Furukawa

Abstract An experimental study has been performed in order to investigate the relationship between the extensional viscosity of polymer solution and the turbulent drag reduction. A flexible polymer and a rigid rod-like polymer were added to the two-dimensional turbulent flow that was visualized by the interference pattern of a flowing soap film and analyzed by a single-image analysis. The power spectra of interference images were obtained, which is related to the water layer fluctuations in turbulence. The power spectra show a scaling behavior and the power components give the information of drag reduction. It was suggested that the energy transfer mechanisms are different in streamwise and normal directions. In the normal direction, the energy transfer was prohibited by the orientation of polymers, while the energy transfer in the streamwise direction was prohibited by extensional viscosity of polymers. The extensional viscosities of polymer solutions were measured by calculating pressure losses at an abrupt contraction flow.

1 Introduction

The addition of very small amounts of flexible polymer to a water flow reduces drag in a turbulent flow, which is known well as drag reduction phenomena, and used in many industries to improve the energy efficiency. Although this effect has been known for more than half a century, the physical mechanism that causes this drag reduction has been still unclarified. The first important work on the drag reduction was made by Lumley [1]. He argued the increase of the apparent viscosity in the

R. Hidema (✉) · H. Suzuki
Kobe University, Kobe, Hyogo, 657-8501, Japan
e-mail: hidema@port.kobe-u.ac.jp

N. Yamada · H. Furukawa
Yamagata University, Yonezawa, Yamagata 992-8510, Japan

Alexander K. Belyaev et al. (eds.), *Mechanics and Model-Based Control of Advanced Engineering Systems*, DOI 10.1007/978-3-7091-1571-8_18,
© Springer-Verlag Wien 2014

flow outside of the viscous sublayer is the key for drag reduction, which is due to the molecular extension of polymers. The extension of polymers occurs only under the rapid strain rates, and gives the anisotropic effects in the fluids. On the other hand, Tabor and de Gennes [2] suggested that the major effect arises only when the elastic energy stored by the partially stretched polymers becomes comparable to the turbulent energy.

However, both analyses have the shortcomings to explain which a strain works for stretching polymers, that is, shear strain or extensional strain. Smith et al. showed that DNA molecules were effectively stretched by extensional flow [3]. Indeed, extensional viscosity of polymers due to extensional strain can reach much higher compared to the intrinsic viscosity that is related to the shear viscosity. Thus, Hidema et al. [4] mentioned that an abrupt increase of the extensional viscosity depending on the extensional strain rate is the key for the turbulent drag reduction. In addition, the extensional viscosity is defined as a viscosity increase in an extensional axis, thus, the increase of the extensional viscosity is due to a flexibility of polymers, which leads anisotropic effects in the flow. In our previous study, we mentioned that a flexible polymer leads abrupt increase of extensional viscosity due to an abrupt extension of polymers in the flow and a rod-like rigid polymer leads slightly increase of the viscosity due to an orientation in extension axis [4, 5].

In this study, we focus on anisotropic effects of polymers on turbulent flow. A flexible polymer and a rod-like rigid polymer were chose to observe how the extensional viscosity works on turbulence drag reduction. In addition, the extensional viscosities of the polymer solution were measured by abrupt contraction flow to confirm our ideas.

In order to observe the turbulent flow affected by stretched polymers, two-dimensional turbulence made by flowing soap films is used (Fig. 1) [6]. As shown in Fig. 1c, both sides of the flowing soap films are free surfaces and the water layer is sandwiched by surfactants. The thickness of water layer is much larger than vanishingly small surfactant molecules, and also the surface area is infinity compared to the thickness. Additionally, surface tension is negligible due to the surfactant in the surface of flowing soap films. Therefore, soap film is considered as a 2D water layer flow. Soap films reflect illumination light, which make interference patterns of the film. Since interferences of the illumination lights are affected by the thickness of the water layer, the interference patterns have information of the dynamics of the water layer as 2D flows. When a grid is inserted to the flow, an extensional strain occurs at the grid, which causes polymer stretching or polymer orientation. Thus, in the case when polymers are added to the flow, the 2D grid turbulence is affected by the polymers. Flows are visualized and analyzed by a single-image processing, which is proposed by our previous work [8]. The extensional strain applied at this apparatus is roughly calculated by Eq. (1).

$$S(t) = S_0 \exp(-\dot{\varepsilon} t) . \quad (1)$$

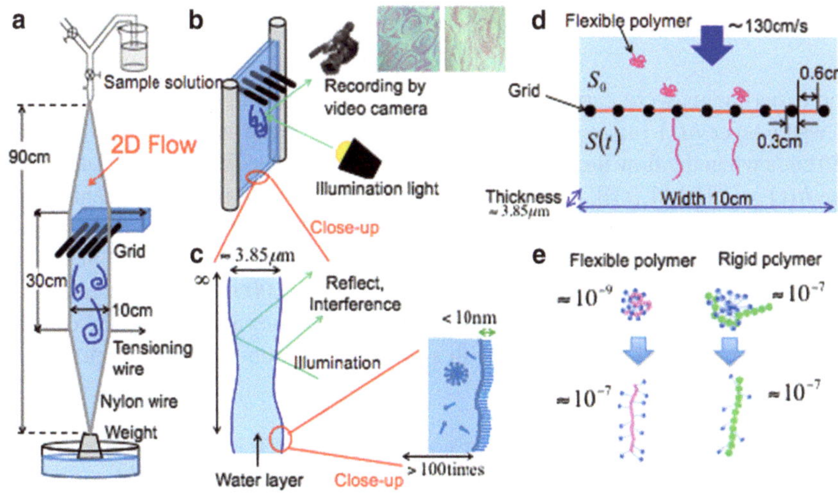

Fig. 1 The experimental set up of flowing soap films. (**a**) Is the whole image, (**b**) is the close up figure of the flowing soap film channel, (**c**) shows the cross section, (**d**) explains how the extensional strain occurs in this flow around the grid, (**e**) shows the expected polymer behavior under extensional strain measured by a dynamic light scattering technique [4]

where S_0 [m^2] is the cross-section area before deformation, $S(t)$ [m^2] is the cross-section area after deformation, t [s] is the time required for deformation and $\dot{\varepsilon}$ [s^{-1}] is the extensional rate. In our case, $\dot{\varepsilon}$ is calculated as 351 s^{-1} by the mean velocity, width of the flow, diameter of the grids and thickness of the film.

An extensional viscosity measured by an abrupt contraction flow is firstly proposed in this paper thus, the technique is described in the experimental section.

2 Experimental

2.1 Materials

For the 2D flow visualization experiments, soap solutions contained sodium dodecylbenzenesulfonate (SDBS) as a surfactant at the concentration of 2 wt%. As the flexible polymers, polyethyleneoxide (PEO, molecular weight of 3.5×10^6) was used at the concentration of 0.25, 0.5, 0.75, 1.0, 1.5 and 2.0×10^{-3} wt%. As the rod-like rigid polymers, hydroxypropyl cellulose (HPC, molecular weight is greater than 1×10^6) was used at the concentration of 0.01, 0.02 and 0.05 wt%. The overlap concentration of PEO was about 0.012 wt% and that of HPC was roughly 0.15 wt%. For the abrupt contraction flow experiments, the solution contains PEO or HPC at the same concentration of 2.0 wt%.

2.2 Turbulence Visualization by Flowing Soap Films

The experiments were carried out in an apparatus shown in Fig. 1. The thickness of water layer $h(t)$ [m] was about 3.85 μm with the mean velocity $V(t)$ [m/s] of 130 cm/s when the flow flux $Q(t)$ [m^3/s] was 0.5 ml/s, since there is the relationship like $h(t) = Q(t)/V(t)W$ where W [m] is the width of the soap film channel [6]. The interference images of soap films were recorded with a digital video camera (Panasonic TM700) at the data acquisition area, which was located at 20 cm behind the grid. The shutter speed of the video camera was 1/3,000 s. A time interval in a series of images was adjusted to 1/60 s. Each of the frames acquired by the camera was converted into RGB form files with a spatial resolution of 640 × 360 pixels.

2.3 Single-Image Processing by Film Interference Flow Imaging

The interference images were analyzed by a 2D-FFT, which calculated the power spectrum of the interference images [7]. Since the RGB pixel intensity of the images is related to the thickness of water layer, the power spectrum of the image shows the dynamics of 2D flows. Here, the power spectrum $\langle I^2(k_x, k_y)\rangle$ was calculated by the pixel intensity G with a hamming window, where the k_x [m^{-1}] and k_y [m^{-1}] are the spatial frequency that are streamwise and normal directions in an interference image. For measurements presented here, the data acquisition area that was clipped from the video frames has 256 × 256 pixels, which correspond to 2.56 × 2.56 cm^2, thus, the k_x and k_y range from 1/2.56 to 1/0.02 cm^{-1}, that is, 0.391–5.00 cm^{-1} as the frequency. $\langle I^2(k_x, k_y)\rangle$ of the pixel intensity G characterizes the strength of the thickness fluctuation of water layer on spatial frequency, k_x and k_y. Here the power spectrum in both frequencies are plotted to discuss energy transfer of turbulence in streamwise and normal directions.

2.4 Extensional Viscosity Measurements by Abrupt Contraction Flow

An abrupt contraction flow was made by connecting a syringe and a glass tube (Fig. 4a). The syringe that has a diameter, D_1 [m], of 28 mm was filled with a polymer solution. The polymer solution was forced out of the syringe into the glass tube that has a diameter of 2.3 mm, D_2 [m]. The forces F [N] to push the solution at the flow rates of 5, 10, 20 and 30 ml/min were measured by a load cell. A pressure P_{EX} [Pa] added to the flow was calculated by F [N], as $P_{EX} = F/A_1$ when A_1 [m^2] is a cross area of the syringe. First, the experimental pressure P_{EX} [Pa] was calculated by water to calculate the friction of the syringe, P_F [Pa].

An energy-balance equation is written as

$$P_{EX} + P_0 + \frac{1}{2}\rho V_1^2 = \Delta P_1 + \Delta P_2 + \Delta P_3 + \Delta P_4 + P_0 + \frac{1}{2}\rho V_2^2 + P_F + P, \quad (2)$$

where P [Pa] is an extensional stress, P_0 [Pa] is the atmosphere pressure, V_1 [m/s] and V_2 [m/s] are velocities in the syringe and the glass tube, ρ [kg/m^3] is the density, and ΔP_1 [Pa], ΔP_2 [Pa], ΔP_3 [Pa] and ΔP_4 [Pa] are the pressure drop in the syringe, in the glass tube, at the abrupt contraction in the middle of the flow and at the abrupt expansion in the end of the glass tube, which is calculated as follows:

$$\begin{aligned} \Delta P_1 &= \frac{64\mu}{V_1 D_1}\frac{l}{D_1}\frac{V_1^2}{2}, \Delta P_2 = \frac{64\mu}{V_2 D_2}\frac{l}{D_2}\frac{V_2^2}{2}, \Delta P_4 = \frac{V_2^2}{2}\rho, \\ \Delta P_3 &= \zeta\frac{V_2^2}{2}\rho, \zeta = 0.04 + \left(1 - \frac{D_2}{D_C}\right)^2, \frac{D_C}{D_2} = 0.582 + \frac{0.0418}{1.1 - \sqrt{D_2 - D_1}} \end{aligned} \quad (3)$$

P_F is calculated by measuring P_{EX} of water. Additionally, only the ΔP_2 should be considered to calculate extensional viscosity since ΔP_1, ΔP_3 and ΔP_4 are much less than P_F. Thus, the extensional viscosity η_e [Pa s] was calculated as follows. Here, extension rate $\dot{\varepsilon}$ is calculated by considering a previous research [8].

$$P = P_{EX} - P_F - \Delta P_2 = \eta_e \dot{\varepsilon}, \dot{\varepsilon} = \frac{0.75 V_2}{D_2/2} \quad (4)$$

By measuring the P_{EX} of polymer solution, the extensional viscosities of polymers were calculated.

3 Results and Discussion

Figure 2 shows the examples of interference patterns of 2D grid turbulence. Turbulence was generated at the comb, and was convected in the observation area. The interference patterns of 2D turbulence visualize the thickness of water layer fluctuations, which is considered as a passive scalar. One of the characteristics of 2D turbulence is an inverse energy cascade from small scales to large scales that is seen through the increase of the apparent size of the eddies compared to injection scales. By adding of polymers, the eddies clearly become long in streamwise direction and narrow in normal direction (Fig. 2b, c). This feature is especially seen in the case of the addition of PEO. This indicates that the polymer inhibits inverse energy transfers in normal direction. In the case of the addition of HPC, the effect can be slightly observed at much higher concentration compared with that in the case of PEO.

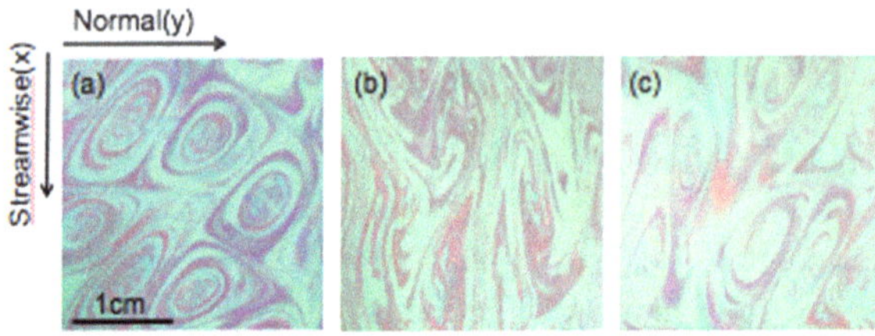

Fig. 2 Interference images of the turbulence in flowing soap films. The images are taken at SDBS 2 wt% solution (**a**), + PEO 2.0 × 10^{-3} wt% solution (**b**), + PEO HPC 0.05 wt% (**c**) for a mean velocity is about 130 cm/s. The *black bars* indicate 1 cm

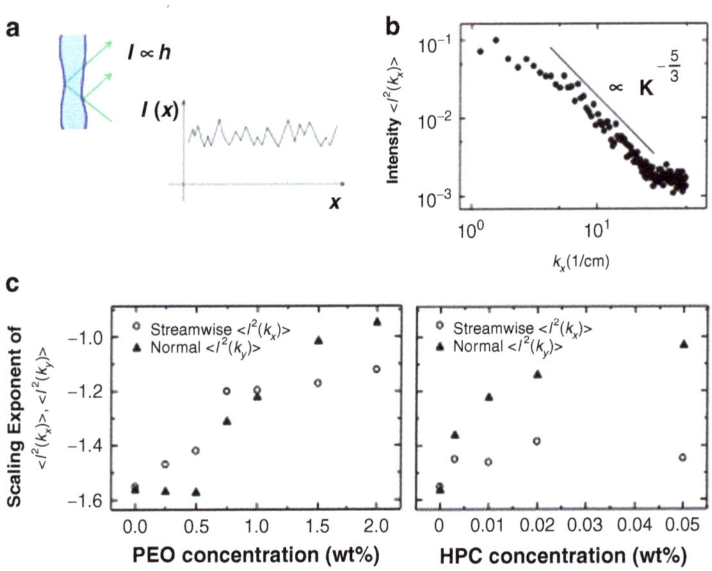

Fig. 3 The example of the power spectra $\langle I^2(k_x, k_y)\rangle$. The k_x and the k_y directions in the image represent the spatial frequencies in the streamwise and the normal directions

Figure 3 shows the example of the power spectra $\langle I^2(k_x, k_y)\rangle$ of the interference pattern. As shown in Fig. 3a, the pixel intensity is related to the thickness. k_x [m^{-1}] and k_y [m^{-1}] in $\langle I^2(k_x, k_y)\rangle$ are the spatial frequencies which have the information of water layer fluctuations directed in streamwise and normal directions. $\langle I^2(k_x, k_y)\rangle$ shows the examples of the scaling behaviors (Fig. 3b). It is found that the power component almost fits with the −5/3 that is theoretically predicted for a 2D turbulence. The variation of the power components is shown in Fig. 3c. In the case of PEO is added, the value was changed from −5/3 to −1 in both streamwise and normal directions. At lower concentrations of PEO, the variation rate of the power component is small but the rate increases rapidly from 0.5 to 0.75×10^{-3} wt%. In the case of HPC is added, the value was changed gradually from −5/3

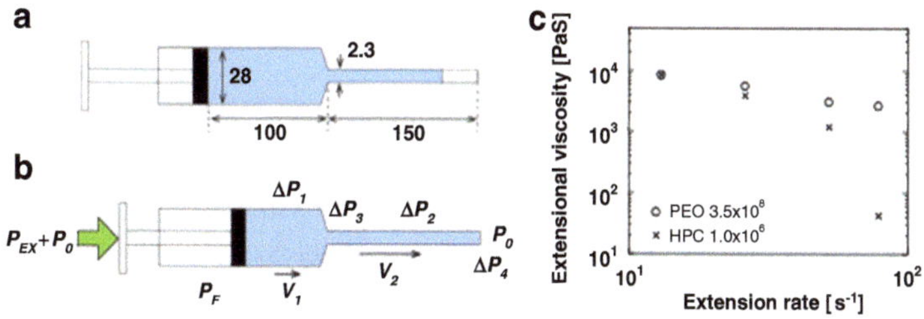

Fig. 4 An abrupt contraction flow and calculated extensional viscosities

to −1 in normal directions. However, the variation of the scaling exponent in the streamwise direction was small and has the absence of a pattern in the case of HPC. As mentioned above, in 2D turbulence, an inverse energy cascade is a characteristic phenomenon. The inverse energy cascade is seen through the vortex mergers in the interference patterns. The scaling exponent detects the prohibition of the inverse energy cascade due to polymers, that is, the variation from −5/3 to −1. Thus, it is considered that PEO prohibit the inverse energy cascade in both directions, while HPC prohibits the inverse energy cascade only in the normal direction. This difference is due to the different mechanism of energy transfer in turbulence when polymers are added. In the normal direction, the orientation of polymers prohibits the energy transfer. Thus it is happen in both PEO and HPC cases. On the other hand, the extensional viscosity prohibits the energy transfer in the streamwise direction. The extensional viscosity of HPC is lower compared to PEO, thus HPC is less effective in the streamwise direction.

Since it is considered that the streamwise direction is affected by extensional viscosity, that of polymer solution was measured by an abrupt contraction flow. The force to push water at 5 ml/min was 10 N, which was used to calculate P_F. The density of water was applied to all solution. By using these values, the forces to push the solution at the flow rates of 5, 10, 20 and 30 ml/min corresponding to the extension rate of 13, 26, 52, 78 s^{-1} were 96, 133, 169 and 228 N for PEO solution, 112, 135, 169 and 190 N for HPC solution from the calculation of PEX, respectively. Shear viscosities of the PEO solution was 0.43 Pa s at applied shear stress in glass tube, and that of the HPC solution was 0.59 Pa s. Other scales which need for calculation are shown in Fig. 4a. Calculated extensional viscosities are shown in Fig. 4c. The extensional viscosity of PEO at the applied extension rate is slightly decreased. On the contrary, that of HPC is suddenly decreased at 78 s^{-1}. This result indicates that the extensional viscosity in 2D turbulence at the extension rate of 351 s^{-1} is expected high in the case of PEO, while that of HPC is already low. Thus, the drag reduction in the streamwise direction occurs effectively by PEO, not by HPC. These increase and decrease of the extensional viscosity were due to the extension of PEO and to the orientation of HPC, but the oriented HPC do not make any entangled structures. Thus, the breakup of the structure occurred at relatively low extension rates.

4 Conclusions

An experimental study has been performed in order to investigate how the extensional viscosity of polymers affects on turbulent drag reduction. PEO as a flexible polymer and HPC as a rod-like rigid polymer were added to the 2D turbulence. The power spectra of interference images were obtained, which is related to the water layer fluctuations in turbulence. The scaling behavior of the power spectra gives the information of energy transfer in turbulence. From these results, it was suggested that the mechanism of drag reduction in turbulence is different in normal and streamwise direction. In the normal direction, the orientation of polymers prohibits the energy transfer. On the other hand, the extensional viscosity prohibits the energy transfer in the streamwise direction. The extensional viscosity of polymer solution in an abrupt contraction flow, which indicates the extensional viscosity of HPC was lower than that of PEO at the applied extensional strain in 2D flow. Thus, HPC was less effective in the streamwise direction.

Acknowledgements This study is supported in part by the Grant-in-Aid for Young Scientists(B) (Project No.: 24760129) and the Grant-in-Aid for Scientific Research(B) (Project No.: 24360319) from the Japan Society for the Promotion of Science (JSPS).

References

1. Lumley, J.: Drag reduction in turbulent flow by polymer additives. J. Polym. Sci. **7**, 263–270 (1973)
2. Tabor, M., de Gennes, P.G.: A cascade theory of drag reduction. Europhys. Lett. **2**, 519–522 (1986)
3. Smith, D.E., Chu, S.: Response of flexible polymers to a sudden elongational flow. Science **281**, 1335–1340 (1998)
4. Hidema, R., Furukawa, H.: Development of film interference flow imaging method (FIFI) studying polymer stretching effects on thin liquid layer. e-J. Surf. Sci. Nanotech. **10**, 335–340 (2012)
5. Hidema, R., Ushiki, H., Furukawa, H.: Polymer effects on turbulence in flowing soap films studied with film interference flow imaging method. J. Solid Mech. Mater. Eng. **5**, 838–848 (2011)
6. Kellay, H., Goldburg, W.I.: Two-dimensional turbulence: a review of some recent experiments. Rep. Prog. Phys. **65**, 845–894 (2002)
7. Hidema, R., Yatabe, Z., Shoji, M., Hashimoto, C., Pansu, R., Sagarzasu, G., Ushiki, H.: Image analysis of thickness in flowing soap films. I: Effects of polymer. Exp. Fluids. **49**, 725–732 (2010)
8. McKinley, G.H., Raidford, W.P., Brown, R.A., Armstrong, R.C.: Nonlinear dynamics of viscoelastic flow in axisymmetric abrupt contractions. J. Fluids. Mech. **223**, 411–456 (1991)

Dynamic Mechanical Properties of Functionally Graded Syntactic Epoxy Foam

Masahiro Higuchi and Tadaharu Adachi

Abstract We investigated dynamic mechanical properties in functionally graded (FG) syntactic foams which had graded distribution of Acrylonitrile micro-balloons in epoxy resin matrix. The density distributions in the FG syntactic epoxy foams were graded by floating phenomenon of the micro-balloons in the matrix resin during curing process. Dynamic viscoelastic measurements and compression tests were conducted to evaluate distributions of mechanical properties in the foams. The dynamic viscoelastic measurements revealed that the thermo-viscoelastic properties of the foams were determined by the properties of the epoxy matrix resin. The static and dynamic compression tests provided relations between the mechanical properties: the compressive Young's modulus and yield stress, and the density in the foams.

1 Introduction

Functionally graded materials (FGMs) [1, 2] were first suggested as materials for relaxing thermal stress in structures. The FGMs can be designed and manufactured to have excellent properties or to suit required conditions of applications by controlling distributions of the phases [2]. Recently, some numerical studies [3–5] have suggested that functionally graded (FG) foams having density distribution are effective for impact energy absorption.

M. Higuchi (✉)
School of Mechanical Engineering, Kanazawa University, Kakuma-machi, Kanazawa, 920-1192, Japan
e-mail: higuchi-m@se.kanazawa-u.ac.jp

T. Adachi
Department of Mechanical Engineering, Toyohashi University of Technology, 1-1 Hibarigaoka, Tempaku, Toyohashi, 441-8580, Japan
e-mail: adachi@me.tut.ac.jp

Alexander K. Belyaev et al. (eds.), *Mechanics and Model-Based Control of Advanced Engineering Systems*, DOI 10.1007/978-3-7091-1571-8_19,
© Springer-Verlag Wien 2014

Concurrently with those numerical studies, syntactic foam [6] having density distribution have been produced for the purpose of impact energy absorption [6–10]. The syntactic foams easily fabricated by filling matrix material with small hollow spherical particles (micro-balloons) have light weight and high absorption of mechanical energy [6]. The mechanical properties of the syntactic foams can be adjusted by the volume fraction of the micro-balloons in the matrix material, namely density of foam as well as properties of the matrix material and the micro-balloons. The authors have developed fabrication process of FG syntactic foams [9–11]. In order to control the distributions of the mechanical properties, the density distributions in the foams are graded by floating phenomenon of the micro-balloons in the matrix resin during curing process [9–11].

In the present study, we investigated distributions of mechanical properties in the FG syntactic epoxy foams which have graded distribution of Acrylonitrile micro-balloons in epoxy resin matrix to apply for impact energy absorption. Dynamic viscoelastic measurements were conducted out to clarify the effect of filling the micro-balloons into the epoxy resin on viscoelastic properties of the FG syntactic epoxy foams. Static and dynamic compression tests were performed to obtain relation between density and mechanical properties including the strain rate effect due to the viscoelastic properties.

2 Fabrication of FG Syntactic Epoxy Foam

Figure 1 shows fabrication process to produce FG syntactic epoxy foams [9–11]. A mixture of micro-balloons and matrix resin was heated up to constant temperature T_F to decrease viscosity of the matrix resin. Then the micro-balloons were floated easily during a retaining time t_F due to density difference between the matrix resin and the micro-balloons, that is, graded distribution of micro-balloons occurs. Here, the distribution of the micro-balloons can be controlled by constant temperature T_F and a retaining time t_F based on temperature dependency on the viscosity of the matrix resin until the gelation of the resin. After the gradating micro-balloons, the mixture was cured based on a standard curing process. In the paper, FG syntactic epoxy foams were produced by adding Acrylonitrile copolymer micro-balloons coated by calcium carbonate powder (MFL-100CA, Matsumoto Yushi-Seiyaku) into epoxy resin. The average diameter of the micro-balloons was 80–110 μm and the density was $130 \pm 30\,\mathrm{kg\ m^{-3}}$. The matrix epoxy resin was bisphenol-A-type epoxide resin (AER2603, Asahi Kasei E-materials) with a curing agent (RIKACID MH-700, New Japan Chemical) which was an ad-mixture of 4-methyl hexahydrophthalic anhydride (4-methyl HHPA) and hexahydrophthalic anhydride (HHPA) and an accelerator (BMI12, Mitsubishi Chemical) which was 1-benzil-2-methylimidazoles. The weight ratio of the main agent, the curing agent and the accelerator was determined as 100:86:0.5 according to stoichiometry. The curing condition for the epoxy resin system was composed of the pre-curing at 373 K for 2 h for the gelling and the post-curing at 403 K for 15 h for cross-linking

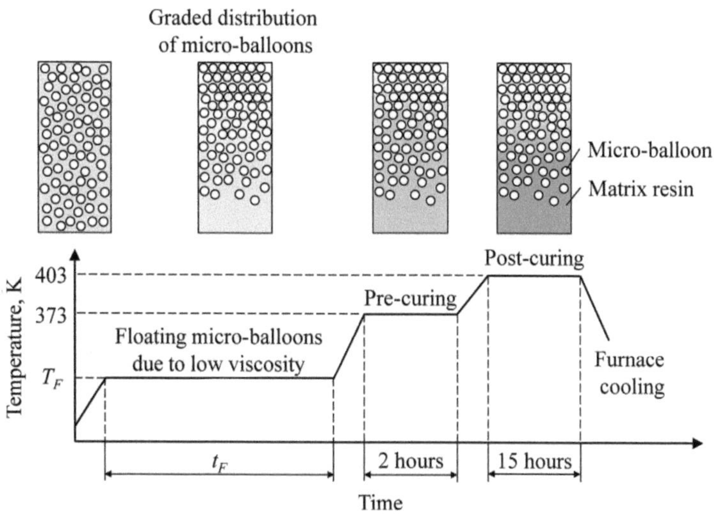

Fig. 1 Fabrication process for FG syntactic epoxy foam

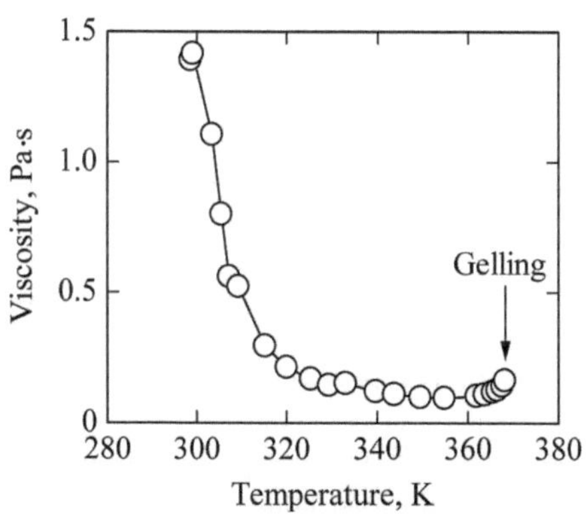

Fig. 2 Relation between viscosity and temperature for epoxy resin

reaction. Figure 2 shows temperature dependency of the viscosity for the epoxy resin before curing measured by a rotational viscometer (DV-1+, Brookfield). The viscosity decreased drastically with temperature in-crease from room temperature to 330 K. The viscosity maintained nearly constant value above 330 K. After that, gelling of the resin occurred at 370 K. After the mixture of the epoxy resin and the micro-balloons was stored in a vacuum vessel to remove voids, the mixture was poured into an aluminum mold. The mold was 260 mm long, 10 mm wide and 130 mm deep. The mixture was kept at $T_F = 323$ K for $t_F = 2$ h to grade the distribution of the micro-balloons as shown in Fig. 1. Then the mixture was cured according to above mentioned curing conditions. After curing, densities ρ

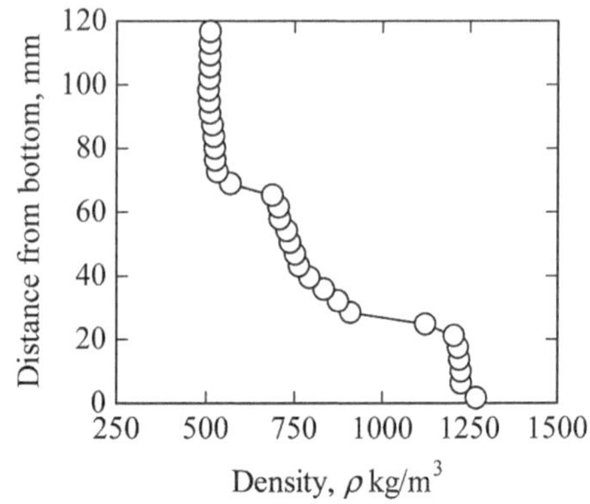

Fig. 3 Density distribution of FG syntactic epoxy foam

at different positions in the foams were measured by the Archimedes method according to JIS Z8807 to obtain the density distributions.

The density distribution of the fabricated FG syntactic epoxy foam having average density 710 kg m^{-3} is shown in Fig. 3. The density in the fabricated FG foam was distributed from minimum density of 500 kg m^{-3} to maximum density of 1,200 kg m^{-3} (density of neat epoxy resin).

3 Dynamic Thermo-Viscoelasticity Measurements

The dynamic thermo-viscoelasticity measurements were conducted out to clarify the effect of filling the micro-balloons into the epoxy resin on the viscoelastic properties in the FG syntactic epoxy foams by using a dynamic viscoelastometer (Rheogel-E4000, UBM) with a non-resonance tensile method. The specimens were 2 mm in thick, 40 mm long and 5 mm wide, which were cut from the fabricated foam (Fig. 3) not to have density distribution in the longitudinal direction of the specimens. The densities of the specimens were 510, 800 and 1,200 kg m^{-3} (epoxy resin density). Storage modulus E', loss modulus E'' and loss factor, $\tan\delta(= E'/E'')$ for tensile oscillation with frequency of 10 Hz were measured at each 5 K in the temperature range from 233 to 493 K. The heating rate was 1 K min^{-1}.

Figure 4 shows temperature dependencies of storage modulus E', loss modulus E''. Regardless of the densities of the specimens, the storage modulus decreased gradually with the temperature increase up to 420 K. After that, the storage modulus decreased rapidly in the temperature range from 420 to 460 K, namely the glass transition occurred. Additionally, the storage and loss moduli in whole temperature range were reduced with decreasing the density from the epoxy resin density to lower densities. In contrast, temperature dependencies of the loss factor defined by $\tan\delta(= E'/E'')$ shown in Fig. 5 were not affected by the densities, and the loss factors showed peak values at temperature of 440 K for all densities. Regardless of the specimen density (namely the volume fraction of the micro-balloons), the glass

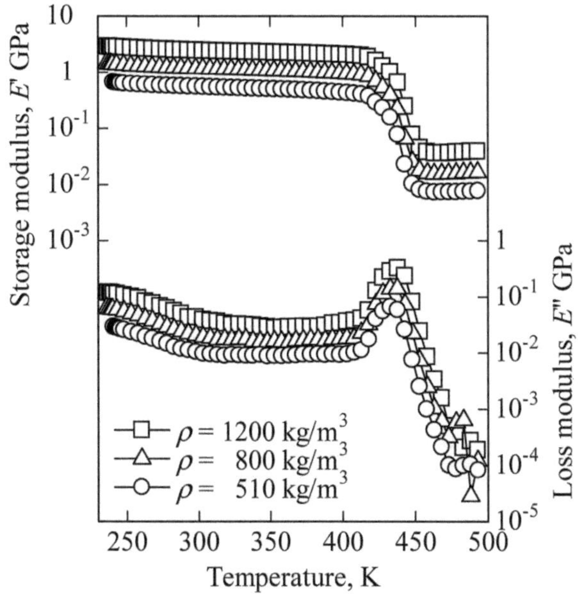

Fig. 4 Temperature dependencies of storage modulus E' and loss modulus E''

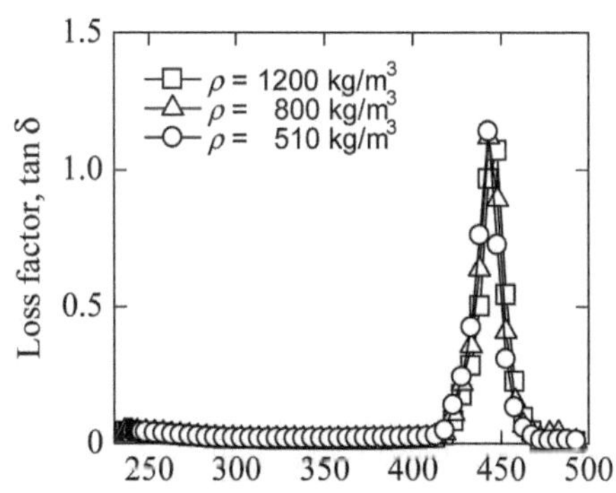

Fig. 5 Temperature dependencies of loss factor $\tan\delta$

transition temperatures were identified to be 440 K. Therefore, absolute values of the viscoelastic properties of the FG syntactic foams depended on the volume fraction of the micro-balloons, meanwhile the temperature dependencies of the viscoelastic properties were found to be governed by the epoxy matrix.

4 Static and Dynamic Compression Tests

Static and dynamic compression tests were performed to evaluate the mechanical properties of the FG syntactic epoxy foams. The specimens were 10 mm long, 10 mm wide and 5 mm high. The average densities of the specimens were 510, 690, 780, 920 and 1,200 kg m^{-3}.

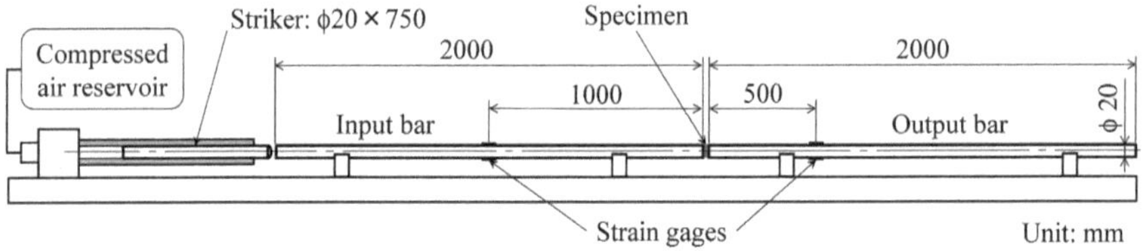

Fig. 6 Schematic diagram of split Hopkinson pressure bar equipment

The static compression tests were carried out under a constant displacement rate of 1 or 10 mm min^{-1} with a universal testing machine (Autograph DSC-25T, Shimadzu). The corresponding strain rates were 3.3×10^{-3} and 3.3×10^{-2} s^{-1}, respectively. The dynamic compression tests were conducted to obtain stress-strain curves under high strain rates by using a split Hopkinson pressure bar (SHPB) equipment. A schematic diagram of the SHPB equipment is shown in Fig. 6. The striker, input and output bars were made of aluminum alloy (JIS A7075-T6) to accurately measure stress-strain response of the low mechanical impedance specimen. The histories of the nominal strain $\varepsilon(t)$, strain rate $\dot{\varepsilon}(t)$ and stress $\sigma(t)$ in the specimen under assumption of force equilibrium between both side of the specimen were determined by the following equations:

$$\varepsilon(t) = \frac{2c_0}{h_s}\int_0^t [\varepsilon_I(t') - \varepsilon_T(t')]dt',$$

$$\dot{\varepsilon}(t) = \frac{2c_0}{h_s}[\varepsilon_I(t) - \varepsilon_T(t)],$$

$$\sigma(t) = \frac{A_0}{A_S}E_0\varepsilon_T(t), \tag{1}$$

where $\varepsilon_I(t)$ and $\varepsilon_T(t)$ denote the incident and transmitted strain pulse, c_0, E_0 and A_0 are the longitudinal elastic wave velocity ($= 5.1 \times 10^3$ ms^{-1}), Young's modulus ($= 73.2$ GPa) and the cross-sectional area of the input and output bars, and h_s and A_s are the height and cross-sectional area of the specimen. The strain rates in the dynamic tests were approximate 2×10^3 ms^{-1}. Figure 7 shows compressive stress-strain curves obtained by the static and dynamic tests for the specimens with the densities of 510 and 920 kg m^{-3}. The static test curves showed typical behavior of conventional foam material; elastic deformation, deformation in plateau region after the yielding, and densitification. The compressive stress-strain relations were found to strongly depend on the density. In the dynamic tests, although the densitification did not occur due to the brittleness of the epoxy resin in high strain rates, the Young's modulus and yield stress increased compared with those of static tests due to the viscoelastic properties of the epoxy matrix as show in Fig. 4. In general, mechanical properties of foam materials depend on the relative density to the one of matrix [12]. Consequently, the relative Young's modulus, $E(\rho, \dot{\varepsilon})/E_0(\dot{\varepsilon})$ and relative yield stress,

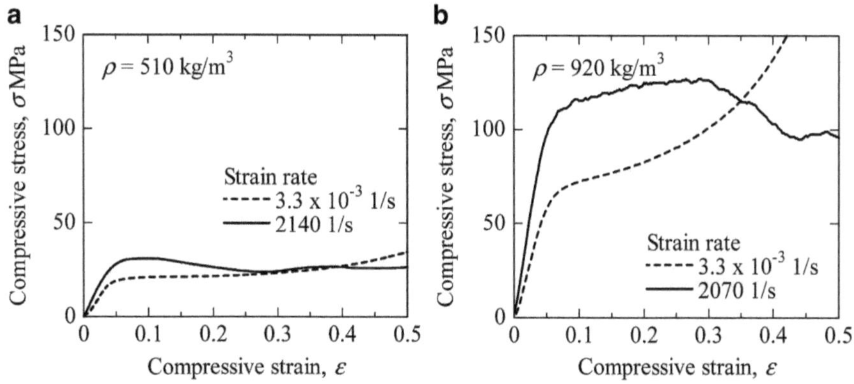

Fig. 7 Compressive stress-strain curves obtained by static and dynamic tests. (**a**) Density, $\rho = 510\,\mathrm{kg\,m^{-3}}$. (**b**) Density, $\rho = 920\,\mathrm{kg\,m^{-3}}$

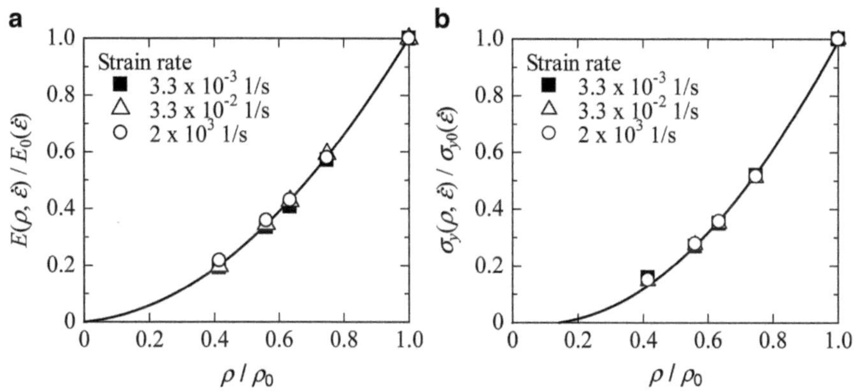

Fig. 8 Relation between viscosity and temperature for epoxy resin. (**a**) Relative Young's modulus. (**b**) Relative yield stress

$\sigma_y(\rho, \dot{\varepsilon})/\sigma_{y0}(\dot{\varepsilon})$ for various strain rates are plotted against the relative density, ρ/ρ_0 in Fig. 8. Here, ρ_0 is the density of the epoxy matrix, $E_0(\dot{\varepsilon})$, and $\sigma_{y0}(\dot{\varepsilon})$ denote the Young's modulus and yield stress including strain rate effect for the epoxy matrix. From Fig. 8, the compressive Young's modulus $E(\rho, \dot{\varepsilon})$ and yield stress $\sigma_y(\rho, \dot{\varepsilon})$ could be expressed in the separation variables composed of the quadratic function of the relative density and the Young's modulus or yield stress for the epoxy matrix including the strain rate effect:

$$E(\rho, \dot{\varepsilon}) = \left[0.89\left(\frac{\rho}{\rho_0}\right)^2 + 0.11\frac{\rho}{\rho_0}\right] E_0(\dot{\varepsilon}),$$

$$\sigma_y(\rho, \dot{\varepsilon}) = \left[1.15\left(\frac{\rho}{\rho_0}\right)^2 - 0.15\frac{\rho}{\rho_0}\right]\sigma_{y0}(\dot{\varepsilon}). \tag{2}$$

Therefore, the graded distributions of the compressive Young's modulus and yield stress can be identified from the density distribution with the relations of Eq. (2).

5 Conclusion

We investigated the dynamic mechanical properties in the FG syntactic foams which have graded distribution of the Acrylonitrile micro-balloons in the epoxy resin matrix. The dynamic viscoelastic measurements and compression tests for low and high strain rates were conducted to evaluate the distributions of mechanical properties in the FG syntactic foams. Because temperature dependences of the FG syntactic epoxy foams are determined by the ones of the epoxy matrix, the compressive Young's modulus and yield stress were expressed in the separation of variables composed of the quadratic function of the relative density and the Young's modulus or yield stress for the epoxy matrix including strain rate effect.

References

1. Tanigawa, Y.: Some basic thermoelastic problems for nonhomogeneous structural materials. Appl. Mech. Rev. **48**, 287–300 (1995)
2. Birman, V., Byrd, L.W.: Modeling and analysis of functionally graded materials and structures. Appl. Mech. Rev. **60**, 195–216 (2007)
3. Cui, L., Kiernan, S., Gilchrist, M.D.: Designing the energy absorption capacity of functionally graded foam materials. Mater. Sci. Eng. A **507**, 215–225 (2009)
4. Kiernan, S., Cui, L., Gilchrist, M.D.: Propagation of a stress wave through a virtual functionally graded foam. Int. J. Nonlinear Mech. **44**, 456–468 (2009)
5. Adachi, T., Yoshigaki, N., Higuchi, M.: Analysis of longitudinal impact problem for functionally graded materials. Trans. Jpn. Soc. Mech. Eng. A **79**, 502–510 (2012)
6. Bunn, P., Mottram, J.T.: Manufacture and compression properties of syntactic foams. Composites **24**, 565–571 (1993)
7. Parameswaran, V., Shukla, A.: Processing and characterization of a model functionally gradient material. J. Mater. Sci. **35**, 21–29 (2000)
8. Rhohatgi, P.K., Matsunaga, T., Gupta, N.: Compressive and ultrasonic properties of polyester/fly ash composites. J. Mater. Sci. **44**, 1485–1493 (2009)
9. Adachi, T., Higuchi, M.: Development of integral molding of functionally-graded syntactic foams. In: Irschik, H., Krommer, M., Belyaev, A.K. (eds.) Advanced Dynamic and Model-Based Control of Structures and Machines, pp. 1–9. Springer, Heidelberg (2012)
10. Higuchi, M., Adachi, T., Yokochi, Y., Fujimoto, K.: Controlling of distribution of mechanical properties in functionally-graded syntactic foams for impact energy absorption. Mater. Sci. Forum. **706–709**, 729–734 (2012)

11. Higuchi, M., Adachi, T., Yoshioka, T., Yokochi, Y.: Evaluation on distributions of mechanical properties in functionally graded syntactic foam. Trans. Jpn. Soc. Mech. Eng. A **78**, 890–901(2012)
12. Gibson, L.J., Ashby, M.F.: Cellular Solids: Structure and Properties, pp. 169–202. Cambridge University Press, Cambridge (1997)

Problems of Describing Phase Transitions in Solids

D.A. Indeitsev, V.N. Naumov, D. Yu. Skubov, and D.S. Vavilov

Abstract This paper is devoted to discrete and continuous models of stress-induced phase transitions in solids. The models are based on non-monotone stress-strain relation. Only mechanical impact is taken into consideration. Thermal effects are omitted. The problem is limited to one space dimension. It is numerically analyzed and the obtained solution allows to demonstrate the penetration of structural conversion inside the material.

1 Introduction: Discrete or Continuous?

Depending on the purpose of the research crystalline body can be treated as a discrete structure consisting of interconnected atoms or as a continuum. In the first case the system with the required number of degrees of freedom is analyzed, and in the end the problem is reduced to a large system of ordinary differential equations. Using the second approach, one assumes that the behavior of material is described by a set of unknown functions with appropriate mathematical properties. Partial differential equations are to be solved in this case.

Not so many years ago the second way seemed to be much more convenient and productive, but nowadays the rapid development of computer techniques allows to study systems with a huge number of particles. However, it is not worth of opposing these approaches to each other. Each of them has its own benefits and drawbacks. Often they supplement each other, especially when it is necessary to describe a microstructure of material.

D.A. Indeitsev (✉) · D. Yu. Skubov · D.S. Vavilov
Institute of Problems of Mechanical Engineering, Saint-Petersburg, Russia
e-mail: Dmitry.Indeitsev@gmail.com; skubov.dsk@yandex.ru; dvs010188@gmail.com

V.N. Naumov

Alexander K. Belyaev et al. (eds.), *Mechanics and Model-Based Control of Advanced Engineering Systems*, DOI 10.1007/978-3-7091-1571-8_20,
© Springer-Verlag Wien 2014

Two methods mainly are used to pass from discrete model to continuous one: long-wave approximation and different averaging procedures. Sometimes, under certain conditions it is possible to demonstrate that discrete and continuous systems are equivalent in some sense. The most obvious example here is a chain of masses connected by elastic links which is often used as the simplest model of crystal lattice. It is quite natural to expect that as far as the number of elements increases, the chain is getting closer to an elastic rod. Its natural frequencies are given by

$$w_s = \frac{s\pi}{L}\sqrt{\frac{E}{\rho}}\,. \tag{1}$$

For the frequencies of the discrete chain one can obtain

$$w_s = 2\sqrt{\frac{c}{m}}\sin\frac{\pi s}{2(n+1)}\,, \tag{2}$$

where n is the number of point-masses. Using the following relations between the parameters of the chain: $c = E/d$, $d = L/(n+1)$, $m = \rho L/(n+1)$, the expression (2) can be rewritten in the form

$$w_s = \frac{2(n+1)}{L}\sqrt{\frac{E}{\rho}}\sin\frac{\pi s}{2(n+1)}\,. \tag{3}$$

It is evident that if n goes to infinity, we get the same result as in the continuum model. Unfortunately, this quick conversion from discrete model to continuous one does not always exist.

For instance, consider the process of phase transition in solids which goes under external impact. The word "phase" refers to a homogeneous domain in the sample separated from other regions with a pronounced border. In order to define the position of the phase boundary it is common to introduce a new function of time making it necessary to come up with additional kinetic equation [1]. The equation relates the velocity of the phase boundary with the constitutive curve, which contains a decreasing segment. It means that the strain energy is the non-convex function in this case. The assumption about the unstable branch is the basic one for different models of phase transforming materials [2, 3]. In terms of the discrete model structural transformation means that some atoms move from their initial equilibriums. The displacement of atoms is accompanied with the rearrangement of the crystal lattice. In case of the one-dimensional chain it is easy to visualize the process of phase transition using a simple mechanical model. For this purpose one should replace the elastic link with some nonlinear element which can abruptly change its configuration as a result of dynamic process. In this case it is convenient to use the two-bar truss, also known as von Mises truss (Fig. 1).

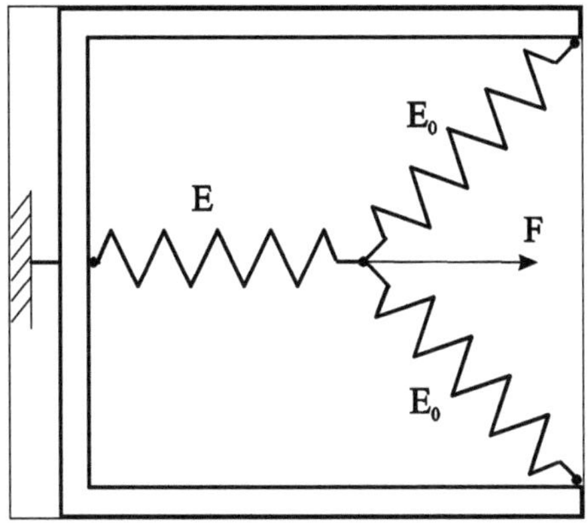

Fig. 1 Mises truss

2 Semi-infinite Rod with Non-monotone Stress-Strain Relation

As soon as the discrete model is constructed, a curious question may arise. We are now investigating the nonlinear system, and he point is that it is not clear whether this chain corresponds to continuous model of material capable of undergoing phase transitions. Perhaps there is contradiction between them. In linear model it is easy to show that the continuous and discrete approaches are rather close to each other, but in this situation the equivalence of two methods seems to be doubtful. The question is significant in case of carrying out numerical computations, because when after replacing derivative expressions with approximately equivalent finite quotients, one actually deals with a discrete representation of the model.

One way to clarify the problem is to look for a simple model, where it is possible to obtain an analytical result and after that compare it with a numerical solution of the problem. If both methods lead to the same conclusions, then we would be able to talk about the conformity of two different models. For this purpose consider a semi-infinite, homogeneous rod with non-monotone constitutive curve, defined with the piece-wise linear function.

$$\sigma(\varepsilon) = \begin{cases} E_1\varepsilon\,, & \varepsilon < \varepsilon_c \\ E_2\varepsilon\,, & \varepsilon > \varepsilon_c \end{cases} \tag{4}$$

We consider the situation, when the slope of the second segment is less than the slope of the first one. Thus, the inequality $E_2 < E_1$ should be satisfied (Fig. 2).

At the initial time a constant force is instantly applied to the left butt of the rod. Let $u(x,t)$ be its longitudinal displacements. In case of small deformations $\varepsilon = \partial u/\partial x$ the governing equation has the form

$$\frac{\partial \sigma}{\partial x} - \rho_0 \frac{\partial^2 u}{\partial t^2} = 0\,. \tag{5}$$

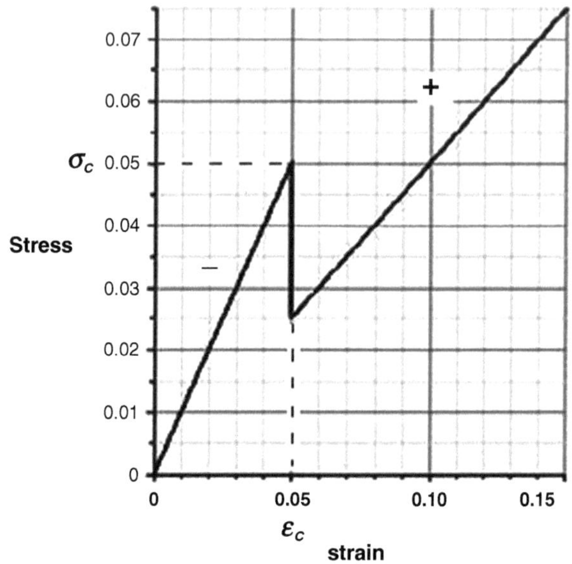

Fig. 2 Piece-wise constitutive curve

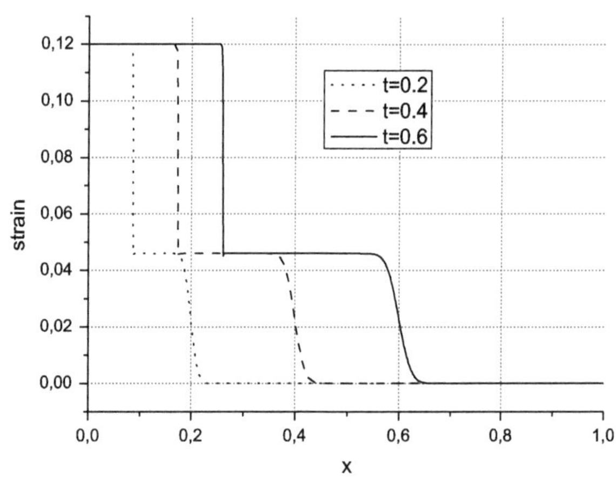

Fig. 3 Strain distribution

Here ρ_0 is the mass density of the rod, which is assumed to be the same for both phases. On the left edge it should satisfy the boundary condition: $\sigma|_{x=0} = \sigma_0 H(t)$, where $H(t)$ is the unit step function. Initial conditions are assumed to be zero. The aim of the research is to determine the stress-strain state of the material at various moments of time. The numerical solution of the problem obtained by using finite difference method is presented in Fig. 3.

Wave process in the medium is characterized with two fronts. The first front is an elastic precursor, moving with the sound velocity. Its emergence is caused by the passing through the first segment of the bilinear stress-strain relation. The next front is responsible for transition of material from phase "–" to phase "+". It's velocity denoted as $\dot{l}(t)$ is not known in advance. Supposing that the structural

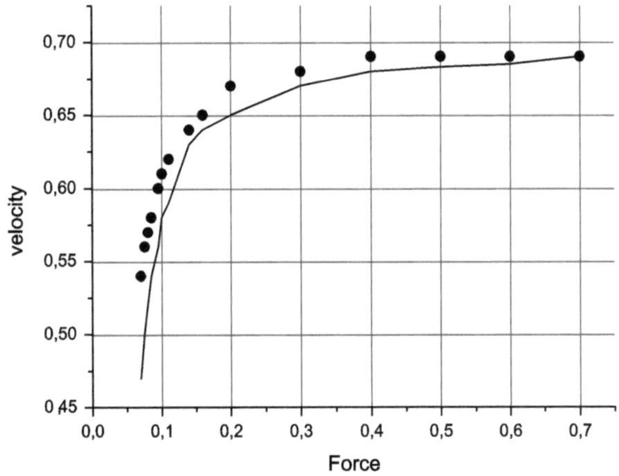

Fig. 4 The velocity of the phase boundary

transformation starts when the deformation reaches its critical value ε_c, it is possible to determine the velocity of the phase boundary using mass and momentum conservation laws.

For small deformations integrating the mass balance equation and the dynamics equation (5) through the phase boundary leads to the following relations, where square brackets denote jump at $x = l(t)$:

$$[\sigma] = -\dot{l}\rho_0[v] \tag{6}$$

$$[\rho]\dot{l} = \rho_0[v] \tag{7}$$

$$\rho^+ = \rho_0 \frac{\partial u^+}{\partial x}, \quad \rho^- = \rho_0 \frac{\partial u^-}{\partial x} \tag{8}$$

Taking into account the jump conditions and assuming that the stress at phase "+" coincide with the stress applied to the left butt of the rod, one can obtain

$$\dot{l}^2 = \frac{\dot{l}^2}{c^{-2}} = \frac{[\sigma]}{\sigma^+/\gamma - \sigma^-}, \tag{9}$$

where $\gamma = c^{+2}/c^{-2} < 1$ is the ratio of sound velocities at different phases [4]. This relation is depicted in Fig. 4. It is clearly seen that the velocity of the phase boundary depends on the external load. Besides, it can't exceed the sound velocity of the second phase. So, it is bounded from above. Dots indicate the result of the numerical solution, which agrees well with the analytical one.

3 Von Mises Truss as a Rheological Model

Good agreement between continuous and discrete models suggests the idea of using von Mises truss as a rheological model of phase transforming material. Then the constitutive law can be written in the form

$$\sigma_M(\varepsilon) = 2E_0\left(1 - \frac{\varepsilon}{a}\tan\alpha\right)\left[\frac{1}{\sqrt{\tan^2\alpha + \left(1 - \frac{\varepsilon}{a}\tan\alpha\right)^2}} - \cos\alpha\right] + E_1\varepsilon, \quad (10)$$

which is a rigorous analogous to the relation between force and displacement in von Mises truss [5]. Depending on the parameters, the stress-strain relation can be presented with one of the curves, depicted in Fig. 5.

Let us return to the model of the semi-infinite rod subjected to the constant load at the end. If the force is not very large, it corresponds to the first increasing branch of the curve. This force is not sufficient enough to transfer the material through the falling segment and initiate the structural conversion.

If the value of the applied load reaches the critical value the emergence of the second front is observed, meaning that the material starts passing to another phase, which penetrates from the left butt of the rod inside the sample with the certain velocity (Fig. 6). If the load is maintained at the same level, i.e. the energy is constantly transmitted to the system, the penetration continues indefinitely.

When the external force disappears, the rod, in accordance with the constitutive law (Fig. 5, curve 1), gradually returns to its initial state, except for the elastic precursor, which keeps on propagating to infinity. The process of phase transition turns to be reversible.

The situation changes when the stress-strain relation (10) has two stable equilibriums (Fig. 5, curve 2). Such curve can be obtained by changing the parameters E_0 and E_1. Now the strain energy possesses two potential wells, and it is not clear which of them the system will take after all.

In this situation it is quite reasonable to expect the emergence of residual strains after the cessation of all dynamic processes. Numerical solution confirms this assumption (Fig. 7). Thereby, this kind of constitutive curve can be used as a model of the irreversible structural transformation.

Since energy is spent during the transition from one increasing branch to another, the velocity of the phase boundary, as it is shown in Fig. 7, gradually decreases. Finally, the propagation of the new phase stops and the system becomes stable. The rod is set into inhomogeneous state, which indicates that the structural transformation has taken place. As a result, two domains with different values of deformations are observed. Consequently, these domains possess different physical qualities, as the internal forces primarily depend on the distance between the particles of continuum.

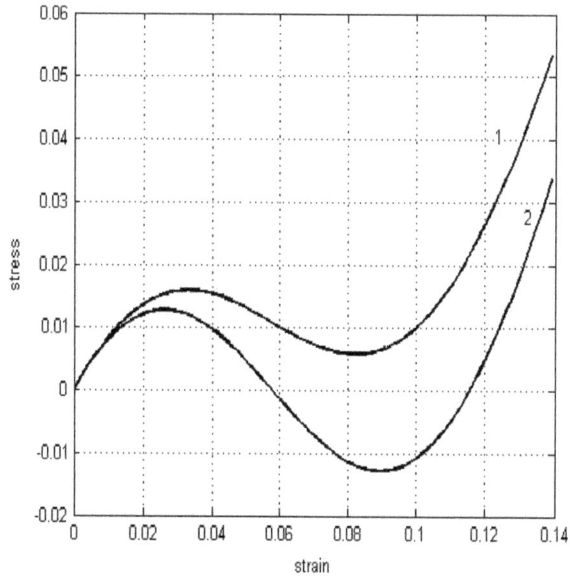

Fig. 5 Stress-strain relation

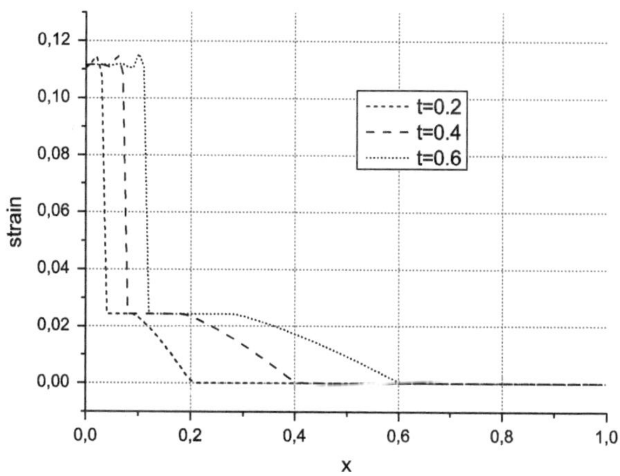

Fig. 6 Penetration of the new phase

4 Conclusion

The present paper deals with the problem of describing phase transitions in solids. The simulation is based on non-monotone stress-strain relation. Discrete and continuous models of material capable of undergoing phase transition are discussed. The point is that the equivalence of discrete to continuous approaches in case of unstable constitutive law is questionable. The main aim of the article is to compare two different representations of the problem and to show their conformity.

At first, a simple model of semi-infinite rod with non-convex strain energy is considered. The non-monotone stress-strain relation is defined with the piece-wise linear function. This assumption allows to study dynamics of the rod and to

Fig. 7 Residual strains

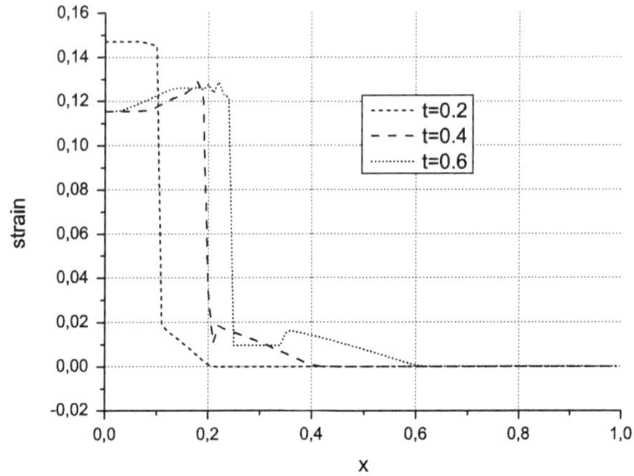

estimate the velocity of the phase boundary which depends on the external load. The analytical result is compared with numerical solution, obtained by using the finite difference method.

The obtained results turn to be rather close to each other, and this fact gives us the possibility to suppose that discrete and continuous models have similar properties. After that we introduce a rheological model for phase transforming material, which is based on von Mises truss. The rod in this case is considered as a chain of these trusses. The suggested model allows to visualize the dynamics of structural transformation. The propagating of deformation wave in the rod is characterized with two fronts. The first one is an elastic precursor, corresponding to the first rising branch of the constitutive law. The second front is responsible for transforming material to the new state.

By changing the parameters of the suggested stress-strain relation we can control the number of equilibriums. If the strain energy has more than one minimum, residual strains may appear in the system. They indicate that the process of phase transition is irreversible in this case.

References

1. Knowles, J.K.: Stress-induced phase transitions in elastic solids. Comput. Mech. **22**, 429–436 (1999)
2. Faciu, C., Molinary, A.: On the longitudinal impact of two phase transforming bars. J. Solids Struct. **43**, 497–522 (2006)
3. Ngan, S.-C., Truskinovsky, L.: Thermo-elastic aspects of dynamic nucleation. J. Mech. Phys. Solids **50**, 1193–1229 (2002)
4. Slepyan, L.I.: Feeding and dissipative waves in fracture and phase transition. J. Mech. Phys. Solids **49**, 513–550 (2001)
5. Eremeyev, V.A., Lebedev, L.P.: On the loss of stability of von Mises truss with the effect of pseudo-elasticity. Matematicas **14**, 111–118 (2006)

A Non-linear Theory for Piezoelectric Beams

Hans Irschik, Alexander Humer, and Johannes Gerstmayr

Abstract An extension of Reissner's geometrically exact theory for the plane deformation of beams allows a consistent constitutive modeling of various types of material behavior. In the present paper, such formulation is used to describe the coupled response of slender piezoelectric structures. Starting from the local equilibrium relations and the geometric description of the cross-sectional deformation, the constitutive equations in the structural mechanics framework are derived from quantities of the non-linear continuum theory in a step-by-step procedure.

1 Introduction

The present paper is motivated by the fact that, even in case of small strains, piezoelectric structures can exhibit a non-linear behavior with respect to the electric field once the field intensity is sufficiently strong, see the contribution of Trinidade and Benjeddou [1]. In the following, we combine this constitutive non-linearity with a theory that can deal with the presence of large strains in structures. For the latter geometric non-linearity, we utilize a theory for plane bending of slender beams by Irschik and Gerstmayr [2, 3]. The Bernoulli-Euler case of an extensible elastica has been treated in [2], while the effect of shear has been additionally included in [3]. The formulations of [2, 3] represent continuum mechanics based extensions of the celebrated large displacement, finite deformation theory by Eric Reissner [5]. In contrast to the theory in [5], which was formulated in terms

H. Irschik (✉) · A. Humer
Institute of Technical Mechanics, Johannes Kepler University, Altenberger Str. 69, A-4040 Linz, Austria
e-mail: hans.irschik@jku.at; alexander.humer@jku.at

J. Gerstmayr
Linz Center of Competence in Mechatronics, Altenberger Str. 69, A-4040 Linz, Austria
e-mail: johannes.gerstmayr@lcm.at

Alexander K. Belyaev et al. (eds.), *Mechanics and Model-Based Control of Advanced Engineering Systems*, DOI 10.1007/978-3-7091-1571-8_21,
© Springer-Verlag Wien 2014

of structural mechanics quantities only, and thus did not utilize the notions of stress and strain, the formulations in [2, 3] do allow the constitutive modeling at the level of non-linear continuum mechanics, as relations between suitable stress and strain measures. Therefore, they appear to be particularly suitable for introducing piezoelectric stress-strain relations into a non-linear theory of structures. More precisely, a non-linear constitutive model for the coupled response to strong electric fields and moderate mechanical strains going back to Tiersten [6] is used in the present considerations. Our contribution is subdivided into the following consecutive steps:

(i) The equilibrium equations that must hold between the static quantities bending moment, normal force and shear force, and
(ii) Geometric relations for certain kinematic quantities of the deformed axis, among them two angles, axial stretch and the displacements of the axis.
(iii) The relations of static equivalence between the second Piola-Kirchhoff stresses and bending moment, normal force and shear force. This brings into the play the deformation of initially plane cross-sections, and thus is also associated with a dependency upon the transverse beam coordinate.
(iv) Introduce the Timoshenko assumption of cross-sections remaining plane and undistorted during the deformation. This allows an interpretation of the aforementioned two angles as cross-sectional rotation and shear angle at the axis. Continuum mechanics then immediately yields a corresponding consequence of (iv), namely
(v) The relations between Green strains and the structural kinematic entities axial stretch, cross-sectional rotation and shear angle. We close the problem by
(vi) Utilizing three-dimensional non-linear piezoelectric stress-strain relations [5] under the Timoshenko assumption, using the strains formulated in (v), which allows a closed cross-sectional integration of the relations of static equivalence (iii).

Due to this combination of continuum mechanics and the structural Bernoulli-Euler-Timoshenko theory, the number of unknown quantities is not only reduced substantially, but all of the remaining unknowns become functions of the coordinate of the undeformed axis only. The resulting theory can thus be considered as an extension of the theory of shear-deformable, extensible elastica theory with respect to a non-linear piezoelectric constitutive behavior.

2 Static and Kinematic Relations of Structural Mechanics: Steps (i) and (ii)

The local equilibrium equations of a differential beam element read, cf. Reissner [5]:

$$M' + \Lambda_{x0} \left(Q \cos\chi - N \sin\chi\right) + m = 0, \tag{1}$$

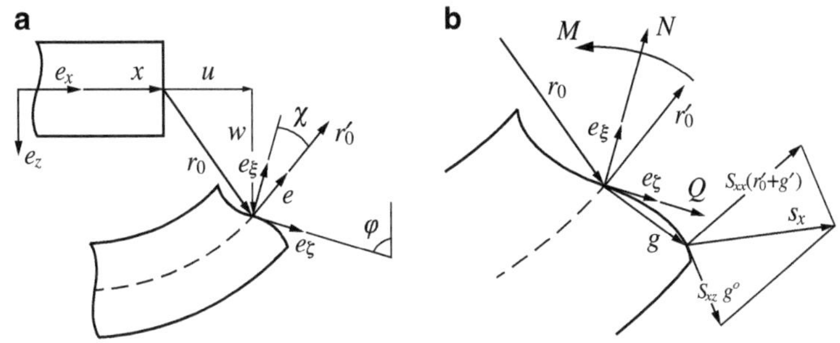

Fig. 1 Coordinate systems and kinematic quantities (**a**); deformed cross-section and static equivalence (**b**)

$$N' - \varphi' Q + n = 0, \tag{2}$$

$$Q' + \varphi' N + q = 0. \tag{3}$$

The bending moment is denoted as M, while Q and N represent the components of the internal force vector in the direction of the unit vectors e_ξ and e_ζ, see Fig. 1a. The latter base vectors are obtained from the global Cartesian (e_x, e_y, e_z)-basis via an elementary rotation about the y-axis with the angle φ, which at the present instant of discussion is not yet given a kinematical meaning. The x-axis of the global system represents the axis of the undeformed beam, which is taken as initially straight, and z is its transverse coordinate in the undeformed configuration. In structural mechanics, the static quantities M, Q and N are considered as a system of inner forces that must be in local equilibrium with the external forces and couples m, q and n per unit axial length in the undeformed configuration, which yields Eqs. (1)–(3). A prime denotes the derivative with respect to the axial coordinate x. While Eq. (1) represents the local equilibrium relation for the axial moment in the y-direction, Eqs. (2) and (3) are the local force equilibrium conditions in the directions of the unit vectors e_ξ and e_ζ. Note that Eqs. (1)–(3) may contain ponderomotive parts, and that dynamic conditions may be accounted for by including inertia forces in m, q and n. The kinematic quantities in Eqs. (1)–(3) have the following meaning: Denoting the position vector of an axis point in the deformed configuration as

$$r_0 = r_0(x) = \left[x + u(x)\right] e_x + w(x)\, e_z, \tag{4}$$

$$r_0' = \frac{\partial r_0}{\partial x} = \Lambda_{x0}\, e, \quad \Lambda_{x0} = \left\| r_0' \right\|. \tag{5}$$

The unit vector in the direction of the deformation gradient vector r_0', which is tangent to the deformed axis, is e, see Fig. 1a. In Eq. (1), the stretch of an element of the beam axis, i.e., the ratio of the length of a differential axis element in the deformed and in the undeformed configuration, is denoted as Λ_{x0}. Moreover, the angle between r_0' and the unit vector e_ξ normal of the cross section in the deformed configuration is χ, see Fig. 1a. From Eqs. (4) and (5), and from Fig. 1a, the following relations result for the derivatives of the axial and the transverse displacements, u and w, respectively:

$$\cos\left(\varphi - \chi\right) = \frac{1 + u'}{\Lambda_{x0}}, \quad \sin\left(\varphi - \chi\right) = -\frac{w'}{\Lambda_{x0}}. \tag{6}$$

3 Static Equivalence of Stresses and Stress Resultants: Step (iii)

In continuum mechanics, the static quantities M, Q and N are considered as stress resultants, where the stresses act in the current configuration upon a deformed cross-section, which has been plane and perpendicular to the x-axis in the undeformed straight reference configuration. Considering a point with coordinates $(x,\ y,\ z)$ in the reference configuration, the position vector of this point in the deformed configuration in the present plane case is written as, see Fig. 1b,

$$r(x, y, z) = r_0(x) + y\, e_y + g(x, z), \quad g = g_\xi e_\xi + g_\zeta\, e_\zeta, \tag{7}$$

where g describes the rotation and deformation of initially plane cross-sections in the actual configuration. The deformation gradient vectors are

$$r' = \frac{\partial r}{\partial x} = r_0' + g' = \Lambda_{x0}\, e + g', \quad r^o = \frac{\partial r}{\partial z} = \frac{\partial g}{\partial z} = g^o. \tag{8}$$

While r' is tangential to the deformed fibres of the beam, the deformation gradient vector g^o—a superimposed "o" denotes the derivative with respect to the transverse coordinate in what follows—is tangential to the deformed cross-section, Fig. 1b. Following, e.g., Washizu [7], the vectorial entity

$$s_x = S_{xx}\left(r_0' + g'\right) + S_{xy}\, e_y + S_{xz}\, g^o \tag{9}$$

represents the Lagrange stress vector needed for the cross-section under consideration, see Fig. 1b. The components of the second Piola-Kirchhoff stresses in the global coordinate system are denoted as S_{xx}, S_{xy} and S_{xz}. The system of inner forces that acts upon the deformed cross-section is formed by the elementary forces $s_x\, dA$, where dA is the elementary cross-sectional area in the undeformed reference configuration with unit normal vector e_x. Noting that

$$e_\xi' = -\varphi' e_\zeta, \quad e_\zeta' = \varphi' e_\xi, \tag{10}$$

the static equivalence of the inner forces with the stress resultants gives, see Fig. 1b:

$$N = \int_A S_{xx}\left(\Lambda_{x0}\cos\chi + g'_\xi + \varphi' g_\zeta\right) + S_{xz}\, g^o_\xi dA, \tag{11}$$

$$Q = \int_A S_{xx}\left(\Lambda_{x0}\sin\chi + g'_\zeta - \varphi' g_\xi\right) + S_{xz}\, g^o_\zeta dA, \tag{12}$$

$$M = \int_A \left[S_{xx}\left(\Lambda_{x0}\cos\chi + g'_\xi + \varphi' g_\zeta\right) + S_{xz}\, g^o_\xi\right] g_\zeta - \left[S_{xx}\left(\Lambda_{x0}\sin\chi + g'_\zeta - \varphi' g_\xi\right) + S_{xz}\, g^o_\zeta\right] g_\xi dA. \tag{13}$$

The relations stated so far are exact, i.e., they must be satisfied by an exact solution of a given problem. However, they are not sufficient for obtaining such solution.

4 The Timoshenko Assumption: Step (iv)

Timoshenko's kinematical assumption states that the deformed cross-section remains plane and undistorted during deformation. In the present non-linear case, this assumption reads

$$g_\xi = 0, \quad g_\zeta = z. \tag{14}$$

This assigns a kinematic meaning to the angle φ, namely as representing the cross-sectional rotation angle about the y-axis. Substituting Eqs. (14) into (11)–(13) gives

$$N = \int_A S_{xx}\left(\Lambda_{x0}\cos\chi + \varphi' z\right) dA, \tag{15}$$

$$Q = \int_A S_{xx}\,\Lambda_{x0}\sin\chi + S_{xz} dA, \tag{16}$$

$$M = \int_A S_{xx}\left(\Lambda_{x0}\cos\chi + \varphi' z\right) z\, dA. \tag{17}$$

These are the relations originally derived in Irschik and Gerstmayr [3] by using Timoshenko's kinematical assumption from the beginning.

5 Green Strains for the Timoshenko Assumption: Step (v)

The deformation gradient F can be computed from the position vector r, see Eq. (7),

$$F = \frac{\partial r}{\partial x} \otimes e_x + \frac{\partial r}{\partial y} \otimes e_y + \frac{\partial r}{\partial z} \otimes e_z. \tag{18}$$

Substituting Timoshenko's kinematical assumption, Eq. (13), into Eq. (7), the deformation gradient F in the common frame attains the following matrix representation:

$$[F] = \begin{bmatrix} \Lambda_{x0} \cos\left(\varphi - \chi\right) + z\varphi' \cos\varphi & 0 & \sin\varphi \\ 0 & 1 & 0 \\ -\left(\Lambda_{x0} \sin\left(\varphi - \chi\right) + z\varphi' \sin\varphi\right) & 0 & \cos\varphi \end{bmatrix}. \tag{19}$$

The Jacobian determinant $J = \det F$ follows to

$$J = \Lambda_{x0} \cos\chi + z\varphi'. \tag{20}$$

The Green strain tensor, which in general is defined by $G = (F^T F - I)/2$, therefore has the following matrix representation:

$$[G] = \frac{1}{2} \begin{bmatrix} J^2 - 1 & 0 & \Lambda_{x0} \sin\chi \\ 0 & 0 & 0 \\ \Lambda_{x0} \sin\chi & 0 & 0 \end{bmatrix}. \tag{21}$$

6 Non-linear Piezoelectric Stress-Strain Relations Under the Timoshenko Assumption: Step (vi)

In order to close the set of Eqs. (1)–(3) provided above, constitutive equations describing the material's stress response to deformation as well as an electric field need to be provided. Following Tiersten's approach [6] for the modeling of structures subjected to strong electric fields but only moderate strains, terms up to the second order in the electric field are regarded, whereas the stress response is assumed to be linear in Green's strain. In particular, the components required for evaluating the stress resultants (11)–(13) are given by

$$S_{xx} = C_{11} G_{xx} + 2C_{15} G_{xz} - e_{11} E_x - e_{31} E_z - \frac{1}{2}\left(b_{111} E_x^2 + 2b_{131} E_x E_z + b_{331} E_z^2\right), \tag{22}$$

$$S_{xz} = C_{15}G_{xx} + 2C_{55}G_{xz} - e_{15}E_x - e_{35}E_z - \frac{1}{2}\left(b_{115}E_x^2 + 2b_{135}E_xE_z + b_{335}E_z^2\right), \tag{23}$$

where C denotes the fourth-order tensor of elastic moduli, e is the third-order tensor of piezoelectric constants, and b is the fourth-order tensor of electrostrictive coefficients. The component notation used for the material tensors follows [4]; otherwise, the indices x, y, z are retained. The electric field vector in the reference configuration E is related to the actual electric field via a proper pull-back operation as explained by Yang [8]. For notational convenience, the stress resultants are decomposed into purely elastic parts, indicated by a superimposed (e), and piezoelectric components (p) accounting for the presence of an electric field:

$$N = N^{(e)} + N^{(p)}, \qquad Q = Q^{(e)} + Q^{(p)}, \qquad M_y = M_y^{(e)} + M_y^{(p)}. \tag{24}$$

Further, the cross-sectional moments of inertia are introduced as

$$A^{(i)} = \int_A z^i \, dA, \qquad i = 0, 1, 2, \ldots \tag{25}$$

with the beam's axis being chosen such that the odd moments vanish identically, i.e.,

$$A^{(i)} = 0, \qquad i = 1, 3, 5, \ldots . \tag{26}$$

In accordance with the above definitions, the elastic part of the normal force is obtained by inserting (22)–(23) into (11) as

$$\begin{aligned} N^{(e)} = {} & \frac{C_{11}A^{(0)}}{2}\Lambda_{x0}\cos\chi\left(\Lambda_{x0}^2\cos^2\chi - 1\right) + \frac{3C_{11}A^{(2)}}{2}(\varphi')^2\Lambda_{x0}\cos\chi \\ & + C_{15}A^{(0)}\Lambda_{x0}^2\cos\chi\sin\chi. \end{aligned} \tag{27}$$

The respective elastic components of the shear force and the bending moment analogously follow as

$$\begin{aligned} Q^{(e)} = {} & \frac{C_{15}A^{(0)}}{2}\left(\Lambda_{x0}^2\cos^2\chi + 2\Lambda_{x0}^2\sin^2\chi - 1\right) + \frac{C_{15}A^{(2)}}{2}(\varphi')^2 + C_{55}A^{(0)}\Lambda_{x0}\sin\chi \\ & + \frac{C_{11}A^{(0)}}{2}\Lambda_{x0}\sin\chi\left(\Lambda_{x0}^2\cos^2\chi - 1\right) + \frac{C_{11}A^{(2)}}{2}(\varphi')^2\Lambda_{x0}\sin\chi, \end{aligned} \tag{28}$$

$$M_y^{(e)} = -\frac{C_{11}A^{(2)}}{2}\varphi' + \frac{3C_{11}A^{(2)}}{2}\varphi'\Lambda_{x0}^2\cos^2\chi + \frac{C_{11}A^{(4)}}{2}(\varphi')^3 + C_{15}A^{(2)}\varphi'\Lambda_{x0}\sin\chi. \tag{29}$$

The piezoelectric parts of the stress resultants, which account for the material response to an electric field, are given by

$$N^{(p)} = -\int_A \left[e_{11} E_x + e_{31} E_z + \frac{1}{2} \left(b_{111} E_x^2 + 2 b_{131} E_x E_z + b_{331} E_z^2 \right) \right] J \, dA, \quad (30)$$

$$Q^{(p)} = -\int_A \left[e_{11} E_x + e_{31} E_z + \frac{1}{2} \left(b_{111} E_x^2 + 2 b_{131} E_x E_z + b_{331} E_z^2 \right) \right] \Lambda_{x0} \sin \chi$$
$$+ e_{15} E_x + e_{55} E_z + \frac{1}{2} \left(b_{115} E_x^2 + 2 b_{135} E_x E_z + b_{335} E_z^2 \right) dA, \quad (31)$$

$$M_y^{(p)} = -\int_A \left[e_{11} E_x + e_{31} E_z + \frac{1}{2} \left(b_{111} E_x^2 + 2 b_{131} E_x E_z + b_{331} E_z^2 \right) \right] J z \, dA. \quad (32)$$

Before the above integrals can be evaluated, either the coupled problem governed by the field equations (1)–(3) and Maxwell's equations needs to be solved, or some simplifying assumptions concerning the distribution of the electric field are required.

7 Conclusion

In the present paper, a theory for large deformation problems of piezoelectric beams, which is based on Reissner's geometrically exact equilibrium relations, has been presented adopting a systematic step-by-step derivation of the governing equations. Not restricting the cross-sectional deformation at first, the kinematic quantities required in the description of the deformation process have been introduced to begin with. Subsequently, the stress resultants within the structural mechanics formulation have been linked to the notion of stresses of non-linear continuum mechanics by inspecting the static equivalence in the beam's deformed configuration. Additionally, Timoshenko's hypothesis regarding the deformation of the cross-sections is introduced, and the kinematic quantities, which then become functions of the beam's axis only, are linked to the components of Green's strain tensor. Having established the connection between quantities of structural and continuum mechanics, various kinds of constitutive models can be consistently included into the beam theory. In particular, the present paper has discussed relations for piezoelectric structures which may be subjected to strong electric fields. The constitutive relations for the normal force, the shear force and the bending moment have been derived for a material model introduced by Tiersten, whose elastic part is of the St. Venant-Kirchhoff type and which accounts for inelastic effects related to piezoelectricity as well as electrostriction.

Acknowledgements The authors gratefully acknowledge the support of the Comet K2 Austrian Center of Competence in Mechatronics (ACCM).

References

1. Trinidade, M.A., Benjeddou, A.: Finite element characterization and parametric analysis of the nonlinear behaviour of an actual d15 shear MFC. Acta Mech. (2013) (online first)
2. Irschik, H., Gerstmayr, J.: A continuum mechanics based derivation of Reissner's large-displacement finite-strain beam theory: the case of plane deformations of originally straight Bernoulli-Euler beams. Acta Mech. **206**(1–2), 1–21 (2009)
3. Irschik, H., Gerstmayr, J.: A continuum-mechanics interpretation of Reissner's non-linear shear-deformable beam theory. Math. Comput. Model. Dyn. Syst. **17**(1), 19–29 (2011)
4. Meitzler, A., Tiersten, H., Warner, A., Berlincourt, D., Couqin, G., Welsh III, F.: IEEE Standard on Piezoelectricity. The Institute of Electrical and Electronics Engineers, Inc., New York (1988)
5. Reissner, E.: On one-dimensional finite-strain beam theory: the plane problem. Zeitschrift für angewandte Mathematik und Physik (ZAMP) **23**(5), 795–804 (1972)
6. Tiersten, H.: Electroelastic equations for electroded thin plates subject to large driving voltages. J. Appl. Phys. **74**(5), 3389–3393 (1993)
7. Washizu, K.: Variational Methods in Elasticity and Plasticity, 2nd edn. Pergamon Press, Oxford (1975)
8. Yang, J.: An Introduction to the Theory of Piezoelectricity. Springer, New York (2004)

Nonlinear Analysis of Phase-Locked Loop (PLL): Global Stability Analysis, Hidden Oscillations and Simulation Problems

G.A. Leonov and N.V. Kuznetsov

Abstract In the middle of last century the problem of analyzing *hidden oscillations* arose in automatic control. In 1956 M. Kapranov considered a two-dimensional dynamical model of phase locked-loop (PLL) and investigated its qualitative behavior. In these investigations Kapranov assumed that oscillations in PLL systems can be *self-excited oscillations* only. However, in 1961, N. Gubar' revealed a gap in Kapranov's work and showed analytically the possibility of the existence of another type of oscillations, called later by the authors *hidden oscillations*, in a phase-locked loop model. From a computational point of view the system considered was globally stable (all the trajectories tended to equilibria), but, in fact, there was a bounded domain of attraction only.

In this review, following ideas of N. Gubar', the qualitative analysis of hidden oscillation bifurcation in a two-dimensional model of phase-locked loop is considered.

1 Introduction: Self-Excited and Hidden Oscillations

From the point of view of the numerical analysis of nonlinear dynamical systems the attractors can be regarded as *self-excited* and *hidden attractors*. Self-excited attractors can be localized numerically by a *standard computational procedure, in which after a transient process a trajectory, started from a point of unstable*

G.A. Leonov
Saint-Petersburg State University, Universitetsky pr. 28, St. Petersburg 198594, Russia
e-mail: leonov@math.spbu.ru

N.V. Kuznetsov (✉)
University of Jyväskylä, P.O. Box 35 (Agora), Jyväskylä FIN-40014, Finland

Saint-Petersburg State University, Universitetsky pr. 28, St. Petersburg 198594, Russia
e-mail: nkuznetsov239@gmail.com

Alexander K. Belyaev et al. (eds.), *Mechanics and Model-Based Control of Advanced Engineering Systems*, DOI 10.1007/978-3-7091-1571-8_22,

© Springer-Verlag Wien 2014

manifold in a neighborhood of equilibrium, reaches a state of oscillation and therefore it can be easily identified. In contrast, for a *hidden attractor, its basin of attraction does not intersect with small neighborhoods of equilibria.* While classical attractors are self-exited attractors and therefore can be obtained numerically by the standard computational procedure, for localization of hidden attractors it is necessary to develop special procedures since there are no similar transient processes leading to such attractors.

At first the problem of investigating hidden oscillations arose in the second part of Hilbert's 16th problem (1900). The first nontrivial results were obtained in Bautin's works, which were devoted to construction of nested limit cycles in quadratic systems and showed the necessity of studying hidden oscillations to solve this problem. Later, the problem of analyzing hidden oscillations arose from engineering problems in automatic control. In 1950–1960s of the last century, the investigations of widely known *Markus-Yamabe's*, *Aizerman's*, and *Kalman's conjectures* on absolute stability have led to the finding of hidden oscillations in automatic control systems with a unique stable stationary point (see, e.g., surveys [1, 10]). In 1961, N. Gubar' revealed [2] a gap in Kapranov's work [3] on phase locked-loops (PLL) and showed the possibility of the existence of hidden oscillations in PLL.

Note that further investigations of hidden oscillations were greatly encouraged by the authors' discovery, in 2010 for the first time, of *chaotic hidden attractor* in Chua's circuit [6, 8, 10, 14, 16].

1.1 Phase-Locked-Loop Circuits: Simulation and Nonlinear Analysis

The phase-locked loop (PLL) systems were invented in 1930s–1940s and were widely used in radio and television (demodulation and recovery, synchronization and frequency synthesis). Nowadays PLL can be produced in the form of single integrated circuit and different modifications of PLL are used in modern electronic applications (radio, telecommunications, computers, and others). Various methods for analysis of phase-locked loops are well developed by engineers and considered in many publications, but the problems of the construction of adequate nonlinear models and the nonlinear analysis of such models are still far from being resolved, turn out difficult, and require the use of special methods of qualitative theory of differential, difference, integral, and integro-differential equations [4, 5, 7, 9, 11–13, 15, 17–20].

Below it will be illustrated some difficulties arising in analysis of comparatively simple nonlinear two-dimensional dynamical model of PLL even in studying piecewise-constant nonlinearity.

In the middle of last century there were begun the investigations of dynamical models of phase synchronization systems. M. Kapranov [3] obtained in his work the conditions of global stability for the following two-dimensional PLL model (with

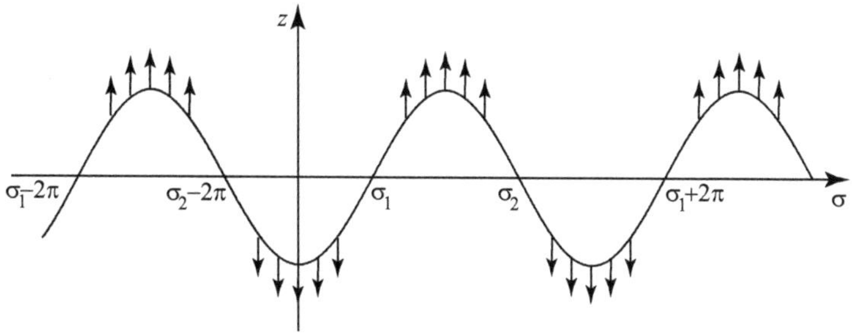

Fig. 1 Vector field on the curve $z = a(\varphi(\sigma) - \gamma)$

the filter of the type $W(p) = \frac{ap+\beta}{p+\alpha}$)

$$\begin{aligned} \dot{z} &= -\alpha z - (1 - a\alpha)\big(\varphi(\sigma) - \gamma\big), \\ \dot{\sigma} &= z - a\big(\varphi(\sigma) - \gamma\big), \end{aligned} \qquad a, \alpha, \gamma \geq 0, \tag{1}$$

where $\varphi(\sigma)$ is 2π-periodic characteristic of phase detector.

In 1961 N. Gubar' [2] revealed a gap in the proof of Kapranov's results and indicated system parameters for which in this system a semistable periodic solution can exist. Such a semistable trajectory cannot be found numerically by the standard computation procedure. From a computational point of view the system considered was globally stable (all the trajectories tend to equilibria), but, in fact, there was a bounded domain of attraction only.

Following [11], the qualitative analysis of system (1) is considered below.

Theorem 1. *Any bounded in* $\mathbb{R}^2$ *for* $t \in [0, +\infty)$ *solution of system* (1) *tends to a certain equilibrium as* $t \to +\infty$.

Further, without loss of generality, it is assumed that $-\min_{\sigma} \varphi(\sigma) = \max_{\sigma} \varphi(\sigma) = 1$. Then from Theorem 1 it follows that for $\gamma > 1$ (i.e. in the absence of equilibria) all the solutions of system (1) are unbounded. In this case the synchronization does not occur under any initial conditions $z(0)$, $\sigma(0)$.

Suppose that on the set $[0, 2\pi)$ there are exactly two zeros of the function $\varphi(\sigma) - \gamma$: $\sigma = \sigma_1$, $\sigma = \sigma_2$. Besides $\varphi'(\sigma_1) > 0$, $\varphi'(\sigma_2) < 0$.

Consider the case $\gamma < 1$. In this case in a phase space an important role plays the curve $z = a(\varphi(\sigma) - \gamma)$ and the placed on this curve equilibria $z = 0$, $\sigma = \sigma_0$, where $\varphi(\sigma_0) = \gamma$ (Fig. 1).

Consider the equilibrium $z = 0$, $\sigma = \sigma_2$. In neighborhood of this point the characteristic polynomial of first approximation linear system is as follows

$$p^2 + \big(\alpha + a\varphi'(\sigma_2)\big)p + \varphi'(\sigma_2). \tag{2}$$

The inequality $\varphi'(\sigma_2) < 0$ implies that characteristic polynomial (2) has one positive and one negative zeros and the stationary point $z = 0$, $\sigma = \sigma_2$ is a saddle.

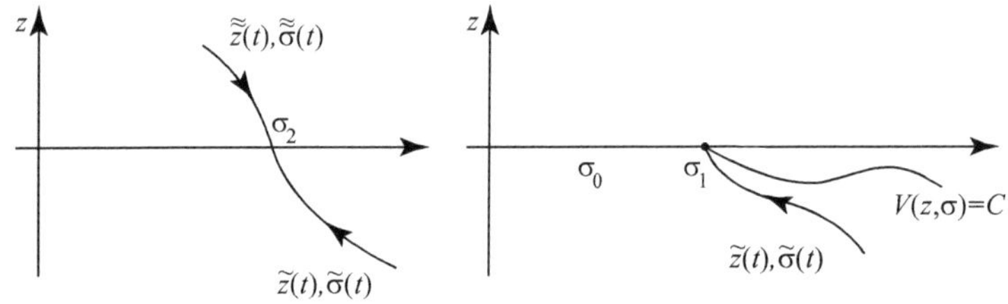

Fig. 2 Separatrices of saddle point

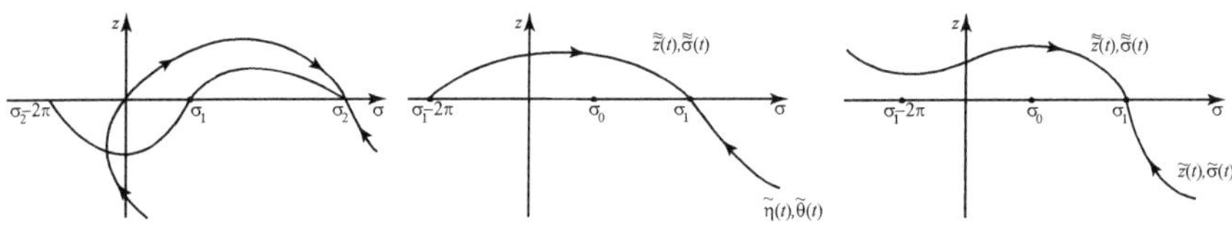

Fig. 3 Behavior of separatrix in Cases 1, 2, 3

In this case only two trajectories of system (1), separatrices of saddle, tend to equilibrium $z = 0, \sigma = \sigma_2$ as $t \to +\infty$. Denote them as $\tilde{z}(t), \tilde{\sigma}(t)$ and $\tilde{\tilde{z}}(t), \tilde{\tilde{\sigma}}(t)$ (Fig. 2).

The following assertion can be proved.

Theorem 2. *The relations*

$$\tilde{z}(t) < 0, \quad \forall\, t, \tag{3}$$

$$\lim_{t\to-\infty} \tilde{z}(t) = -\infty, \quad \lim_{t\to-\infty} \tilde{\sigma}(t) = +\infty \tag{4}$$

are satisfied.

The proof of this theorem is based on consideration of Lyapunov function in the form

$$V(z,\sigma) = \frac{z^2}{2} + (1 + a\alpha)\int_0^\sigma \big(\varphi(\theta) - \gamma\big)\, d\theta\,.$$

Consider now the behavior of separatrix $\tilde{\tilde{z}}(t)$, $\tilde{\tilde{\sigma}}(t)$. In this case there are more opportunities of qualitative behavior, than those given by relations (3) and (4) for the separatrix $\tilde{z}(t)$, $\tilde{\sigma}(t)$. Three cases are possible:

1. There exists a number τ such that $\tilde{\tilde{z}}(\tau) = a(\varphi(\tilde{\sigma}(\tau)) - \gamma)$, $\tilde{\tilde{\sigma}}(\tau) \in (\sigma_2 - 2\pi, \sigma_1)$, $\tilde{\tilde{z}}(t) > a(\varphi(\tilde{\tilde{\sigma}}(t)) - \gamma)$, $\forall\, t \in (\tau, +\infty)$. In Fig. 3 are shown the separatrices of saddle $z = 0$, $\sigma = \sigma_2$ and the curve $z = a(\varphi(\sigma) - \gamma)$.

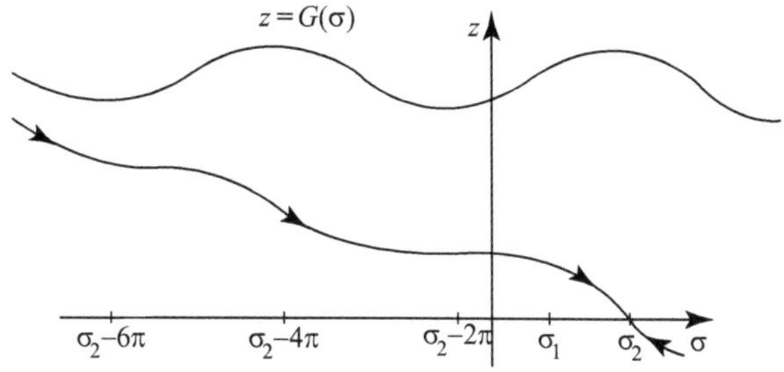

Fig. 4 Separatrices and the unstable cycle $z = G(\sigma)$ in Case 3b

2. For all $t \in \mathbb{R}^1$ the relation $\tilde{\tilde{z}}(t) > 0$ is satisfied (see Fig. 3, middle) and

$$\lim_{t \to -\infty} \tilde{\tilde{z}}(t) = 0, \qquad \lim_{t \to -\infty} \tilde{\tilde{\sigma}}(t) = \sigma_2 - 2\pi.$$

3. For all $t \in \mathbb{R}^1$ the relation $\tilde{\tilde{z}}(t) > 0$ is satisfied (see Fig. 3, right) and

$$\lim_{t \to -\infty} \tilde{\tilde{\sigma}}(t) = -\infty.$$

In this case there are two possibilities:

(3a) As $t \to -\infty$, the separatrix tends to infinity as the coordinate z tends to infinity:

$$\lim_{t \to -\infty} \tilde{\tilde{z}}(t) = +\infty. \tag{5}$$

As was shown first by F. Tricomi [21], this case always occurs for $a = 0$.

(3b) As $t \to -\infty$, the separatrix remains bounded with respect to a coordinate $z \left(\overline{\lim_{t \to -\infty}} \tilde{\tilde{z}}(t) \le \text{const} \right)$ and tends to – graph of a certain periodic function $\{z = G(\sigma),\ G(\sigma + 2\pi) = G(\sigma)\}$ (Fig. 4). The curve $z = G(\sigma)$ corresponds to an unstable trajectory, to which the separatrices of saddles tend (from below) as $t \to -\infty$.

For $\varphi(\sigma) = \operatorname{sign}\sin(\sigma)$ this effect was discovered first in the work [2]. In this case the piecewise-linearity property of nonlinearity allows one to integrate a system on linearity intervals and then to apply Andronov's point-transformation method to the investigation of limit cycle existence.

From Theorems 1 and 2 and the stated above properties of the separatrix of saddle $\tilde{\tilde{z}}(t), \tilde{\tilde{\sigma}}(t)$ a conclusion can be made that for $\gamma \in [0, 1]$ the following topologies of the phase space are possible:

1. The separatrices $\tilde{z}(t), \tilde{\tilde{\sigma}}(t)$ and $\tilde{\tilde{z}}(t), \tilde{\tilde{\sigma}}(t)$ are boundaries of domains of attraction of locally asymptotically stable equilibrium $z = 0,\ \sigma = \sigma_1$ (Fig. 5). In the displacement along σ by 2π these domains become the domains of attraction of stationary solutions $z = 0, \sigma = \sigma_1 + 2k\pi$.

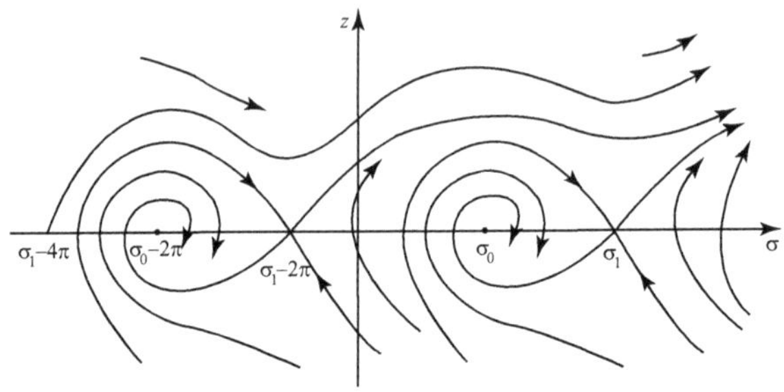

Fig. 5 The attraction domains bounded by separatrices

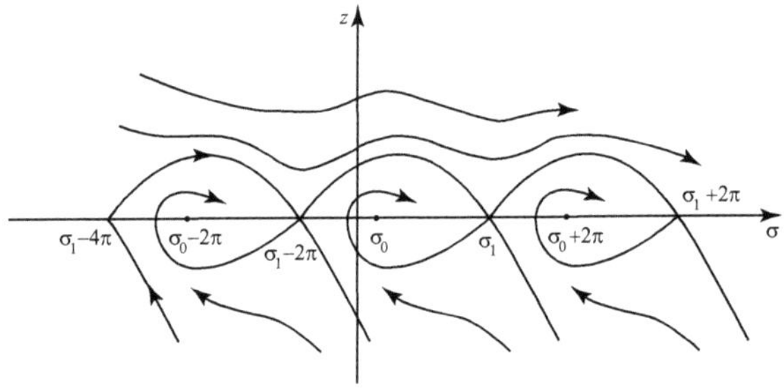

Fig. 6 Bifurcation of heteroclinic trajectory

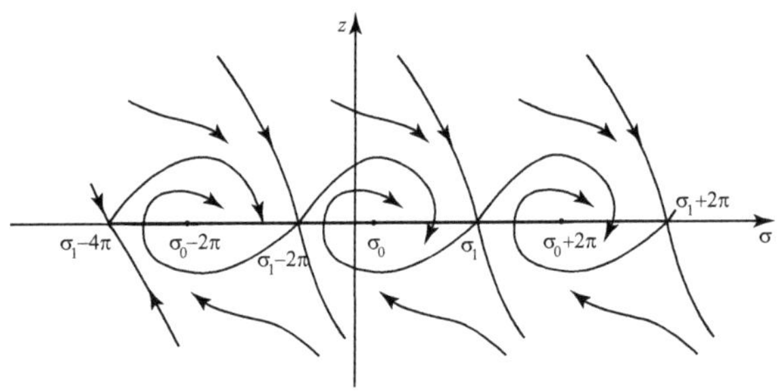

Fig. 7 Global asymptotic stability

The trajectories, placed outside these domains bounded by separatrices, tend to infinity as $t \to +\infty$.

2. The separatrix $\tilde{\tilde{z}}(t)$, $\tilde{\tilde{\sigma}}(t)$ is a heteroclinic trajectory in $\mathbb{R}^2$, i.e.

$$\lim_{t\to-\infty} \tilde{\tilde{z}}(t) = \lim_{t\to+\infty} \tilde{\tilde{z}}(t) = 0,$$
$$\lim_{t\to+\infty} \tilde{\tilde{\sigma}}(t) = \sigma_2, \quad \lim_{t\to-\infty} \tilde{\tilde{\sigma}}(t) = \sigma_2 - \pi.$$

In this case the attraction domains of asymptotically stable equilibria are also bounded by separatrices but in the semiplane $\{z \leq 0\}$ the unstable "corridors" are absent (Fig. 5).

3. The phase space is partitioned into domains of attraction of stable equilibria (this corresponds to Case 3a – see Fig. 6). The boundaries of these domains are separatrices of saddle equilibria (Fig. 7).

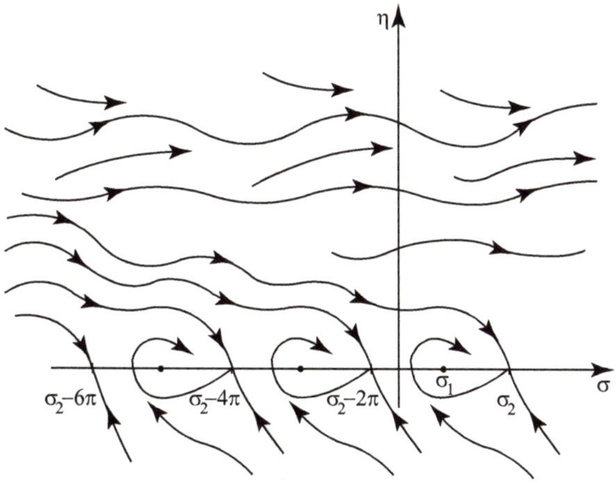

Fig. 8 Stability domains bounded by the cycle $z = G(\sigma)$

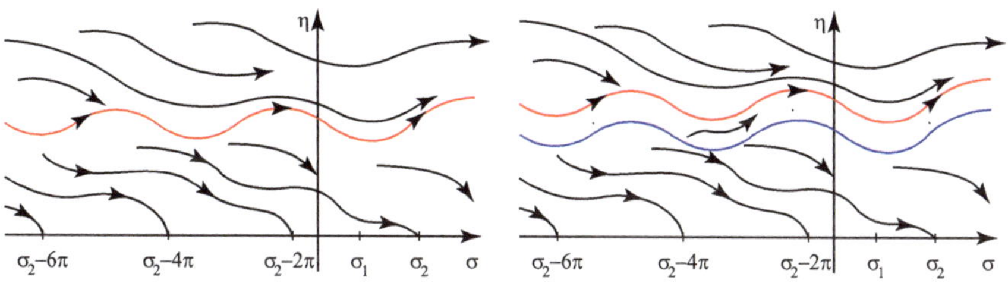

Fig. 9 Stable (hidden oscillation) and unstable periodic trajectories are bifurcated from semistable periodic trajectories. If stable and unstable periodic solutions are very close one another, then from a computational point of view, all the trajectories tend to equilibria, but, in fact, there is a bounded domain of attraction only

4. The phase space is partitioned into domains of attraction of stable equilibria with the boundaries, consisting of separatrices of saddles, and into a domain, placed above the curve $\{z = G(\sigma)\}$. In this domain all trajectories tend to infinity as $t \to +\infty$. This corresponds to Case 3b (Fig. 8).

Obviously, that under any initial data $z(0)$, $\sigma(0)$ the synchronization corresponds to global asymptotic stability of system (1). This property corresponds to Case 3 (Fig. 7).

The loss of stability because of continuous dependence of trajectories with respect to system parameters can be caused only by two global bifurcations: the appearance of heteroclinic trajectory (Case 2, Fig. 3) or the appearance of semistable trajectory of the form $\{z = G(\sigma)\}$ (so called a second-order cycle) (Case 4, Fig. 9, left).

2 Conclusions

In this review it is demonstrated that oscillations in dynamical models of PLL can be not only self-excited but hidden. For investigation of hidden oscillations it is necessity to develop special analytical-numerical methods.

Acknowledgements This work was partly supported by Academy of Finland, Ministry of Education and Science of the Russian Federation, RFBR and Saint-Petersburg State University.

References

1. Bragin, V.O., Vagaitsev, V.I., Kuznetsov, N.V., Leonov, G.A.: Algorithms for finding hidden oscillations in nonlinear systems. The Aizerman and Kalman conjectures and Chua's circuits. J. Comput. Syst. Sci. Int. **50**(4), 511–543 (2011). doi:10.1134/S106423071104006X
2. Gubar', N.A.: Investigation of a piecewise linear dynamical system with three parameters. J. Appl. Math. Mech. **25**(6), 1011–1023 (1961)
3. Kapranov, M.: Locking band for phase-locked loop. Radiofizika (in Russian) **2**(12), 37–52 (1956)
4. Kudrewicz, J., Wasowicz, S.: Equations of Phase-Locked Loop. Dynamics on Circle, Torus and Cylinder, A, vol. 59. World Scientific, Singapore (2007)
5. Kuznetsov, N.V., Leonov, G.A., Seledzhi, S.S.: Phase locked loops design and analysis. In: ICINCO 2008 – Proceedings of the 5th International Conference on Informatics in Control, Automation and Robotics, Funchal, vol. SPSMC, pp. 114–118 (2008). doi:10.5220/0001485401140118
6. Kuznetsov, N.V., Leonov, G.A., Vagaitsev, V.I.: Analytical-numerical method for attractor localization of generalized Chua's system. IFAC Proc. Vol. (IFAC-PapersOnline) **4**(1), 29–33 (2010). doi:10.3182/20100826-3-TR-4016.00009
7. Kuznetsov, N.V., Leonov, G.A., Yuldashev, M.V., Yuldashev, R.V.: Analytical methods for computation of phase-detector characteristics and pll design. In: ISSCS 2011 – Proceedings of the International Symposium on Signals, Circuits and Systems, Iasi, pp. 7–10 (2011). doi: 10.1109/ISSCS.2011.5978639
8. Kuznetsov, N., Kuznetsova, O., Leonov, G., Vagaitsev, V.: Analytical-numerical localization of hidden attractor in electrical Chua's circuit. In: Informatics in Control, Automation and Robotics. Lecture Notes in Electrical Engineering, vol. 174, part 4, pp. 149–158. Springer, Berlin (2013). doi:10.1007/978-3-642-31353-0_11
9. Leonov, G.A.: Sets of transversal curves for two-dimensional systems of differential equations. Vestnik St. Petersburg Univ. **39**(4), 219–245 (2006)
10. Leonov, G., Kuznetsov, N.V.: Analytical-numerical methods for hidden attractors localization: the 16th Hilbert problem, Aizerman and Kalman conjectures, and Chua circuits. In: Numerical Methods for Differential Equations, Optimization, and Technological Problems, Computational Methods in Applied Sciences, vol. 27, part 1, pp. 41–64. Springer, Dordrecht (2013). doi:10.1007/978-94-007-5288-7_3
11. Leonov, G.A., Ponomarenko, D.V., Smirnova, V.B.: Frequency-Domain Methods for Nonlinear Analysis. Theory and Applications. World Scientific, Singapore (1996)
12. Leonov, G.A., Kuznetsov, N.V., Seledzhi, S.M.: Analysis of phase-locked systems with discontinuous characteristics. IFAC Proc. Vol. (IFAC-Papers Online) **1**, 107–112 (2006). doi: 10.3182/20060628-3-FR-3903.00021

13. Leonov, G.A., Kuznetsov, N.V., Seledzhi, S.M.: Nonlinear analysis and design of phase-locked loops. In: Automation Control – Theory and Practice, pp. 89–114. In-Tech (2009). doi:10.5772/7900
14. Leonov, G.A., Kuznetsov, N.V., Vagaitsev, V.I.: Localization of hidden Chua's attractors. Phys. Lett. A **375**(23), 2230–2233 (2011). doi:10.1016/j.physleta.2011.04.037
15. Leonov, G.A., Kuznetsov, N.V., Yuldahsev, M.V., Yuldashev, R.V.: Computation of phase detector characteristics in synchronization systems. Dokl. Math. **84**(1), 586–590 (2011). doi:10.1134/S1064562411040223
16. Leonov, G.A., Kuznetsov, N.V., Vagaitsev, V.I.: Hidden attractor in smooth Chua systems. Physica D **241**(18), 1482–1486 (2012). doi:10.1016/j.physd.2012.05.016
17. Leonov, G.A., Kuznetsov, N.V., Yuldashev, M.V., Yuldashev, R.V.: Differential equations of Costas loop. Dokl. Math. **86**(2), 723–728 (2012). doi:10.1134/S1064562412050080
18. Leonov, G.A., Kuznetsov, N.V., Yuldahsev, M.V., Yuldashev, R.V.: Analytical method for computation of phase-detector characteristic. IEEE Trans. Circuits Syst. II Express Briefs **59**(10), 633–637 (2012). doi:10.1109/TCSII.2012.2213362
19. Margaris, W.: Theory of the Non-linear Analog Phase Locked Loop. Springer, New York (2004)
20. Suarez, A., Quere, R.: Stability Analysis of Nonlinear Microwave Circuits. Artech House, Boston (2003)
21. Tricomi, F.: Integrazione di unequazione differenziale presentatasi in elettrotechnica. Annali della R. Shcuola Normale Superiore di Pisa **2**(2), 1–20 (1933)

Constitutive Models for Anisotropic Materials Susceptible to Loading Conditions

E.V. Lomakin, B.N. Fedulov, and A.M. Melnikov

Abstract The mechanical behavior of many structural materials displays anisotropic elastic and plastic properties in different degree. Along with anisotropy, their behavior may demonstrate the asymmetry in the mechanical properties. The simplest example of this asymmetry is the difference of the deformation characteristics or the yield limits under conditions of tension and compression in the same directions in a material. Generally the mechanical properties of materials may be expressed as the functions of direction of the applied load and some parameters of the stress state type. In some of recently proposed theories of elasticity and plasticity different from the classic isotropic theories based on the Hooke's law and Tresca or von Mises yield criteria, the anisotropy and asymmetry of the mechanical properties in some degree are involved, but the formulation of general equations that may be easily used in engineering is still a challenged problem. In this paper, generalized constitutive equations for the description of elastic and plastic behavior of anisotropic materials with stress state dependent properties are considered.

1 Introduction

Mechanical properties of many structural materials exhibit dependence on the type of stress state and the direction of the applied loads. While this difference of the strength properties may be negligible in some cases, it can play the significant role for anisotropic materials such as composites, polycrystalline materials and other materials. The sources of the difference of the properties under different types of stress state are the special processes, which accompany the deformation of the materials under corresponding loadings. Under conditions of compression

E.V. Lomakin (✉) · B.N. Fedulov · A.M. Melnikov
Faculty of Mechanics and Mathematics, Moscow State Lomonosov University, 119992 Moscow, Russia
e-mail: lomakin@mech.math.msu.su; fedulov.b@mail.ru; m_andrew_m@mail.ru

Alexander K. Belyaev et al. (eds.), *Mechanics and Model-Based Control of Advanced Engineering Systems*, DOI 10.1007/978-3-7091-1571-8_23,
© Springer-Verlag Wien 2014

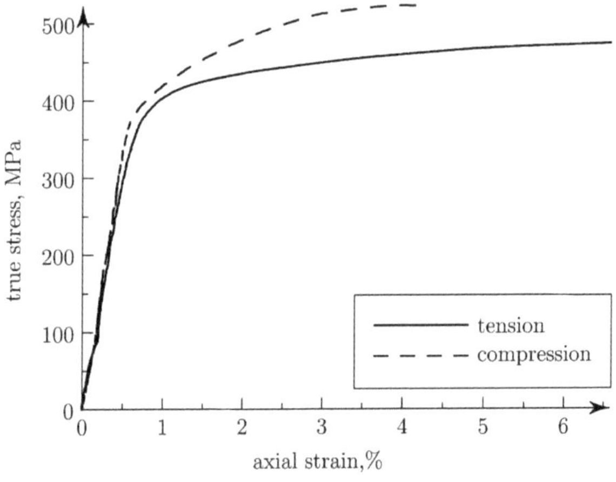

Fig. 1 The stress-strain diagrams for aluminum D16 rod

the closure of the microcracks in the material may occur, that lead to the contact interaction on their faces. This interaction involves mechanisms of deformation and fracture corresponding to the action of pressure and compressive loads rather than classic slip one. In general case, it should lead to the increased strength of a material under compression in comparison with its properties under tension. The opposite effect sometimes can be observed in composites. The fibers of the fiber-reinforced composites can tighten up under tension and buckle under compression that can result in the opposite behavior than described above. These processes can take place in the inelastic deformation of the isotropic materials and result in different plastic limits under different stress states. The anisotropy of materials causes the increasing of complexity of the mathematical models describing their properties and behavior. Some elastically isotropic materials such as metallic alloys may express the anisotropy under plasticity conditions. This anisotropy may be the result of manufacture and forming processes of materials.

Some anisotropy of elastic and plastic properties can be observed in the tests of materials without initial structural asymmetries under some special conditions of the forming process. Numerical values of parameters of this anisotropy and the differences of the yield limits for large number of metallic alloys may be found in [6]. The variation of tensile properties depending on the orientation with respect to the rolling direction for a certain aluminum-lithium alloys is described in [1, 2]. The stress-strain curves for aluminum extruded rod under tension and compression along the same direction of extrusion are plotted on the Fig. 1. The aluminum alloy D16 had been used in this experiment. It's clear from this figure that there's essential difference between the yield limits and diagrams in the plastic zone for different stress states.

There are many theories tending to describe the behavior of this kind of materials especially in the framework of the mechanics of porous media and the mechanics of soils, but the development of the general theory that can be easily applied for engineering proposes is still a challenging problem. In this paper, an approach to the formulation of the generalized anisotropic constitutive equations taking into account the dependence of material properties on the stress state type is proposed.

2 Constitutive Relations for Anisotropic Elasticity

Several parameters characterizing the stress state are developed in the framework of solid mechanics to describe the mechanical properties of solids. The most well-known of them are the Lode parameter and Lode angle that are functions of the ratio between the second and the third invariants of the deviator of stress tensor. In general case the stress state type can be characterized by two parameters: $\xi = \sigma/\sigma_0$ and S_{III}/σ_0^3, where $\sigma = \sigma_{ii}/3$ is the hydrostatic component of the stress tensor, $\sigma_0 = \sqrt{3S_{ij}S_{ij}/2}$ is the effective von Mises stress, $S_{ij} = \sigma_{ij} - \sigma\delta_{ij}$ is the deviator of the stress tensor and $S_{III} = S_{ij}S_{jk}S_{ik}$ is the third invariant of the deviator of the stress tensor. The first parameter ξ is the ratio of triaxiality of the stress tensor. This parameter characterizes the average ratio between normal and shear stresses. The using of this parameter may be used to distinguish different stress states. For example: $\xi = 1/3$ for pure tension, $\xi = -1/3$ for pure compression and $\xi = 0$ for torsion or pure shear loads and so on.

The second parameter S_{III}/σ_0^3 characterizes the deviation of the stresses from this averaged value. The second parameter in many cases has no significant influence on the mechanical properties of materials therefore the first parameter only could be used for the description of the stress state in the practical purposes. The isotropic and anisotropic elastic models based on the ratio of triaxiality were proposed in [5].

The classical models of linear-elastic materials may be extended by replacing the constant coefficients with the functions dependent on ξ. The elastic potential for isotropic material in this modeling can be represented in the following form:

$$\Phi = \frac{1}{2}[1 + \zeta(\xi)]\left(A + B\xi^2\right)\sigma_0^2 \tag{1}$$

The function $\zeta(\xi)$ may be determined on the base of a series of effective stress-strain curves for different values of the parameter ξ. More complex model is proposed for anisotropic materials. In this case, all the coefficients of the anisotropic matrix of stiffness may be considered as functions of ξ. The elastic potential for anisotropic material can be represented in the following form:

$$\Phi = \frac{1}{2}A_{ijkl}(\xi)\sigma_{ij}\sigma_{kl} \tag{2}$$

The stress-strain relations may be obtained by differentiating of (2) with respect to the stresses:

$$\varepsilon_{ij} = A_{ijkl}(\xi)\sigma_{kl} + \frac{1}{2} \cdot \frac{dA_{ijkl}(\xi)}{d\xi}\left[\left(\frac{1}{3} + \frac{3}{2}\xi^2\right)\delta_{ij} - \frac{3}{2}\xi\sigma_{ij}\sigma_0^{-1}\right]\sigma_0^{-1} \tag{3}$$

The relation (3) has more complex form than the relations for linear anisotropic elastic material. Determination of the coefficients $A_{ijkl}(\xi)$ in the elastic potential (2) in common case is more difficult task than in the case of isotropic material and requires more experimental data. In order to simplify relations (3) some additional hypotheses about nature of the dependence of coefficients $A_{ijkl}(\xi)$ on the parameter ξ can be applied. For example, if all the coefficients $A_{ijkl}(\xi)$ are considered to depend on the ratio of triaxiality in a similar way then the simplified form of the elastic potential may be used:

$$\Phi = \frac{1}{2}\left[1 + \zeta(\xi)\right]A_{ijkl}\sigma_{ij}\sigma_{kl} \tag{4}$$

where the coefficients A_{ijkl} are the material constants.

3 Constitutive Relations for Anisotropic Plasticity

The yield criterion for stress state dependent material can be formulated similar to the construction of stress state dependent elastic potential. The yield criterion for the isotropic material described in [4] is represented in the form:

$$f(\xi)\sigma_0 = k\ . \tag{5}$$

Here $f(\xi)$ is a function of the ratio of triaxiality that can be determined in the series of experiments under different loading conditions. The function $f(\xi)$ can be regarded as a particular scale factor for the classic von Mises criterion. The anisotropic yield criterion can be constructed in a similar way. Following to the paper [5], one of the most general forms of criterion can be obtained with the yield criterion in the quadratic form:

$$\frac{1}{2}A_{ijkl}(\xi)\sigma_{ij}\sigma_{kl} = k\ . \tag{6}$$

The coefficients $A_{ijkl}(\xi)$ are the components of fourth rank tensor and they should have the same symmetries as they have in the elastic potential (2) represented in [5]: $A_{ijkl} = A_{jikl} = A_{ijlk} = A_{jilk} = A_{klij} = A_{lkij} = A_{klji}$. The yield criterion in this form can be used to describe plastic properties of the anisotropic material with different yield limits in tension and compression and under other conditions of loading, too. The asymmetry of the plastic properties can be handled by the proper determination of the coefficients $A_{ijkl}(\xi)$. The criterion proposed above can be used to describe inelastic behavior of the material in general case. The main flaw of this form is its complexity. The functions $A_{ijkl}(\xi)$ should be determined experimentally therefore one should define 13 independent functions of the ratio of triaxiality. The determination of these functions requires enormous number of experiments. Analytical study of the stress-strain relations constituted by the yield criterion in this form is quite challenging, too. The particular forms of this criterion may be constructed on the base of the Hill's yield criterion [3]:

$$f(\xi)\sqrt{F(\sigma_{22}-\sigma_{33})^2+G(\sigma_{11}-\sigma_{33})^2+H(\sigma_{11}-\sigma_{22})^2+2(L\sigma_{23}^2+M\sigma_{13}^2+N\sigma_{12}^2)}=k$$

$$\sqrt{F(\sigma_{22}-\sigma_{33})^2+G(\sigma_{11}-\sigma_{33})^2+H(\sigma_{11}-\sigma_{22})^2+2(L\sigma_{23}^2+M\sigma_{13}^2+N\sigma_{12}^2)}=k$$

where F, G, H, L, M, N are the functions of ξ.

The first formulation is easier to implement but it's less flexible than the second one. The value of parameter k could be determined for chosen type of experiment and all other coefficients have to be related to this value. The coefficients F, G, H, L, M, N could be expressed using the coefficients R_{ij} with simple physical meaning:

$$\begin{aligned}
F(\xi) &= \frac{1}{2}\left(\frac{1}{R_{22}(\xi)^2}+\frac{1}{R_{33}(\xi)^2}-\frac{1}{R_{11}(\xi)^2}\right),\\
G(\xi) &= \frac{1}{2}\left(\frac{1}{R_{11}(\xi)^2}+\frac{1}{R_{33}(\xi)^2}-\frac{1}{R_{22}(\xi)^2}\right),\\
H(\xi) &= \frac{1}{2}\left(\frac{1}{R_{11}(\xi)^2}+\frac{1}{R_{22}(\xi)^2}-\frac{1}{R_{33}(\xi)^2}\right),\\
L(\xi) &= \frac{3}{2}\frac{1}{R_{23}(\xi)^2}, \quad M(\xi)=\frac{3}{2}\frac{1}{R_{13}(\xi)^2}, \quad N(\xi)=\frac{3}{2}\frac{1}{R_{12}(\xi)^2}.
\end{aligned} \tag{7}$$

Here R_{ij} can be considered as the ratios of following quantities: $R_{ii}=\sigma_{ii}^y/k$ without summation over i, $R_{ij}=\sigma_{ij}^y\sqrt{3}/k$ for $i\neq j$, where σ_{ij}^y is the yield stress for uniaxial tension, uniaxial compression, pure shear or other types of tests. The coefficients in any of these formulations can be found in a series of experiments with different stress state types. The more precise determination of the coefficients requires the larger amount of tests. The piecewise linear approximation for the functions $R_{ij}(\xi)$ can be used for practical purposes. While piecewise linear approximation is the natural result of the analysis of the experimental data based on the limited set of stress states with different values of ξ, the approximation of the coefficients $R_{ij}(\xi)$ with smooth functions seems to be more advantageous in terms of analytic description of the material behavior.

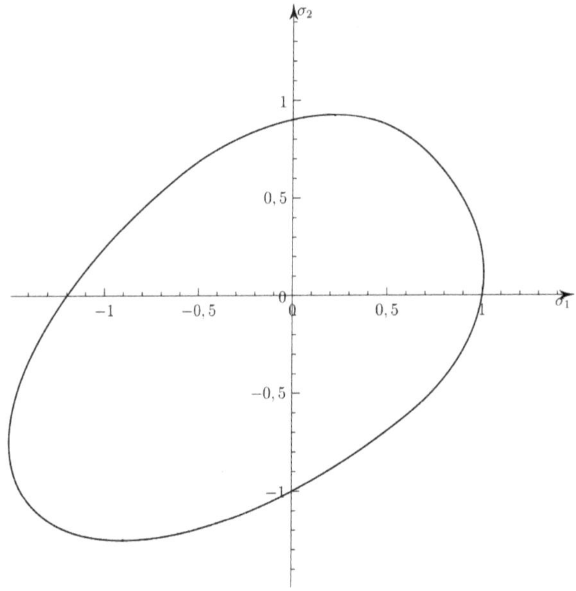

Fig. 2 The yield surface of the anisotropic stress-state dependent yield criterion in the $\sigma_1\sigma_2$ plane

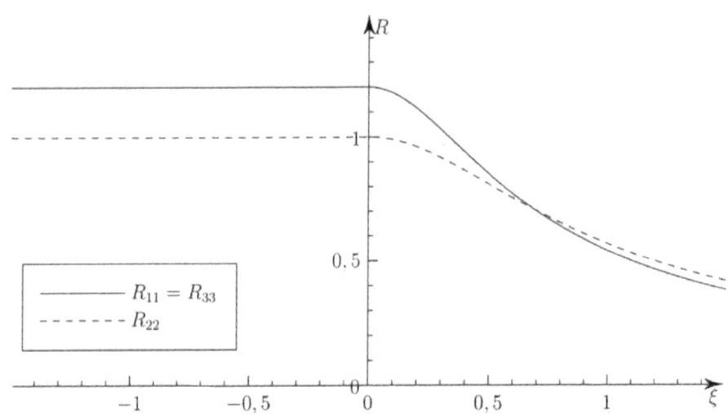

Fig. 3 The stress ratios used in the definition of the yield surface in example

One of the possible ways of this approximation is the use of monotonic functions ξ. These functions have smooth transition between asymptotic values and can be adjusted to the real values of the experiments.

An example of anisotropic stress-state dependent yield condition is shown in the Figs. 2 and 3.

The surface in the Fig. 2 is plotted in principal stresses plane (σ_1, σ_2), with $\sigma_3 = 0$. The yield stresses in tension are less than yield stresses in compression in this example. Only three coefficients R_{ij} should be defined in case of principal stresses consideration under condition of $\sigma_3 = 0$. The coefficients R_{ij} used in this example are:

$$R_{11}(\xi) = R_{33}(\xi) = \begin{cases} 1.2 & , \quad \textit{if} \quad \xi \le 0 \\ 1.2/\sqrt{1 + 9 \cdot 0.44\xi^2} & , \quad \textit{if} \quad \xi \ge 0 \end{cases}$$

$$R_{22}(\xi) = \begin{cases} 1 & , \quad \textit{if} \quad \xi \le 0 \\ 1/\sqrt{1 + 9 \cdot (1/0.9^2 - 1)\xi^2} & , \quad \textit{if} \quad \xi \ge 0 \end{cases} \tag{8}$$

The use of the smooth functions in the definition of the yield criterion makes the application of the associated flow rule quite simple.

The quadratic yield criteria presented above are the possible variants of the general formulation:

$$A_{ijkl}(\xi)S_{ij}S_{kl} = k \tag{9}$$

with the same symmetries as in Eq. (7) and additional conditions: $A_{iikl} = 0$ for $k \neq l$ and $A_{ijkl} = 0$ for $k \neq l$, $i \neq k$ and $j \neq l$. The stress-strain relation for the associated plastic flow rule has the form:

$$d\varepsilon_{ij}^p = d\lambda \frac{\partial \left(A_{klmn}S_{kl}S_{mn}\right)}{\partial \sigma_{ij}} =$$

$$= d\lambda \left[\left(\frac{dA_{klmn}(\xi)}{d\xi} \cdot \frac{\frac{1}{3}\sigma_0^2\delta_{ij} - \frac{3}{2}\sigma S_{ij}}{\sigma_0^3} \right) S_{kl}S_{mn} + 2A_{klmn} \left(\delta_{ik}\delta_{jl} - \frac{1}{3}\delta_{ij}\delta_{kl} \right) S_{mn} \right] \tag{10}$$

where $d\lambda$ depends on the form of the hardening rule. The increment of inelastic volumetric deformation can be expressed in the form:

$$d\varepsilon^p = \frac{1}{3} d\varepsilon_{ii}^p = d\lambda \frac{dA_{klmn}(\xi)}{d\xi} \cdot \frac{S_{kl}S_{mn}}{3\sigma_0} \tag{11}$$

The increment of plastic volumetric deformation in general is non-zero for variable coefficients A_{ijkl}.

The increment $d\lambda$ may be found for a given hardening rule. Let the yield limit k be the function of plastic strains ε_{ij}^p, stresses σ_{ij} and some other hardening parameters χ_i. In this case:

$$\frac{\partial \left(A_{klmn}S_{kl}S_{mn}\right)}{\partial \sigma_{ij}} d\sigma_{ij} - \frac{\partial k}{\partial \varepsilon_{ij}^p} d\varepsilon_{ij}^p - \frac{\partial k}{\partial \sigma_{ij}} d\sigma_{ij} - \frac{\partial k}{\partial \chi_i} d\chi_i = 0 \tag{12}$$

$d\lambda$ may be found after substitution of $d\varepsilon_{ij}^p$ in (12) with its value from (10):

$$d\lambda = \frac{\dfrac{\partial\left(A_{klmn}S_{kl}S_{mn}\right)}{\partial\sigma_{ij}}d\sigma_{ij} - \dfrac{\partial k}{\partial\sigma_{ij}}d\sigma_{ij} - \dfrac{\partial k}{\partial\chi_i}d\chi_i}{\dfrac{\partial k}{\partial\varepsilon_{ij}^p}\dfrac{\partial\left(A_{klmn}S_{kl}S_{mn}\right)}{\partial\sigma_{ij}}} \tag{13}$$

The constitutive relations described above show that proposed formulation of the yield criterion may be used to describe behavior of large variety of anisotropic stress state dependent materials and takes into account inelastic volumetric deformation.

4 Conclusions

In the present paper the generalized models of the elastic and plastic behavior of anisotropic materials are discussed. The presented constitutive relations give the possibilities to describe rather complex elastic behavior of a material and complex forms of the yield surface. Despite the complexity of the proposed equations it can help to increase accuracy of the theoretical predictions of the material behavior. The determination of the coefficients included in the formulations of the constitutive relations in general case requires more efforts than in the case of less universal models. The proposed relations may be used in the cases where anisotropy and asymmetry of the material properties become important for the characterization of the behavior of materials under critical conditions or under critical loads and for materials with strongly anisotropic properties.

Acknowledgements The work was supported by the Russian Foundation for Basic Research and the National Science Council, Taiwan (grants 12-01-31184 and 11-01-92001).

References

1. Eswara Prasad, N., Gokhale, A.A., Rama Rao, P.: Mechanical behaviour of aluminium-lithium alloys. Sadhana **28**(1 & 2), 209–246 (2003)
2. Gregson, P.J., Flower, H.M.: Microstructural control of toughness in aluminium-lithium alloys. Acta Metall. **33**, 527–537 (1985)
3. Hill, R.: A theory of the yielding and plastic flow of anisotropic materials. Proc. R. Soc. Lond. **193**, 281–297 (1948)
4. Lomakin, E.V.: Plastic flow of dilatants solids with stress-state dependent material properties. In: Gupta, N.K., Manzhirov, A.V. (eds.) Topical Problems in Solid Mechanics, pp. 122–132. SBS Publishers & Distributors, Bangalore (2009)
5. Lomakin, E.V.: Constitutive models of mechanical behavior of media with stress state dependent material properties. In: Altenbach, H., Maugin, G.A., Erofeev, V. (eds.) Mechanics of Generalized Continua, pp. 339–350. Springer, Heidelberg (2011)
6. MIL-HDBK-5J (2003)

Applicability of Various Fracture Mechanics Approaches for Short Fiber Reinforced Injection Molded Polymer Composites and Components

Z. Major, M. Miron, M. Reiter, and Tadaharu Adachi

Abstract The applicability and constraints of two fracture mechanics approaches for designing and dimensioning of short fiber reinforced injection molded components are described in the paper. For simplicity, a 3D random short fiber orientation of the component was assumed in the modeling. Based on injection molding and preliminary finite element simulations, the area of interest was selected and mode I stress intensity factors were calculated along the crack front of a corner crack in the component. Furthermore, fiber orientation dependent fracture toughness values were determined at moderate loading rates and at room temperature and compared with the mode I stress intensity factor values. Based on this comparison the occurrence of the crack initiation was predicted. Moreover, welding line formation was observed in the injection molding simulations of this model component and these simulations were verified by computer tomography images. Cohesive elements were embedded at the vicinity of the welding line in the model and the failure evolution was predicted. The existing model assumes constant fracture toughness in the entire ligament and hence constant values for the cohesive zone model. The real fracture toughness, however, depends on the fiber orientation which may vary with the distance and it requires the application of more complex cohesive zone model parameter functions. This is described in a follow-up paper of the authors.

Z. Major (✉) · M. Miron · M. Reiter
Institute of Polymer Product Engineering, Johannes Kepler University Linz, Linz, Austria
e-mail: Zoltan.Major@jku.at

T. Adachi
Department of Mechanical Engineering, Toyohashi University of Technology, 1-1 Hibarigaoka, Tempaku, 441-8580 Toyohashi, Japan
e-mail: adachi@me.tut.ac.jp

Alexander K. Belyaev et al. (eds.), *Mechanics and Model-Based Control of Advanced Engineering Systems*, DOI 10.1007/978-3-7091-1571-8_24,

© Springer-Verlag Wien 2014

1 Introduction

Short fiber reinforced polymers are frequently used in injection molded polymeric components for many demanding engineering applications (e.g. automotive, oilfield, medical). These components are often exposed to a complex combination of dynamic mechanical and thermal loading over a wide loading rate/frequency and temperature range. As these materials are potential candidates to substitute metals in various light-weight design structures, it is of prime practical importance for polymer engineers to understand the local failure mechanism and to develop proper models to further improve the strength and durability of these materials. This amendment includes in addition to the improvement of the mechanical properties of the polymer matrix, the application of various reinforcing fibers with varying lengths, the compatibilization of the polymer matrix/fiber interface and tailor made fiber orientation by proper processing techniques. Furthermore, short fiber reinforced injection molded components reveal a highly heterogeneous microstructure and a component specific fiber orientation (FOD) and fiber length distribution (FLD). Due to this fiber orientation, both the deformation and the failure behavior are anisotropic and the properties depend on the actual values of the fiber length distribution (FLD) and the fiber orientation distribution (FOD) [4, 5].

Moreover, in addition to the global stiffness, the practical performance of these components depends on the local failure behavior. Failure prediction of such complex systems involving multi-scale, nonlinearity, and uncertainties is one of most critical engineering challenges. By using mean-field data for heterogeneous materials, homogeneous materials-based models might result in prediction of unrealistic smooth crack trajectories and unreliable load-carrying capacity [25]. As a local approach, fracture mechanics offers further options for predicting the failure behavior and for dimensioning against fracture of these components. Fracture mechanics techniques are also frequently used for characterizing the fracture behavior of short fiber reinforced polymeric composites [9, 11, 12, 23, 24, 32, 33]. While the characterization techniques are well developed, partly standardized and widely used for supporting material development efforts, hardly any description was found for consequently applying fracture mechanics methods for dimensioning of real components [3, 27].

The essential prerequisite of the applicability of various fracture mechanics approaches is the presence of crack or crack-like defects in the specific structure. It is well known that injection molded parts may contain various imperfections. These imperfections can be located both on the surface and in the bulk. The surface defects are visible, the bulk imperfections can be detected by X-Ray computer tomography experiments. While many of these defects can be consider as smooth stress concentrators, others may generate a highly singular stress field and may behave as crack-like defect. Crack-like defects in short fiber reinforced and injection molded polymer components (sfrimc) are micro-cracks, inclusions, voids and welding lines. Special kinds of these imperfections are the welding lines which are frequently observed in injection molded short fiber reinforced polymeric composite parts. The welding lines are locations where the continuity of the fiber orientation flow is distorted and this high degree of local inhomogeneity may cause a significant reduction of both the global and local stiffness and the strength of such components. In spite of all efforts, the presence of the welding lines in complex injection molded parts cannot completely be avoided [14, 35]. Welding lines are assumed as crack-like defects in the second part of this paper and the fracture behavior is modeled by Cohesive Zone Models [1]. In many practical loading situations, however, due to the low and non-severe loading situation these defects do not influence the component behavior. Hence, to apply complex and time consuming fracture mechanics models and experiments is reasonable only for components which are exposed to a complex loading situation and for which the failure behavior is of prime practical importance (e.g. crash relevant automotive parts).

Hence, the main objective of this paper is the definition of a proper methodology for applying the critical stress intensity factor concept and the cohesive zone model concept for short fiber reinforced injection molded composites. The methodology involves modeling approaches along with appropriate modeling parameters and material models along with proper characterization techniques.

2 Methodology

The fundamental feature of sfrim composites is their anisotropy. This anisotropy is generated by the fiber orientation distribution (FOD) during the injection molding. The FOD can be predicted by proper injection molding simulation. The essential prerequisites of accurate estimation of the FOD are the adequate model (2D or 3D [2, 10]) considering fiber interactions for the specific compound and proper material data (viscosity, thermal properties). A selected result of an injection molding simulation for a model component is shown Fig. 1. Fiber orientation distribution (color intensity) and the main directions of the fibers (indicated by the arrows) are visible. Furthermore, welding line formation was observed in the mid line of the component at the corner of the hole and it was defined as area of interest (AoI) for further investigations. The accuracy of the predicted fiber orientation was verified by X-ray computer tomography experiments. For more details of these technique and the detailed results for various specimen configurations and component geometries please refer to Jerabek et al. [21]. The accurate distribution of the fiber orientation and the relevant fiber orientation tensors are analyzed and used and described elsewhere [36, 37].

The stiffness highly depends on the FOD of the component and can be predicted by the application of micromechanics based software tools. These tools either use the main field homogenization technique [7, 8, 22] or finite element method. Based on all above, the stress distribution for an arbitrary loading situation was predicted for the model component and the von Mises stress values are shown in Fig. 2.

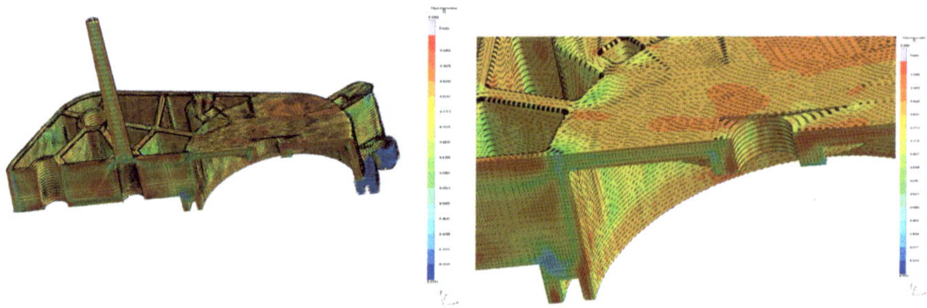

Fig. 1 Injection molding simulation of the model component fiber orientation distribution and the main direction (indicated by the *arrows*) are observed. The welding line formation was observed at the corner of the hole (area of interest (AoI))

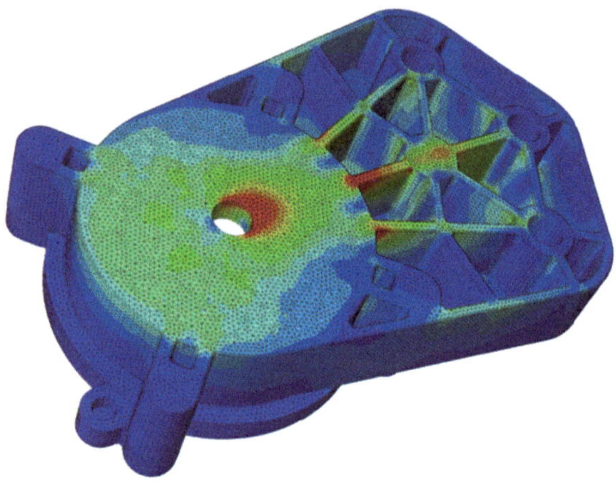

Fig. 2 Determination of the stress state of the component for an arbitrary loading using specific tensile modulus and Poissons ratio values for a 3D random fiber orientation

The sharp corner around the hole of the component was selected as area of interest (AoI). An evenly distributed constant pressure on the surface of the hole was applied as the driving force for the crack initiation and propagation. The definition of the finite elements meshes with deformation elements (four nodes tetrahedron in the component body far field) and with crack elements (degenerated hexahedron second degree (Abaqus 6.11)) at the vicinity of the crack are shown in Fig. 3. It was assumed that due to the high stress concentration, the crack starts from this area and growths to the welding line direction. During the fill-in process the melt fronts must circuit the hole and the filling grade of the sharp corner is lower compared to regions with less geometric constraint. A linear elastic material model with 3D random fiber orientation was applied to model the deformation behavior of the PP-GF30 used. All stress intensity factors were determined but fur further dimensioning only the mode I considered. The spatial distribution of the mode I stress intensity factor along the crack front of the corner crack applying linear elastic material model is shown

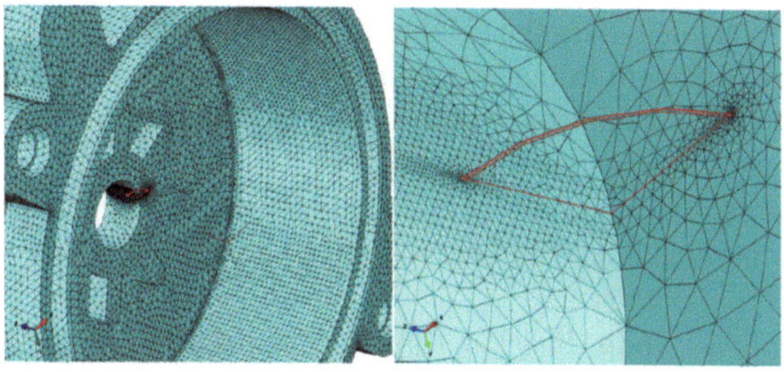

Fig. 3 The definition of the finite elements mesh with deformation elements in the component body (far field) and with crack elements at the vicinity of the crack assumed (**a**) global view and (**b**) local view of the corner crack defined

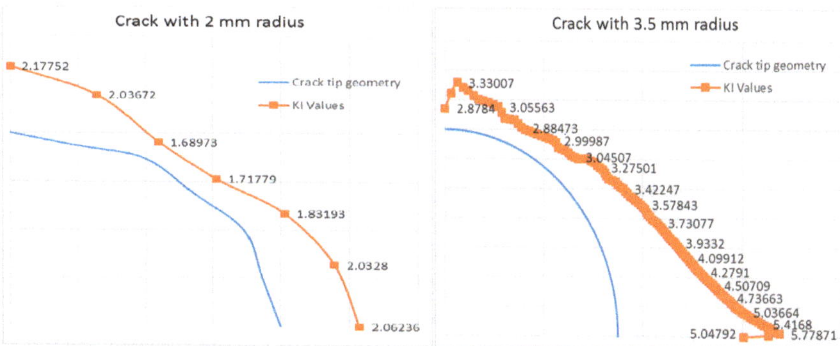

Fig. 4 Determination of the spatial distribution of the mode I stress intensity factor along the crack front of the corner crack applying linear elastic material model

in Fig. 4 for 2 mm and for 3.5 mm crack radius. Small scale yielding condition with a predominantly linear elastic material behavior was assumed. The size of the plastic zone at the crack tip is significantly smaller than the crack length or the relevant dimension of the component. Due to the constraint, the crack front is not regular and the location of the maximum mode I stress intensity factor, K_{Imax} was determined 2.8–5.8 MPa $m^{-1/2}$. These values are used in the dimensioning process and compared with the relevant fracture toughness values. While the estimation of the size of the plastic zone is straightforward for homogenous materials and widely used [28] additional considerations are needed for anisotropic materials [32]. In spite of the complex processes observed in the process zone around the crack tip, the applicability of LEFM for sfrim composites under cyclic loading situation was clearly assured by Lang [26] and Nowotny [34]. The fracture criteria is defined for linear, isotropic solids neglecting time/rate and temperature dependence (conventional approach)

$$K(a, W, \sigma) < K_{IC}(B), \tag{1}$$

Where a is the crack length in the component, W is the relevant dimension (width or the remaining ligament in the crack plane $W - a$) of the component, σ is the stress, B is the specimen thickness and K_{IC} is the plane strain fracture toughness of the material used for the component.

This criteria can be modified for rate and temperature dependent solids (e.g., unreinforced or particle filled and fiber reinforced polymers)

$$K(a, W, \sigma) < K_{IC}(B, T, strainrate), \tag{2}$$

where K_{IC} is the plane strain fracture toughness for materials with distinct rate and temperature dependence. It was shown that the fracture toughness of unreinforced PP decreases with increasing loading rate and reach a material dependent minimum value of about 1.5–2.5 MPa $m^{-1/2}$. Due to dynamic effects the fracture toughness can reliably be determined up to 1 m/s. Hence, for a conservative dimensioning the lowest value with reliable experimental background has to be considered [29].

Additional considerations are required, however, for heterogeneous and anisotropic materials [31].

$$K(a, W, FOD, \sigma(FOD)) < K_{IC}(B, T, FOD, strainrate), \tag{3}$$

The local stress at the vicinity of the crack tip is a function of the microstructure (FOD) and in addition the SIF may depend also on the FOD. On the other hand, the fracture toughness depends also on the local FOD value.

The fracture toughness of short fiber reinforced injection molded polymers can be determined by using the relevant standards [18–20]. Fiber orientation dependent fracture toughness, K_{IC} values are shown in Fig. 5 for three different fiber orientations. Representative load-displacement curves are shown in the small diagrams for both fiber orientation/crack growth direction combinations. The fracture toughness was calculated using the peak load values. As expected, lower fracture toughness values were observed for the parallel crack growth. Furthermore, while in the perpendicular case a pop-in crack initiation and quasi-brittle crack propagation was obtained (energy absorption after the peak load), a rather brittle fracture (sudden drop of the load) was obtained in the parallel situation. Hence, based on mode I stress intensity factor values at the corner crack tip, $KI(r)$ and fiber orientation dependent fracture toughness values, K_{Ic}^{FO} the dimensioning of the component regarding the fracture initiation can be performed. In the specific model with the arbitrary loading stress intensity factor values slightly lower and higher than the fracture toughness values were observed. Hence, as it was expected the AoI is a possible site of the crack initiation.

In the second part of the paper the fracture mechanics based assessment of the welding line is described and the applicability of the cohesive zone model is discussed. The preliminary injection molding simulation clearly indicated the presence of a pronounced welding line on the component. Cohesive elements (hexahedral) were embedded at the vicinity of the welding line in the finite element model. The loading of the component was realized by a cylindrical intender which

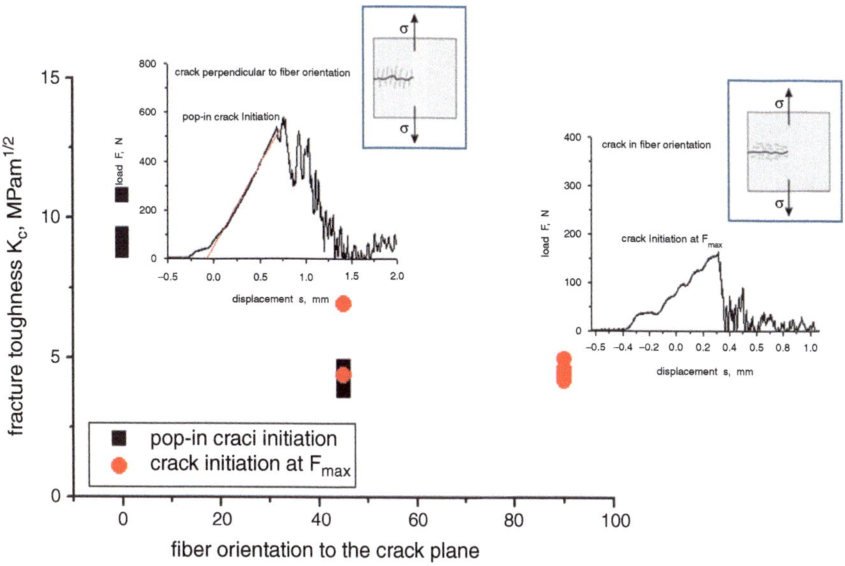

Fig. 5 Determination of fiber orientation dependent fracture toughness KIC values. Representative load-displacement diagrams are shown for both fiber orientation/crack growth direction combinations Pop.in crack initiation was frequently for perpendicular case

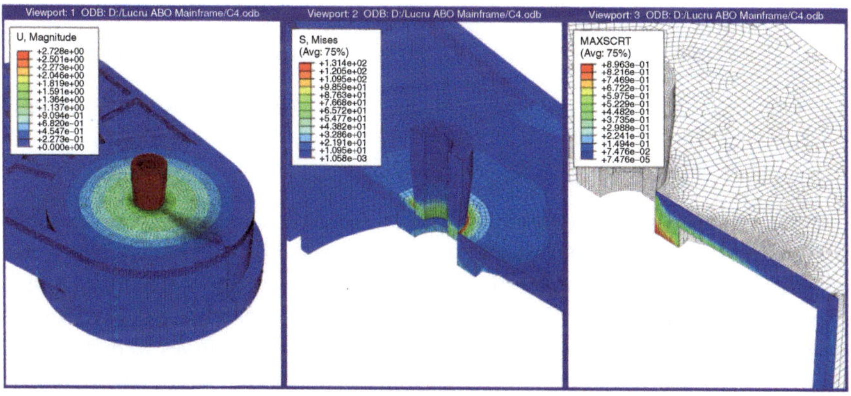

Fig. 6 Implementation of cohesive elements into the component investigated. The injection molding situation clearly indicated the presence of a pronounced welding line on the component cohesive elements were embedded at the vicinity of the welding line in the model

was pushed into the hole. The loading situation, the von Mises stress distribution around the welding line and the damage initiation at a selected stage is shown in Fig. 6.

To prove this model, novel experiments were conceived and carried out. The details are described by Major et al. [30]. A 4 mm thick specimen containing welding line was injection molded. A novel fracture specimen for characterizing the

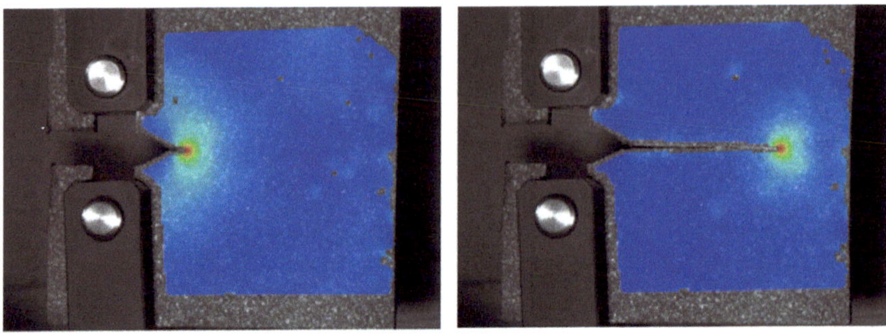

Fig. 7 Various stages of the fracture process (crack initiation and growth) and the strain distribution at the vicinity (near field) and in the ligament (far field) of the crack tip measured by digital image correlation

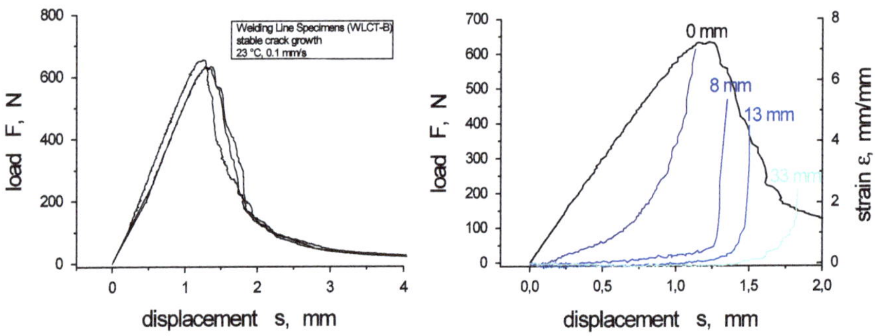

Fig. 8 (**a**) Load-displacement curves of the welding line fracture specimens and (**b**) the strain gradients at various crack lengths

fracture toughness of the welding line was designed. A sharp notch was machined into the welding line and sharpened by a fresh razor blade. Similar to a conventional compact type (CT) fracture specimen, this specimen was loaded by two loading pins as it is shown in Fig. 7. The crack plane is identical with the welding line and straight crack propagation was observed. Load-displacement curves were measured (Fig. 8a) and fracture energy values were calculated. Critical strain energy release rate, G_{Ic} values were than estimated and were found in the range from 4.1 to 4.4 kJ m^{-2}. To gain more insight into the fracture process of the welding line, full-field strain analysis by digital image correlation method was also conducted. This method yielded the spatial and temporal strain distribution both at the vicinity of the crack tip as well in the ligament (see in Fig. 8). The strain curves at various crack lengths are shown in Fig. 8b. Based on these measurements, parameters for a linear cohesive zone model of the welding line were defined and used in subsequent component simulation. The common procedure for the simulation of crack propagation using cohesive elements is based on a priori considered cohesive surfaces [17]. It was assumed that the crack exactly follows the welding line in the component. Conventional cohesive zone fracture models assume, however, isotropic material behavior and do not consider a spatial variation of the fracture toughness [6, 13, 15, 16, 27, 38].

Due to the melt flow direction and due to the geometrical constraints, however, the fiber orientation is not constant along the welding line. As the fracture toughness clearly depends on the fiber orientation, the fracture toughness may vary over the entire ligament length. Hence, the development of a new model which considers these effects was started and the results of the implementation will be reported in a follow-up paper.

3 Summary, Conclusions and Future Work

The applicability and constraints of two fracture mechanics approaches for dimensioning of a short fiber reinforced injection molded component are described in the paper. It was assumed that short fiber reinforced components may contain crack-like defects and the area-of interest was defined at a sharp corner for this specific component. Stress intensity factors were calculated for a corner crack and for a 3D random fiber orientation state and compared with fiber orientation dependent fracture toughness values. It was shown that for a specific crack size the SIF values are higher than the relevant fracture toughness and thus fracture instability may occur. In addition, a welding line formation with highly anisotropic fiber orientation was predicted in the component and the welding line was assumed as a crack like defect. The failure behavior was modeled by cohesive zone element. In the first model constant values for the cohesive zone model parameters were used and the damage evolution in the welding line was predicted. The parameters for the cohesive zone model were determined in novel fracture experiments using a fracture specimen with welding line. Based on the variation of the fiber orientation along the length of the welding line, this model was further developed and a linear function of the fracture toughness along with the ligament was assumed. The initial stiffness, the peak stress and the fracture toughness in terms of G_{Ic} is increasing along the ligament in the welding line in the real component. To gain more insight into this complex failure process, the fiber orientation mapping will be refined and fracture toughness functions will be determined using specimens containing welding lines. Finally, instrumented impact tests will be performed and the simulations verified by the experimental results.

Acknowledgements The research work of this paper was performed in the ACCM and KAPMT projects. The ACCM and KAPMT are funded by the Austrian Government and the State Government of Upper Austria.

References

1. Abaqus User's Manual, Abaqus 6.11: Simulia, Dassault Systems (2012)
2. Advani, S., Tucker, C.L., III: The use of tensors to describe and predict fiber orientation in short fibre composites. J. Rheol. **31**(8), 751–784 (1987)
3. Andriyana, A., Billon, N., Silva, L.: Mechanical response of a short fiber-reinforced thermoplastic: experimental investigation and continuum mechanical modeling. Eur. J. Mech. A/Solids **29**, 1065–1077 (2010)
4. Bernasconi, A.P., Davoli, A., Basile, B., Filippi, A.: Effect of fibre orientation on the fatigue behaviour of a short glass fibre reinforced polyamide-6. Int. J. Fatigue **29**, 199–208 (2007)
5. Bouaziz, A., Zairi, F., Nait-Abdelaziz, M., Gloaguen, J.M., Lefebvre, J.M.: Mi-cromechanical modelling and experimental investigation of random discontinuous glass fiber polymer-matrix composites. Compos. Sci. Technol. **67**, 3278–3285 (2007)
6. Cornec, A., Scheider, I., Schwalbe, K.H.: On the practical application of the co-hesive model. Eng. Fract. Mech. **70**, 1963–1987 (2003)
7. Doghri, I., Tinel, L.: Micromechanics of inelastic composites with misaligned inclusions: numerical treatment of orientation. Comput. Methods Appl. Mech. Eng. **195**, 1387–1406 (2006)
8. Doghri, I., Adam, L., Bilger, N.: Mean-field homogenization of elasto-viscoplastic composites based on a general incrementally affine linearization method. Int. J. Plast. **26**, 219–238 (2010)
9. Fara, S., Pavan, A.: Fibre orientation effects on the fracture of short fibre polymer composites: on the existence of a critical fibre orientation on varying in-ternal material variables. J. Mater. Sci. **39**, 3619–3628 (2004)
10. Folgar, F., Tucker, C.L., III: Orientation behaviour of fibers in concentrated suspensions. J. Reinf. Plast. Compos. **3**, 98–119 (1984)
11. Fu, S.Y., Lauke, B.: Effects of fiber length and fiber orientation distributions on the tensile strength of short-fiber-reinforced polymers. Compos. Sci. Technol. **56**, 1179–1190 (1996)
12. Fu, S.Y., Lauke, B., Mai, Y.W.: Science and Engineering of Short Fibre Rein-Forced Polymer Composites. Woohead Publishing in Materials. CRC, Boca Raton (2009)
13. Fuchs, P., Major, Z.: Experimental determination of cohesive zone models for epoxy composites. Exp. Mech. **51**, 779–786 (2011)
14. Gamba, M.M., Pouzada, A.S., Frontini, P.M.: Impact properties and microhard-ness of double-gated glass-reinforced polypropylene injection moldings. Polym. Eng. Sci. (2009). doi:10.1002/pen.21393
15. Geers, M.G.D., Kouznetsova, V.G., Aydemir, A., Brekelmans, W.A.M.: Multi-scale modelling of failure through delamination and decohesion. In: Onate, E., Owen, D.R.J. (eds.) VIII International Conference on Computational Plasticity COMPLAS VIII, CIMNE, Barcelona (2005)
16. Geers, M.G.D., Kouznetsova, V.G., Brekelmans, W.A.M.: Multi-scale computational homogenization: trends and challenges. J. Comput. Appl. Math. **234**, 2175–2182 (2010)
17. Geißler, G., Netzker, C., Kaliske, M.: Discrete crack path prediction by an adaptive cohesive crack model. Eng. Fract. Mech. **77**, 3541–3557 (2010)
18. ISO 13586: Plastics – determination of fracture toughness (Gc and Kc) – linear elastic fracture mechanics (LEFM) approach (1998)
19. ISO 13586: Plastics – determination of fracture toughness (Gc and Kc) – linear elastic fracture mechanics (LEFM) approach, Amendment 1: guidelines for the testing of injection molded plastics containing discontinuous reinforcing fibres (2001)
20. ISO/FDIS 17281: Plastics – determination of fracture toughness (GIC and KIC) at moderately high loading rates (1 m/s)
21. Jerabek, M., Reiter, M., Plank, D., Salaberger, D., Major, Z.: Characterization of short fibre reinforecd composites. In: European Conference on Composite Materials ECCM15, Poster 3.82, Venice, 24–28 June 2012

22. Kammoun, S., Doghri, I., Adam, L., Robert, G., Delannay, L.: First pseudo-grain failure model for inelastic composites with misaligned short fibers. Compos. A Appl. Sci. Manuf. **42**(12), 1892–1902 (2011)
23. Karger Kocsis, J.: Structure and fracture mechanics of injection-molded composites. In: Wiley Encyclopedia of Composites (2012). doi:10.1002/9781118097298.weoc240
24. Kim, J.K., Mai, Y.W.: Engineering Interfaces in Fiber Reinforced Composites. Elsevier, London (1998)
25. Krajconovic, D., Vujosevic, M.: Strain localization-short to long correlation length transition. Int. J. Solids Struct. **35**, 4147–4166 (1998)
26. Lang, R.W.: Applicability of linear elastic fracture mechanics to fracture in polymers and short-fiber composites. Dissertation, Lehigh University Bethlehem (1984)
27. Li, S., Thouless, M.D., Waas, A.M., Schroeder, J.A., Zavattieri, P.D.: Use of a cohe-sive-zone model to analyze the fracture of a fiber-reinforced polymer-matrix composite. Compos. Sci. Technol. **65**, 537–549 (2005)
28. Major, Z.: A fracture mechanics approach to characterize rate dependent fracture behavior of engineering polymers. Dissertation, Montanuniversität Leoben (2002)
29. Major, Z., Lang, R.W.: Rate dependent fracture toughness of plastics. In: Williams, J.G., Pavan, A. (eds.) 3rd ESIS Conference on Polymers and Composites, Les Diablerets, 15–18 Sept 2002, ESIS Technical Publication. Elsevier, London (2002)
30. Major, Z., Miron, M., Reiter, M., Adachi, T.: Real and virtual fracture tests on welding lines of injection molded short fiber reinforced polymers. In: European Conference of Fracture, ECF19, Kazan, 26–31 Aug 2012
31. Miyazaki, N., Ikeda, T.: Computational fracture mechanics of heterogeneous materials and structures. In: 8th World Congress on Computational Mechanics (Wccm8), Venice, 30 June – 5 July 2008
32. Mower, T.M., Li, V.C.: Fracture characterization of random short fibre rein-forced thermoset resin composites. Eng. Fract. Mech. **26**, 593–603 (1987)
33. Norman, D.A., Robertson, R.E.: The effect of fiber orientation on the toughen-ing of short fiber-reinforced polymers. J. Appl. Polym. Sci. **90**, 2740–2751 (2003)
34. Nowotny, M.: Fatigue crack propagation in engineering thermoplastics – effects of temperature and short fibre reinforcement. Dissertation, Montanuniversität Leoben (1997)
35. Patcharaphun, S., Zhang, Y.B., Mennig, G.: Simulation of three-dimensional fiber orientation in weldline areas during push-pull-processing. J. Reinf. Plast. Compos. **26**, 977 (2007)
36. Reiter, M., Jerabek, M., Machado, M., Herbst, H., Major, Z.: A Combined ex-perimental numerical characterization of the damage behavior for discontinuous fiber reinforced polymers considering various model scales. In: 6th International Conference on Multiscale Materials Modeling (MMM 2012), Singapore (2012)
37. Reiter, M., Hartl, A., Jerabek, M.: Evaluation of the applicability of the first pseudo-grain failure model for short glass fiber reinforced polypropylene materials. In: International conference on composite materials (ICCM), Montreal (2013)
38. Schwalbe, K.H.: Integrity of lightweight structures, Universita degli studi di Trieste, Trieste (2010).

The Deformation–Diffusion Coupling in Thin Elastomeric Gels

Takuya Morimoto and Hiroshi Iizuka

Abstract Elastomeric gel is a three-dimensional network of cross-linked polymer which are able to swell in an environment containing solvent molecules. When a gel undergoes external load or stimuli such as solvent concentration and temperature, the gel evolves over time in which solvent diffusion is coupled to the elastic deformation of the network. We first present a mixture formulation of elastomeric gels and derive the linearized deformation–diffusion coupling model based on the incremental motion and the Onsager's variational principle. Then, we derive the Föppl–von Kármán equations for gel sheets by making dimensional reduction of the deformation–diffusion coupling model.

1 Introduction

Gel is a three-dimensional network of cross-linked polymer which are able to swell in an environment containing solvent molecules. When a gel undergoes external load and/or stimuli such as solvent concentration and temperature, the gel evolves over time in which solvent diffusion is coupled to the elastic deformation of the network. In order to describe the mechanics of gels, three-dimensional nonlinear field theories have been developed. Many biological structures, however, are characterized as low-dimensional continua such as one-dimensional filaments or two-dimensional membranes. A deformation–diffusion coupling model

T. Morimoto (✉)
Department of Mechanical, Electrical and Electronic Engineering, Shimane University, 1060 Nishikawatsu, Matsue, Shimane 690-8504, Japan
e-mail: morimoto@riko.shimane-u.ac.jp

H. Iizuka
Department of Mechanical Systems Engineering, Yamagata University, 4-3-16 Jonan, Yonezawa, Yamagata 992–8510, Japan
e-mail: h-iizuka@yz.yamagata-u.ac.jp

Alexander K. Belyaev et al. (eds.), *Mechanics and Model-Based Control of Advanced Engineering Systems*, DOI 10.1007/978-3-7091-1571-8_25,
© Springer-Verlag Wien 2014

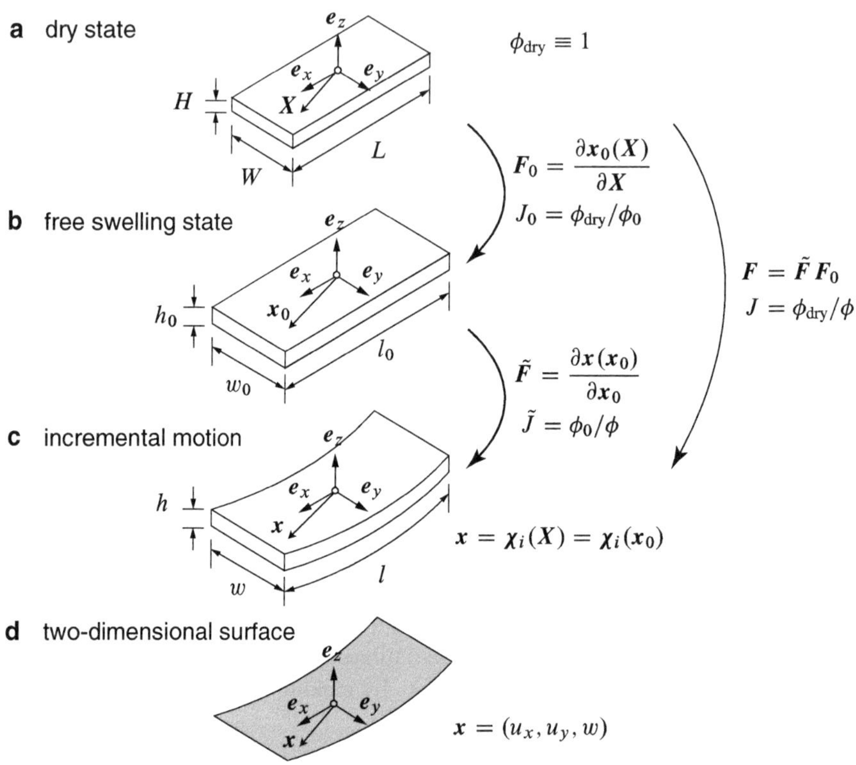

Fig. 1 (**a**) Dry state without solvent. (**b**) Free swelling state. (**c**) Incremental state superimposed on the swelling state. (**d**) A two-dimensional surface by dimensional reduction under appropriate assumptions

for low-dimensional continua would be useful for biological applications and other related mechanics problems.

In this paper, we first present a mixture formulation of elastomeric gels and derive the linearized deformation–diffusion coupling model based on the incremental motion and the Onsager's variational principle [3]. We then derive the Föppl–von Kármán (FvK) equations for gel sheets by making dimensional reduction of the deformation–diffusion coupling model.

2 The Deformation–Diffusion Coupling Theory

We briefly formulate a mixture theory (e.g. Refs. [1, 2]) to describe the dynamics of elastomeric gels. As shown in Fig. 1, let us imagine that the gel sheet is reached in an isotropically swelling equilibrium state from a dry state, and describe the incremental state superimposed to the free swelling equilibrium state.

Gel is a homogeneous mixture of elastomeric polymer and solvent. In the dry state, each component of polymer and solvent occupy a region Ω_i with $i = (\mathrm{p}, \mathrm{s})$, respectively, in which the subscript (p, s) denotes *polymer* or *solvent*. Polymer and solvent are occupied different space $(\Omega_\mathrm{p}, \Omega_\mathrm{s})$ before mixturing the components, while in the swelling equilibrium state or the incremental state, those share the same region in space, Ω. Let $\boldsymbol{X} \in \Omega_i$ and $\boldsymbol{x}_0 \in \Omega$ be position vectors in the dry state and the free swelling state, respectively. The motion of each component is represented by the mapping $\boldsymbol{\chi}_i : \Omega_i \mapsto \Omega$ which takes points $\boldsymbol{X}$ in Ω_i to points $\boldsymbol{x}_0$ in Ω: $\boldsymbol{x}_0 = \boldsymbol{\chi}_i(\boldsymbol{X}, t)$, $\boldsymbol{X} \in \Omega_i$. In addition, let $\boldsymbol{x} = \boldsymbol{x}_0 + \epsilon \boldsymbol{u}$ be position vector in the incremental state from the swelling state, where $\boldsymbol{u}$ is the displacement field and ϵ is a small scale parameter. The incremental motion is represented by the mapping $\boldsymbol{x} = \boldsymbol{\chi}_i(\boldsymbol{x}_0, t)$, $\boldsymbol{x}_0 \in \Omega$. The deformation gradient from the dry state $\boldsymbol{F} = \partial \boldsymbol{x}(\boldsymbol{X}, t)/\partial \boldsymbol{X}$ is defined as

$$\boldsymbol{F} = \tilde{\boldsymbol{F}} \boldsymbol{F}_0, \quad \boldsymbol{F}_0 = \lambda_0 \boldsymbol{I}, \quad \tilde{\boldsymbol{F}} = \boldsymbol{I} + \epsilon \nabla_{\boldsymbol{x}_0} \boldsymbol{u}, \tag{1}$$

where $\tilde{\boldsymbol{F}} = \partial \boldsymbol{x}/\partial \boldsymbol{x}_0$ is the incremental deformation gradient from the free swelling equilibrium state, $\boldsymbol{F}_0 = \partial \boldsymbol{x}_0/\partial \boldsymbol{X}$ is the deformation gradient from the dry state to the free swelling equilibrium state with the stretch λ_0, and $\nabla_{\boldsymbol{x}_0} = \partial/\partial \boldsymbol{x}_0$. The velocity fields of the polymer network and solvent in material description, $\boldsymbol{V}_i$, and in spatial description, $\boldsymbol{v}_i$, are given by

$$\boldsymbol{V}_i(\boldsymbol{X}, t) = \frac{\partial \boldsymbol{\chi}_i(\boldsymbol{X}, t)}{\partial t}, \quad \boldsymbol{v}_i(\boldsymbol{X}, t) = \boldsymbol{V}_i[\boldsymbol{\chi}_i^{-1}(\boldsymbol{x}, t), t], \quad \boldsymbol{X} \in \Omega_i. \tag{2}$$

For elastomeric gels, we assume that the true density of polymer network component ρ_p^* is equal to that of the solvent ρ_s^*, and is a constant independent of temperature and pressure. In addition, we assume both the polymer and solvent are *incompressible*; we set the true densities to unity $\rho_\mathrm{p}^* = \rho_\mathrm{s}^* = 1$. With the volume fraction of polymer network $\phi_\mathrm{p}(\boldsymbol{x}, t) \equiv \phi$ in the current configuration, the mass densities are equal to the volume fractions of polymer and solvent, respectively,

$$\rho_\mathrm{p} = \phi_\mathrm{p} \equiv \phi, \quad \rho_\mathrm{s} = \phi_\mathrm{s} = 1 - \phi, \tag{3}$$

where we have used the mixture law that the sum of volume fractions of polymer and solvent equals to unity,

$$\phi_\mathrm{p}(\boldsymbol{x}, t) + \phi_\mathrm{s}(\boldsymbol{x}, t) = 1. \tag{4}$$

Let ϕ_dry, ϕ_0, and ϕ be the volume fractions of the polymer network in the dry, free swelling, and incremental configurations, respectively. Each local form of mass conservation is expressed as

$$J \equiv \det \boldsymbol{F} = \frac{\phi_\mathrm{dry}}{\phi}, \quad J_0 \equiv \det \boldsymbol{F}_0 = \frac{\phi_\mathrm{dry}}{\phi_0}, \quad \tilde{J} \equiv \det \tilde{\boldsymbol{F}} = \frac{\phi_0}{\phi}, \tag{5}$$

from which we have the relation $\tilde{J} = J/J_0$. Taking the material time derivative to Eq. (5)$_1$, we obtain the local form of mass conservation in terms of the volume fraction for polymer network ϕ,

$$\frac{\partial \phi}{\partial t} + \nabla \cdot (\phi \boldsymbol{v}_{\mathrm{p}}) = 0, \tag{6}$$

where $\nabla = \partial/\partial \boldsymbol{x}$. In addition, from the mixture law given by Eq. (4), the above equation can be written in terms of the solvent volume fraction $\phi_{\mathrm{s}} = 1 - \phi$:

$$\frac{\partial (1-\phi)}{\partial t} + \nabla \cdot [(1-\phi)\boldsymbol{v}_{\mathrm{s}}] = 0. \tag{7}$$

Adding Eqs. (6) and (7) gives the incompressible condition of the gel,

$$\nabla \cdot [\phi\, \boldsymbol{v}_{\mathrm{p}} + (1-\phi)\boldsymbol{v}_{\mathrm{s}}] = 0. \tag{8}$$

The constitutive behavior of a gel is described by the Flory–Rehner free energy per unit dry volume which is expressed as the sum of two terms:

$$W(\boldsymbol{F}, \phi) = W_e(\boldsymbol{F}) + J\, W_m(\phi), \tag{9}$$

where W_e is associated with the elastic deformation of the polymer and W_m is the mixing free energy; $W_e(\boldsymbol{F}) = \frac{1}{2} G_{\mathrm{dry}}(|\boldsymbol{F}|^2 - 3)$ and $W_m(\phi) = \frac{k_{\mathrm{B}}T}{v_{\mathrm{s}}}[(1-\phi)\ln(1-\phi) + \chi(T)\phi(1-\phi)]$ in which we have set $\phi_{\mathrm{dry}} = 1$, and $G_{\mathrm{dry}} = N k_{\mathrm{B}} T$ the shear modulus of the dry polymer, v_{s} the volume per solvent molecule, k_{B} Boltzmann's constant, N the cross density, and $k_{\mathrm{B}}T$ the thermal energy. In order to consider the incremental deformation of the gel, let us rewrite the free energy (9) to that with reference to the swelling equilibrium state. The free energy per unit volume in the swelling equilibrium sate, $\tilde{W}(\tilde{\boldsymbol{F}})$, can be rewritten as

$$\tilde{W}(\tilde{\boldsymbol{F}}, \phi) = \frac{1}{J_0} W(\tilde{\boldsymbol{F}} \boldsymbol{F}_0, \phi). \tag{10}$$

To derive the mechanical equilibrium equation and the kinetic equation of the diffusion of solvent, we use the Onsager's variational principle [3] which is an extension of Rayleigh's least energy dissipation principle to that in irreversible processes. The energy dissipation in a solid-fluid mixture has mainly one origin, which arises from the relative motion [2]. Taking the incompressible condition (8) into account, the Rayleighian functional $R = D + \dot{W}$, which is sum of the energy dissipation function D and the rate of elastic energy $\dot{W}$, to be minimized can be expressed as

$$R = \int \left\{ \frac{1}{2} \xi(\phi)(\boldsymbol{v}_{\mathrm{p}} - \boldsymbol{v}_{\mathrm{s}})^2 + \boldsymbol{\sigma} : \nabla \boldsymbol{v}_{\mathrm{p}} - p\, \nabla \cdot [\phi\, \boldsymbol{v}_{\mathrm{p}} + (1-\phi)\boldsymbol{v}_{\mathrm{s}}] \right\} \mathrm{d}\boldsymbol{x}, \tag{11}$$

where $\xi(\phi)$ is the friction constant per unit swelling equilibrium volume, p is a Lagrangian multiplier which comes from the constraint (8), and $\boldsymbol{\sigma}$ is the incremental stress derived as

$$\boldsymbol{\sigma} = \left(K - \frac{2}{3}G\right)(\nabla \cdot \boldsymbol{u})\,\boldsymbol{I} + G\left[\nabla \boldsymbol{u} + (\nabla \boldsymbol{u})^{\mathrm{T}}\right] - p\,\boldsymbol{I} \tag{12}$$

where G and K are called the shear modulus, and the osmotic bulk modulus respectively [4, 5]:

$$G = \frac{G_0}{\lambda_0}, \quad K = \frac{1}{3}\frac{G_0}{\lambda_0} + \frac{k_{\mathrm{B}}T}{v_{\mathrm{s}}}\frac{J^{(0)} - 2\chi(J^{(0)} - 1)}{(J^{(0)})^2(J^{(0)} - 1)} \tag{13}$$

The variational calculations $\delta R/\delta \boldsymbol{v}_{\mathrm{p}} = 0$ and $\delta R/\delta \boldsymbol{v}_{\mathrm{s}} = 0$ give $\xi(\phi)(\boldsymbol{v}_{\mathrm{p}} - \boldsymbol{v}_{\mathrm{s}}) = \nabla \cdot \boldsymbol{\sigma} - \phi \nabla p$ and $\xi(\phi)(\boldsymbol{v}_{\mathrm{s}} - \boldsymbol{v}_{\mathrm{p}}) = -(1-\phi)\nabla p$, respectively. Thus, we have

$$\nabla \cdot (\boldsymbol{\sigma} - p\,\boldsymbol{I}) = 0, \quad \boldsymbol{v}_{\mathrm{p}} - \boldsymbol{v}_{\mathrm{s}} = \kappa \nabla p, \tag{14}$$

where $\kappa = (1-\phi)^2/\xi$. The velocity of polymer network $\boldsymbol{v}_{\mathrm{p}}$ can be expressed as $\boldsymbol{v}_{\mathrm{p}} = \dot{\boldsymbol{u}}$ and the volume average velocity $\boldsymbol{v} = \phi \boldsymbol{v}_{\mathrm{p}} + (1-\phi)\boldsymbol{v}_{\mathrm{s}}$ is introduced in the incremental motion. Applying the incompressible condition given by Eqs. (8)–$(14)_2$ and substituting Eq. (12) into Eq. $(14)_1$, thus, the incremental motion of gels may be written explicitly in terms of $\boldsymbol{u}$ and p:

$$\left(K + \frac{1}{3}G\right)\nabla(\nabla \cdot \boldsymbol{u}) + +G\nabla^2 \boldsymbol{u} = \nabla p, \tag{15}$$

$$\nabla \dot{\boldsymbol{u}} = \kappa \nabla^2 p. \tag{16}$$

These are the linearized form of the gel dynamics which is called as the stress–diffusion coupling model [2]. We prefer to call it deformation–diffusion coupling throughout this paper. Equation (16) is well-known as Darcy's law in poroelasticity.

3 Thin Gel Sheets

The theories of gel sheets or poroelastic plates have been addressed in few papers. Cederbaum et al. [6] have analyzed the fluid-saturated poroelastic plates using Biot's constitutive law for transversely isotropic poroelastic materials and Darcy's law for describing the fluid flow. They assumed the pressure gradient depends only on the in-plane directions. On the other hand, Taber [7] has analyzed an isotropic poroelastic plates in which the pressure gradient depends only on the thickness direction. These works have formulated the so-called Kirchhoff–Love equations for plates. In this paper, we relax the main assumption of the membrane theory and nevertheless remain in the frame work of small strain. This requires that the two-dimensional strain along the middle surface remains small, and the radii of bending curvature of the sheet are moderately large compared with the thickness h. These conditions lead the FvK equations.

3.1 Kinematics and Constitutive Equations

Since the displacement $\boldsymbol{u}$ is small in the incremental motion, we assume that the small strain tensor $\epsilon_{\alpha\beta}(x, y)$ keeps the component of the displacement gradient that depends on the deflection $w(x, y)$:

$$\epsilon_{\alpha\beta}(x, y) = \epsilon^0_{\alpha\beta} + z\kappa_{\alpha\beta} \tag{17}$$

where $\epsilon^0_{\alpha\beta} = \frac{1}{2}(u_{\alpha,\beta}+u_{\beta,\alpha})+\frac{1}{2}w_{,\alpha}w_{,\beta}$ and $\kappa_{\alpha\beta} = -w_{,\alpha\beta}$. The Greek indices run over the directions x and y tangent to the plate, and the commas in the indices denote partial derivatives.

Since we only consider moderate deflections of the gel sheet, the curvature of the sheet is small and we assume that the normal $\boldsymbol{n}$ does not deviate too much from $\boldsymbol{e}_z$, i.e., $\boldsymbol{\sigma} \cdot \boldsymbol{n} = \boldsymbol{0}$. This hypothesis is called the membrane assumption. Applying to Eq. (12), we have

$$\epsilon_{zz} = -\frac{K - \frac{2}{3}G}{K + \frac{4}{3}G}(\epsilon_{xx} + \epsilon_{yy}) + \frac{p}{K + \frac{4}{3}G}. \tag{18}$$

Substituting Eq. (18) into (12),

$$\left.\begin{aligned} \sigma_{xx} &= -\frac{6G}{4G + 3K}p + \frac{4G(G + 3K)}{4G + 3K}\epsilon_{xx} + \frac{2G(-2G + 3K)}{4G + 3K}\epsilon_{yy}, \\ \sigma_{yy} &= -\frac{6G}{4G + 3K}p + \frac{4G(G + 3K)}{4G + 3K}\epsilon_{yy} + \frac{2G(-2G + 3K)}{4G + 3K}\epsilon_{xx}, \\ \sigma_{xy} &= \sigma_{yx} = 2G\epsilon_{xy}. \end{aligned}\right\} \tag{19}$$

Here we define the resultant forces and moments (total stress components per unit area in bulk material) by integration through the thickness h as

$$(N_{\alpha\beta}, M_{\alpha\beta}) = \int \sigma_{\alpha\beta}\,(1, z)\,\mathrm{d}z, \quad (N_p, M_p) = -\int p\,(1, z)\,\mathrm{d}z, \tag{20}$$

where $\int \equiv \int_{-h/2}^{h/2} \mathrm{d}z$. Substituting Eq. (20) into Eqs. (17) and (12), the resultant forces are obtained as

$$N_{xx} = A(\epsilon^0_{xx} + \nu\epsilon^0_{yy}) + \eta N_p, \quad N_{yy} = A(\epsilon^0_{yy} + \nu\epsilon^0_{xx}) + \eta N_p,$$
$$N_{xy} = N_{yx} = A(1-\nu)\epsilon^0_{xy}, \tag{21}$$

and the resultant moments are also obtained as

$$M_{xx} = D(\kappa_{xx} + \nu\kappa_{yy}) + \eta M_p, \quad M_{yy} = D(\kappa_{yy} + \nu\kappa_{xx}) + \eta M_p$$
$$M_{xy} = N_{yx} = D(1-\nu)\kappa_{xy}, \tag{22}$$

where

$$\left.\begin{aligned} A &= \frac{4G(3K+G)}{3K+4G}h, \quad D = \frac{4G(3K+G)}{3K+4G}\frac{h^3}{12}, \\ \eta &= \frac{6G}{3K+4G}, \quad \nu = \frac{3K-2G}{2(3K+G)}. \end{aligned}\right\} \tag{23}$$

3.2 *Balance Equations*

The Föppl–von Kármán Equations

From Eqs. (15), (17), (21) and (22), we can derive the compatibility condition for the in-plane strain and the equilibrium equation of the transverse deflection $w(x, y)$ in terms of the Airy potential $F(x, y)$:

$$\frac{1}{A(1-\nu^2)}\nabla^4 F - \eta\frac{1}{1+\nu}\nabla^2 N_p + \frac{1}{2}[w, w] = 0, \tag{24}$$

$$D\nabla^4 w - [F, w] - \eta\nabla^2 M_p = 0, \tag{25}$$

where the Airy's potential F is defined by $N_x = F_{,yy}$, $N_y = F_{,xx}$ and $N_{xy} = -F_{,xy}$. $[U, V] \equiv (U_{,xx}V_{,yy} + V_{,xx}U_{,yy} - 2U_{,xy}V_{,xy})$ is the differential operator, where $U(x, y)$ and $V(x, y)$ are two arbitrary functions. For completeness, when the transverse load p_0 and a body force f_z (gravity) are applied, Eq. (25) is replaced with

$$D\nabla^4 w - [F, w] - \eta\nabla^2 M_p = p_0 + f_z \tag{26}$$

where $p_0 = p_l - p_u$ in which p_l and p_u are uniform pressure at lower and upper surface, respectively.

Solvent Diffusion

Substitution of Eq. (18) into the Darcy's law given by Eq. (16) gives

$$2G(\dot{\epsilon}_{xx} + \dot{\epsilon}_{yy}) + \dot{p} = \kappa \left(\frac{3K + 4G}{3} \right) \nabla^2 p \tag{27}$$

Now we make the assumption that in-plane fluid-velocity gradients relative to the solid are small compared to the transverse fluid-velocity gradient [7]: $\nabla p = p_{,z}$. Thus, substituting Eq. (17) into Eq. (27) with Eq. (23) and the Airy's potential F, Eq. (27) yields

$$\frac{1}{2G}\frac{\partial p}{\partial t} - \frac{\kappa}{\eta}\frac{\partial^2 p}{\partial z^2} = \left(z\nabla^2 \dot{w} - \frac{1}{A(1+\nu)}\nabla^2 \dot{F} + \frac{2\eta}{A(1+\nu)}\dot{N}_p \right) \tag{28}$$

Eqs. (24), (25) (or Eq. (26)), and (28) constitute the deformation–diffusion coupling model described by the FvK equations in which variables $\{w, F, p\}$ are coupled.

The Kirchhoff–Love Equations

Further, linearizing Eqs. (24), (25) and (28), the in-plane and out-of-plane deformations can be decoupled:

$$D\nabla^4 w - \eta\nabla^2 M_p = 0, \quad \frac{1}{2G}\frac{\partial p}{\partial t} - \frac{\kappa}{\eta}\frac{\partial^2 p}{\partial z^2} = z\nabla^2 \dot{w}. \tag{29}$$

These constitute the Kirchhoff–Love equations for plates while the linearized equations are still coupled for variables $\{w, p\}$. These are the same form with that derived by Taber [7], who has solved the boundary value problems using Laplace transform and perturbation techniques. Since Eq. (29) are analogous to coupled thermoelasticity, as pointed out by Biot [8], we can also be solved using the methods developed in that discipline.

4 Conclusions

We have briefly formulated a mixture theory for elastomeric gels and derive the linearized deformation–diffusion coupling model based on the incremental motion and the Onsager's variational principle. Then, we have derived the FvK equations for gel sheets under appropriate assumptions by making dimensional reduction of the deformation–diffusion coupling model. We have illustrated that the FvK equations are coupled with the in-plane and out-of-plane deformations as well as the solvent diffusion.

References

1. Rajagopal, K.R., Tao, L.: Mechanics of Mixtures. World Scientific, Singapore (1996)
2. Doi, M.: Gel dynamics. J. Phys. Soc. Jpn. **78**, 052001 (2009)
3. Doi, M.: Onsager's variational principle in soft matter. J. Phys. Condens. Matter **23**, 284118 (2011)
4. Bouklas, N., Huang, R.: Swelling kinetics of polymer gels: comparison of linear and nonlinear theories. Soft Matter **8**, 8194–8203 (2012)
5. Lucantonio, A., Nardinocchi, P.: Reduced models of swelling-induced bending of gel bars. Int. J. Solids Struct. **49**, 1399–1405 (2012)
6. Cederbaum, G., Li, L., Schulgasser, K.: Poroelastic Structures. Elsevier, Oxford (2000)
7. Taber, L.A.: A theory for transverse deflection of poroelastic plates. J. Appl. Mech. **59**, 628–634 (1992)
8. Biot, M.A.: Theory of buckling of a porous slab and its thermoelastic analogy. J. Appl. Mech. **31**, 194–198 (1964)

Elastoplastic Analogy Constitutive Model for Rate-Dependent Frictional Sliding

Shingo Ozaki and Koichi Hashiguchi

Abstract The aim of this study is to propose a numerical approach for analyzing stick-slip motion; the approach is based on the finite element method implemented using an elastoplastic analogy constitutive model for rate-dependent frictional sliding. The present rate-dependent friction model can rationally describe the reciprocal transition between the static friction and the kinetic friction by a unified formulation. The rate-dependent friction model is implemented to FEM by using the user subroutine of the commercial software package. Then, the typical FE analysis of stick-slip motion is carried out. From the results of the FE analysis, we report that the present FE approach is applicable to the practical contact boundary value problems, including stick-slip motion.

1 Introduction

In various fields of engineering, it is important to clarify friction-induced vibration, such as stick-slip motion, for a wide range of scales from microscopic elements to continental plates. Stick-slip motion in a low-velocity regime is known to be caused by the time dependence of static friction, the transition between static and kinetic frictions, and the velocity-weakening of the friction coefficient (cf. [1]). Furthermore, stick-slip motion is induced by elastic interactions. In numerical simulation of stick-slip motion, Oden and Martins [2] have noted that "an acceptable theory for the stick-slip motion must explain the complex relationship between the

S. Ozaki (✉)
Yokohama National University, Yokohama, 240-8501 Kanagawa, Japan
e-mail: s-ozaki@ynu.ac.jp

K. Hashiguchi
Daiichi University, Kirishima, 899-4395 Kagoshima, Japan

Alexander K. Belyaev et al. (eds.), *Mechanics and Model-Based Control of Advanced Engineering Systems*, DOI 10.1007/978-3-7091-1571-8_26,
© Springer-Verlag Wien 2014

friction force and the sliding velocity, rather than assuming a simplified relationship from which the experimental evidence will deviate."

To solve the contact boundary value problems of finite-degree-of-freedom systems by the FEM, numerous researchers have described the friction phenomenon as elastoplastic constitutive equations, i.e., the elastoplastic relationship between the traction rate and the sliding velocity (e.g., [3, 4]). In particular, the present authors have proposed a rate-dependent friction model [5–7]. The model describes the smooth transition from static friction to kinetic friction and has a high robustness in numerical analysis, even for dynamic cyclic frictional behavior because it does not require switching between the "sticking state" and "sliding state" equations.

In the present study, the rate-dependent friction model is implemented to FEM by using the user subroutine of the commercial software package. Then, the typical FE analysis of stick-slip motion is carried out. The results from the analyses of typical stick-slip motion obtained using the current FEM approach are presented.

2 Formulation of the Rate-Dependent Friction Model

In this section, we briefly describe the formulation of the rate-dependent friction model with Coulomb's condition. Refer to Hashiguchi and Ozaki [5], Ozaki and Hashiguchi [6] and Ozaki et al. [7] for the detailed formulation.

The sliding velocity $\bar{\mathbf{v}}$ between contact surfaces is additively decomposed into the normal component $\bar{\mathbf{v}}_n$ and the tangential component $\bar{\mathbf{v}}_t$. In addition, $\bar{\mathbf{v}}$ is assumed to be a additively decomposed into the elastic-sliding velocity $\bar{\mathbf{v}}^e$ and the plastic-sliding velocity $\bar{\mathbf{v}}^p$ as follows:

$$\bar{\mathbf{v}} = \bar{\mathbf{v}}^e + \bar{\mathbf{v}}^p = (\bar{\mathbf{v}}_n^e + \bar{\mathbf{v}}_n^p) + (\bar{\mathbf{v}}_t^e + \bar{\mathbf{v}}_t^p) \tag{1}$$

The elastic part is given by

$$\overset{\circ}{\mathbf{f}} = \mathbf{C}^e \bar{\mathbf{v}}^e, \quad \mathbf{C}^e = -\alpha_n \mathbf{n} \otimes \mathbf{n} - \alpha_t (\mathbf{I} - \mathbf{n} \otimes \mathbf{n}), \tag{2}$$

where $\mathbf{C}^e$ is the contact elastic tensor and α is the contact elastic modulus, which is, at times, referred to as the penalty coefficient because the sliding velocity is equivalent to the gap velocity in numerical analyses based on the penalty method (e.g., [4]). The variable $\mathbf{f}$ is the traction vector applied to a unit area of the contact surface, and $\circ$ denotes the co-rotational rate with objectivity. The unit outward-normal vector $\mathbf{n}$ is defined by the normal to the contact surface, and it satisfies the equation $\mathbf{n} = \mathbf{f}_n / ||\mathbf{f}_n||$. Here, $\mathbf{I}$ is the unit tensor. The symbols $\otimes$ and $||\ ||$ denote the tensor product and the magnitude, respectively.

We assume Coulomb's friction condition for the sliding surface, as shown below.

$$||\mathbf{f}_t|| = \mu f_n, \tag{3}$$

where the equation $f_n = \mathbf{f}_n \cdot \mathbf{n} > 0$ govern, and μ is the friction coefficient. The symbol $\cdot$ denotes the scalar product. In this model, we assume that the interior of the sliding surface is not purely elastic domain, but that plastic-sliding velocity is induced depending on the rate of traction inside that surface. Therefore, the conical surface described by Eq. (3) is called the normal-sliding surface.

Next, we introduce the subloading-sliding surface, which always passes through the current traction $\mathbf{f}$ and maintains a shape similar to the normal-sliding surface in the tangential traction plane. The sizes ratio of the subloading-sliding surface to the normal-sliding surface in the tangential traction plane is called the normal-sliding ratio and is denoted by R. Here, $R = 0$ corresponds to the null traction state ($||\mathbf{f}_t||/f_n = 0$) as the most elastic state, $0 < R < 1$ to the sub-sliding state ($0 < ||\mathbf{f}_t||/f_n < \mu$), and $R = 1$ to the normal-sliding state in which the traction lies on the normal-sliding surface ($||\mathbf{f}_t||/f_n = \mu$). Therefore, the normal-sliding ratio R acts as a measure of the degree of approach to the normal-sliding state. The subloading-sliding surface is described by the following equation:

$$||\mathbf{f}_t|| = R\mu f_n \tag{4}$$

In this model, the friction coefficient is not assumed to be a function of time and velocity, rather the friction coefficient is assumed to be a function of the plastic-sliding velocity and the friction coefficient itself. Based on this concept, we adopt the following function for the evolution rule of the friction coefficient.

$$\dot{\mu} = -\kappa\left(\frac{\mu}{\mu_k} - 1\right)||\bar{\mathbf{v}}^p|| + \xi\left(1 - \frac{\mu}{\mu_s}\right), \tag{5}$$

where μ_s and μ_k are the maximum and minimum values of μ for static friction and kinetic friction, respectively, and κ and ξ are the material constants that influence the rates of decrease and recovery of the friction coefficient, respectively. Equation (5) is composed of two terms with contributions as follows:

1. The first term in Eq. (5) contributes to the reduction in static friction leading to a state of kinetic friction as a softening behavior occurs owing to plastic-sliding, which is caused by deterioration of the contact junction.
2. The second term in Eq. (5) contributes to the recovery of static friction as a hardening behavior occurs owing to creep deformation of the surface asperities under normal compression, i.e., aging of the contact junction.

Thus, Eq. (5) describes the rate-dependent competition between the deterioration and formation of adhesions at asperities. Hence, the velocity-weakening of the friction coefficient is naturally described.

It has been observed in experiments that when the tangential traction increases under constant normal traction, the tangential traction increases almost elastically from zero and thereafter gradually increases to approach the normal-sliding surface with increasing microscopic sliding; however, the tangential traction does not increase further after reaching the normal-sliding surface. We then assume the following simplest function for the evolution rule of the normal-sliding ratio.

$$\dot{R} = -r(\ln R)||\bar{\mathbf{v}}^p|| \quad \text{for } \bar{\mathbf{v}}^p \neq \mathbf{0}, \tag{6}$$

where r is the material constant.

Based on elastoplastic theory, we assume the following sliding-flow rule for the plastic-sliding velocity:

$$\bar{\mathbf{v}}^p = -\lambda \mathbf{t}, \quad \mathbf{t} = \frac{\mathbf{f}_t}{||\mathbf{f}_t||}, \tag{7}$$

where $\lambda(> 0)$ is a positive proportionality factor. The proportionality factor λ expressed in terms of the sliding velocity is obtained from Eqs. (1), (2), and (4)–(7) as follows:

$$\lambda = \frac{(\alpha_n R\mu\mathbf{n} - \alpha_t \mathbf{t}) \cdot \bar{\mathbf{v}} - \xi R\left(1 - \frac{\mu}{\mu_s}\right) f_n}{\alpha_t - r(\ln R)\mu f_n - R\kappa\left(\frac{\mu}{\mu_k} - 1\right) f_n}. \tag{8}$$

Consequently, the contact traction rate is obtained from Eqs. (1), (3), (7), and (8), as follows:

$$\overset{\circ}{\mathbf{f}} = \mathbf{C}^{ep}\bar{\mathbf{v}} + C^c \mathbf{t}, \tag{9}$$

where

$$\mathbf{C}^{ep} \equiv \mathbf{C}^e - \frac{\alpha_t \mathbf{t} \otimes (\alpha_n R\mu\mathbf{n} - \alpha_t \mathbf{t})}{\alpha_t - r(\ln R)\mu f_n - R\kappa\left(\frac{\mu}{\mu_k} - 1\right) f_n}, \tag{10}$$

$$C^c \equiv \frac{\alpha_t \xi R\left(1 - \frac{\mu}{\mu_s}\right) f_n}{\alpha_t - r(\ln R)\mu f_n - R\kappa\left(\frac{\mu}{\mu_k} - 1\right) f_n}. \tag{11}$$

3 Numerical Analysis

For analyzing boundary value problems, the experimental verification of constitutive equations is essential. In this section, first, we validate the rate-dependent friction model for the analysis of stick-slip motion by comparison with experimental results under a constant driving velocity V using a typical one-degree-of-freedom system [6]. The equation of motion is given as follows:

$$M\bar{a} = K(U - \bar{u}) + Sf_t, \tag{12}$$

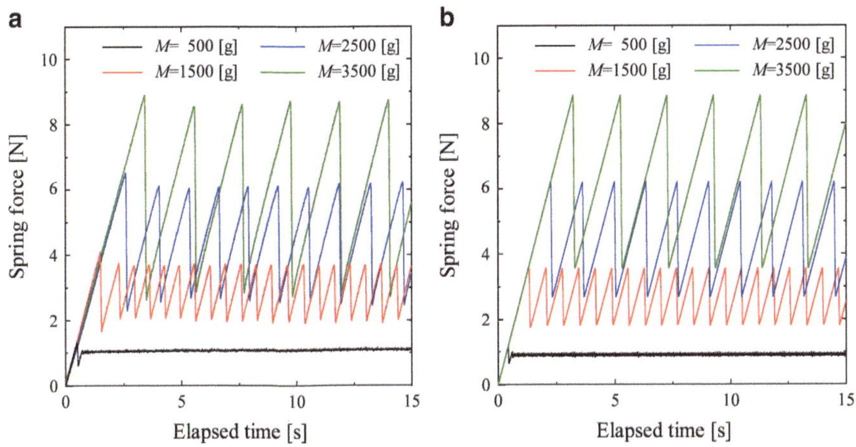

Fig. 1 Stick-slip motion of POM-POM contact: (**a**) experimental result; (**b**) numerical analysis

Table 1 Friction model parameters for one-degree-of-freedom system

$\mu_s = \mu^0$	α	κ	μ_k	r	ξ
0.25	1,000 N mm^{-3}	50 mm^{-1}	0.18	1,000 mm^{-1}	0.08 s^{-1}

where M is the mass of the slider, K is the spring stiffness, S is the nominal contact area, and $\bar{a}$ and $\bar{u}$ are the acceleration and displacement, respectively, of the slider, which are relative values with respect to the fixed base. For simplicity, we ignore the damping effect. The tangential contact traction f_t is estimated by Eq. (9).

In the present study, the Newmark β method is adopted for the time discretization of the equation of motion. We adopted a β of 1/4 (trapezoidal rule), which ensures second-order accuracy and unconditional stability in the linear regime. Thus, we obtain the stick-slip response of the present spring-mass system by executing iterative calculations.

Figure 1 show the comparison of variations of spring force with the elapsed time, as obtained in experiment and numerical analysis; the combination of test pieces is POM-POM. The spring stiffness K is 6.87 N/mm, the driving velocity V is 0.4 mm/s, and the mass of slider M are 500, 1,500, 2,500, and 3,500 g. The parameters of friction model are listed in Table 1. The figures confirm that the experimental results for various combinations of materials are simulated well by the present approach using the rate-dependent friction model.

Next, we demonstrate the analysis of stick-slip motion within the framework of continuum theory using the FEM. In this study, the rate-dependent subloading-friction model is implemented in the commercial FEM software package LS-DYNA Ver.971 [8]. The penalty method is adopted for the treatment of the frictional contact, which is useful and allows constrained conditions to be easily incorporated. We then employ the user subroutine “usrfrc” to introduce Eq. (9) into the FEM. A simple contact problem is analyzed under the coordinate system

Fig. 2 FE model for the analysis of stick-slip motion

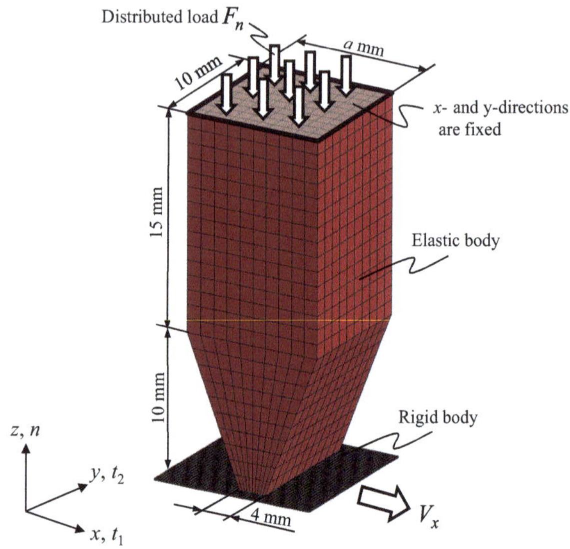

Table 2 Friction model parameters for FE analyses

$\mu_s = \mu^0$	α	κ	μ_k	r	ξ
0.6	Default	$10\,\text{mm}^{-1}$	0.3	$1{,}000\,\text{mm}^{-1}$	$0.1\,\text{s}^{-1}$

defined by the equality $(x, y, z) = (t_2, T_2, n)$, in which (x, y, z) refers to the global coordinate system. In the calculation, a dynamic implicit scheme based on the time discretization of the Newmark β method is adopted, and the updated Lagrangian method is used to formulate the geometric nonlinear behavior, including the large sliding phenomena.

To analyze stick-slip motion, an FE model consisted of two bodies is created as shown in Fig. 2. The model is discretized by eight-node solid elements. The upper elastic body has a thickness of 10 mm, a width of 10 mm, and a height of 25 mm. The elastic characteristic of the Young's modulus E is 200 MPa and Poisson's ratio ν is constant at 0.2. In the calculation, we apply the normal load $F = 5\,\text{N}$ to the top surface of the upper body and the two bodies come into contact. We then apply the forced velocity $V_x = 0.01$ mm/s in the x-direction to the lower plate, which is a rigid body, while the displacements in other directions are fixed. The displacements in x- and y-directions are also fixed on the top surface of the upper body. For the analyses, the parameters of the rate-dependent friction model are shown in Table 2. The default value in LS-DYNA is adopted as the penalty coefficient $\alpha(= \alpha_n = \alpha_t)$. Figure 3 shows the variations of contact forces with the elapsed time. In addition, Fig. 4 shows the distributions of shear stress and velocity fields at the sticking state (point A) and the sliding state (point B) as shown in Fig. 3. These figures show that the present FE approach can analyze the stick-slip motion during elapsed time under a constant imposed velocity. Furthermore, as confirmed from Figs. 3 and 4, the shear

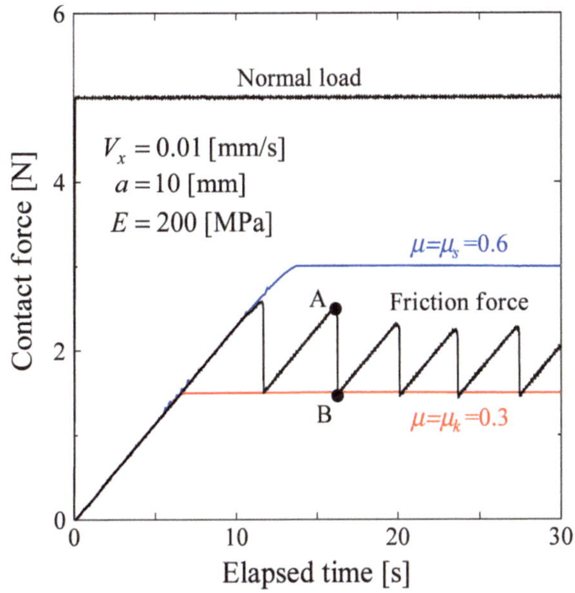

Fig. 3 Relations of contact forces with elapsed time

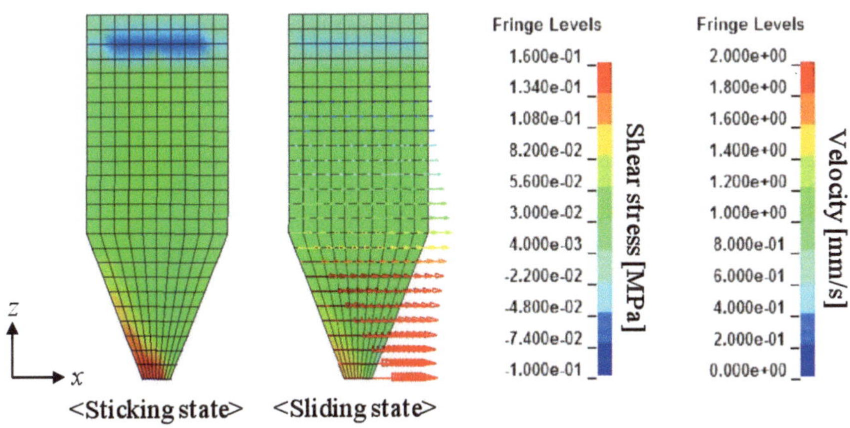

Fig. 4 Distributions of shear stress and nodal velocity in x-direction

stress concentrates at the sticking state and is released at the sliding state. Therefore, the characteristics of bulk stress distribution and the variation in the velocity field of a finite-degree-of-freedom body subjected to stick-slip motion can be investigated by the present FE approach. Furthermore, the present FE approach can also be used to examine the transition behaviors between stick-slip motion and nearly stable sliding owing to changes in loading conditions.

4 Conclusions

The present study demonstrates the capabilities of FEM introducing a rate-dependent subloading-friction model [5, 6] with Coulomb's condition for the analysis of stick-slip motion. A typical FE analysis of stick-slip motion is conducted. The present FE approach using the rate-dependent friction model considers not only the properties of friction and materials but also variations in boundary conditions. In addition, the responses of the friction force and displacement of a target body but also the stress and strain states in bulk can be grasped. From the results of the FE analyses, we suggest that the present FE approach is applicable to the practical contact boundary value problems.

References

1. Persson, B.N.J.: Sliding Friction: Physical Principles and Application, 2nd edn. Springer, Berlin/New York (2000)
2. Oden, J.T., Martins, J.A.C.: Models and computational methods for dynamic friction phenomena. Comput. Methods Appl. Mech. Eng. **52**, 527–634 (1985)
3. Peric, D., Owen, R.J.: Computational model for 3-D contact problems with friction based on the penalty method. Int. J. Numer. Methods Eng. **35**, 1289–1309 (1992)
4. Wriggers, P.: Computational Contact Mechanics. Wiley, Chichester (2003)
5. Hashiguchi, K., Ozaki, S.: Constitutive equation for friction with transition from static to kinetic friction and recovery of static friction. Int. J. Plast. **24**, 2102–2124 (2008)
6. Ozaki, S., Hashiguchi, K.: Numerical analysis of stick-slip instability by a rate-dependent elastoplastic formulation for friction. Tribol. Int. **43**, 2120–2133 (2010)
7. Ozaki, S., Hikida, K., Hashiguchi, K.: Elastoplastic formulation for friction with orthotropic anisotropy and rotational hardening. Int. J. Solids Struct. **49**, 648–657 (2012)
8. LSTC: LS-DYNA Ver.971, Keywords Manual & Theory Manual (2011)

Improved Position Control of a Mechanical System Using Terminal Attractors

Markus Reichhartinger and Martin Horn

Abstract In this contribution a control law for nonlinear uncertain plants is discussed. The synthesis of the controller combines ideas of the backstepping algorithm and terminal sliding-mode control. Latter one is introduced to improve the convergence properties of the closed-loop system. The application of the proposed concept is demonstrated on a real world experiment. A comparison of a conventional sliding-mode controller and the proposed approach demonstrates the improved performance.

1 Introduction

The synthesis of controllers stabilizing nonlinear systems which are affected by e.g. matched disturbances and uncertain plant parameters is a challenging task. A considerable number of applications (see e.g. [7–9, 11]) and theoretical contributions (e.g. [3, 12]) illustrate that sliding-mode control is a promising approach. For single-input single-output plants the design of conventional sliding-mode control laws (also referred to as "1-sliding") consists of two steps [10]: Firstly the plant model is transformed into a so-called regular form such that secondly the desired closed-loop behavior is achieved by a straightforward definition of the sliding-surface and its dynamics. The design technique discussed in this article is based on the regular form as well. In contrast to conventional approaches the specified dynamics describing the motion of the closed-loop system in sliding-mode is modified significantly.

M. Reichhartinger (✉) · M. Horn
Control and Mechatronic Systems Group, Alpen-Adria Universität Klagenfurt, Universitätsstraße 65–67, 9020 Klagenfurt, Austria
e-mail: markus.reichhartinger@aau.at; martin.horn@aau.at

Alexander K. Belyaev et al. (eds.), *Mechanics and Model-Based Control of Advanced Engineering Systems*, DOI 10.1007/978-3-7091-1571-8_27,
© Springer-Verlag Wien 2014

The remainder of this chapter is organized as follows: Sect. 2 introduces the class of considered systems and outlines the controller design goal. A detailed development of the proposed control law is given in Sect. 3. Results of numerical simulation and real world experiments of a mechanical system are presented in Sect. 4, a conclusion is given in Sect. 5.

2 Problem Formulation

It is assumed that the dynamic behavior of the process to be controlled is appropriately described by the differential equations

$$\frac{dx_1}{dt} = x_2, \tag{1a}$$

$$\frac{dx_2}{dt} = f(x_1, x_2) + g(x_1, x_2)u + \Delta(x_1, x_2, t), \tag{1b}$$

where $x_1, x_2 \in \mathbb{R}$ are measurable state variables and the scalar functions $f(x_1, x_2)$ and $g(x_1, x_2)$ represent the nominal behavior of the process. Additionally the function $g(x_1, x_2)$ satisfies the inequality

$$0 < g(x_1, x_2) < \infty \quad \forall x_1, x_2 \in \mathbb{R}. \tag{2}$$

The second order system (1) is a mathematical model of a real process and therefore shows an approximate behavior. In order to consider e.g. unmodelled effects or variations of process parameters the function $\Delta(x_1, x_2, t)$ is introduced into (1b). This matched uncertainty has to satisfy

$$|\Delta(x_1, x_2, t)| \leq \Delta_{\max} < \infty, \qquad \forall x_1, x_2 \in \mathbb{R} \text{ and } \forall t \geq 0, \tag{3}$$

where $\Delta_{\max}$ is a known positive constant. The design of the control signal $u = u(x_1, x_2)$ has to guarantee that the closed-loop system has a globally asymptotically stable origin $(x_1, x_2) = \mathbf{0}^T$ and furthermore the relation

$$x_1(t) = x_2(t) \stackrel{!}{=} 0 \quad \forall t \geq T < \infty \tag{4}$$

holds.[1]

[1]Throughout this article the solutions of a differential equation with discontinuous right hand side are understood in the sense of Filippov, see e.g. [1].

3 Controller Design

The synthesis of the control law $u(x_1, x_2)$ is based on the well-known backstepping-approach, see e.g. [2]. A desired dynamic behavior for (1a) is specified by

$$\frac{dx_1}{dt} = x_2 \stackrel{!}{=} w(x_1) := -\alpha \, |x_1|^\beta \, sign(x_1). \tag{5}$$

The choice of the function $w(x_1)$ is motivated by the ideas of a so-called terminal attractor, see e.g. [13]. In order to ensure an asymptotically stable equilibrium point $x_1 = 0$ and a non-singular solution of the desired dynamics (5), the constant parameters α and β have to be selected according to the inequalities

$$\alpha > 0 \quad \text{and} \quad \frac{1}{2} \leq \beta < 1. \tag{6}$$

Introducing the deviation ε of the desired dynamics (5) to the actual value x_2 as

$$\varepsilon := x_2 + \alpha \, |x_1|^\beta \, sign(x_1), \tag{7}$$

system (1) represented in (x_1, ε)-coordinates reads as

$$\frac{dx_1}{dt} = \varepsilon - \alpha \, |x_1|^\beta \, sign(x_1), \tag{8a}$$

$$\frac{d\varepsilon}{dt} = f(x_1, \varepsilon + w) + g(x_1, \varepsilon + w) \, u + \Delta(x_1, \varepsilon + w, t) + d(x_1, \varepsilon), \tag{8b}$$

where the argument of the function w is skipped. The function

$$d(x_1, \varepsilon) = \frac{\alpha\beta}{|x_1|^{1-\beta}} \varepsilon - \alpha^2 \beta \, |x_1|^{2\beta-1} \, sign(x_1) \tag{9}$$

is the time-derivative of the specified dynamics $w(x_1)$ given in[2] (5). Exploiting the quadratic control-Lyapunov-function

$$V(x_1, \varepsilon) = \frac{1}{2} x_1^2 + \frac{1}{2} \varepsilon^2 \tag{10}$$

and the control law

$$u(x_1, \varepsilon) = -\frac{1}{g(x_1, \varepsilon + w)} \left[x_1 + f(x_1, \varepsilon + w) + d(x_1, \varepsilon) + \gamma \, sign(\varepsilon) \right] \tag{11}$$

[2] A detailed computation of $d(x_1, \varepsilon)$ is given in the Appendix, see Sect. 5.

the time-derivative of (10) is globally negative definite, i.e.

$$\frac{V(x_1, \varepsilon)}{dt} = -\alpha\,|x_1|^{\beta+1} + \varepsilon\Delta(x_1, \varepsilon + w, t) - \gamma\,|\varepsilon|\,, \tag{12}$$

where the constant controller parameter has to satisfy $\gamma > \Delta_{\max}$. As shown in [6] the convergence time T is estimated with the help of Lyapunov-function (10) as

$$T \leq \frac{V(x_1, \varepsilon)}{c\,(1-d)} \quad \forall (x_1, \varepsilon) \in \mathbb{D} := \left\{ (x_1, \varepsilon) \;\middle|\; |x_1| \leq 1, |\varepsilon| \leq 1 \right\}, \tag{13}$$

where the constants c and d are chosen such that the conditions

$$d \geq \frac{\beta+1}{2} \quad \text{and} \quad c \leq 2^d \min\left(\gamma - \Delta_{\max}, \alpha\right) \tag{14}$$

hold.

4 Application

The proposed control law is implemented on the real world system depicted in Fig. 1. Introducing a position error x_1 and its corresponding velocity error x_2 w.r. to a smooth reference signal r as

$$x_1 := z_1 - r \quad \text{and} \quad x_2 := z_2 - \frac{dr}{dt}, \tag{15}$$

the dynamics of the considered plant are described by system (1) where the functions $f(x_1, x_2)$, $g(x_1, x_2)$ and $\Delta(x_1, x_2, t)$ are given by

$$f = -\frac{1}{I_0}\left(k_f + \frac{k_t\,k_i\,\eta^2}{R}\right)\left(x_2 + \frac{dr}{dt}\right) - \frac{d^2 r}{dt^2}, \tag{16a}$$

$$g = \frac{k_t \eta}{I_0\,R} \quad \text{and} \tag{16b}$$

$$\Delta = \frac{\delta}{I_0}\left[\frac{k_t \eta}{R} u - \left(k_f + \frac{k_t\,k_i\,\eta^2}{R}\right)\left(x_2 + \frac{dr}{dt}\right)\right]. \tag{16c}$$

The uncertain moment of inertia I w.r.t. the load side is expressed as

$$I(\delta) = \frac{I_0}{1+\delta} \quad \text{with} \quad \delta \in [\delta_{\min}, \delta_{\max}]\,, \tag{17}$$

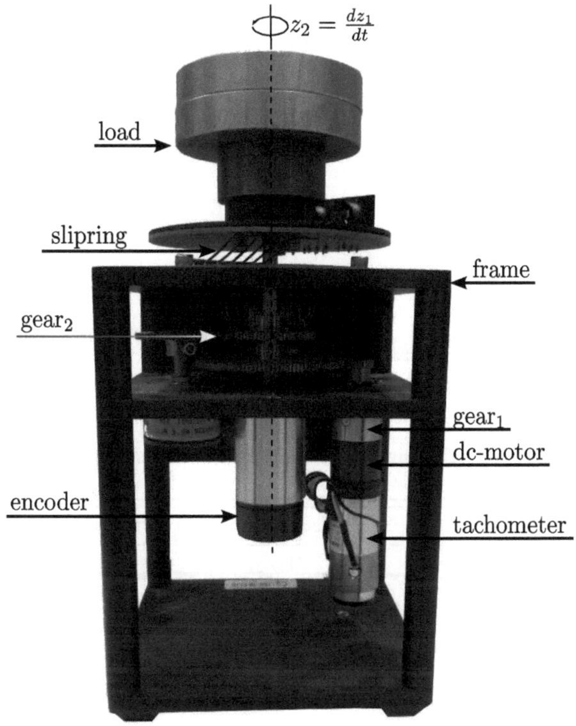

Fig. 1 The illustration shows the pilot plant considered in the present application. The dc-motor drives the load via a multi-stage gear unit. A slipring device allows to transmit measurements from the rotating load to the stationary frame of the pilot plant. Although the signal transmission is not used in the current application the slipring introduces a high amount of friction and considerably influences the system behavior. The measurements evaluated by the discussed controller are the position z_1 of the load measured by the encoder and the velocity z_2 of the dc-motor measured by the tachometer

Table 1 Parameters of the plant depicted in Fig. 1 and of the implemented control law (11)

	Plant parameters		Controller setting	
Parameter	Value	Unit	Parameter	Value
k_f	0.1916	Nm s rad^{-1}	α	35
k_i	$7.67 \cdot 10^{-3}$	V s	β	0.65
k_t	$7.67 \cdot 10^{-3}$	Nm/A	γ	85
R	2.44	Ω		
I	$[3.9 \cdot 10^{-3}, 4.9 \cdot 10^{-3}]$	kg m^2		
I_0	$4.4 \cdot 10^{-3}$	kg m^2		
δ	$[-0.1, 0.13]$	kg m^2		
η	70	1		

where I_0 is the entire nominal moment of inertia. The positive constants k_t, k_i, R, k_f and η are the dc-motor torque constant, the current constant, the amature resistor, the viscous friction constant and the overall gear ratio respectively. Effects like stiction mainly caused by the slipring remain unmodelled. The parameters of the setup shown in Fig. 1 are listed in Table 1. Considering transformation (7) the implementation of control law (11) is straightforward and is realized on a control unit operating at 1 ms sampling time. So-called robust exact differentiators (see e.g. [3, 4]) are employed to realize the required time-differentiation of the reference signal r. The parameter setting of control law (11) and of the differentiators is either computed with the help of linear matrix inequalities or is based on numerical

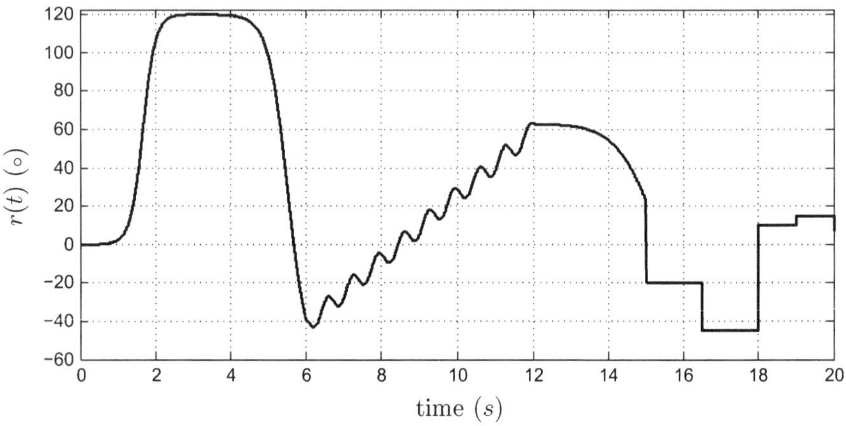

Fig. 2 Reference signal $r(t)$ used for numerical simulation and real world experiments

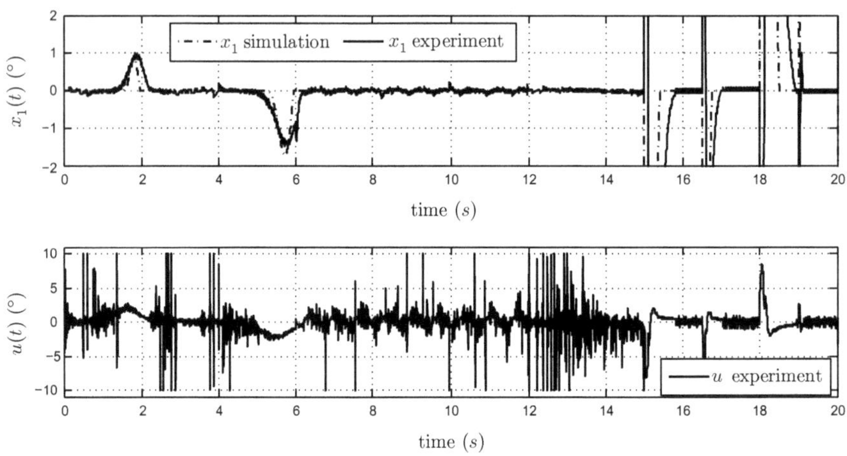

Fig. 3 The *upper plot* compares the position error x_1 achieved by numerical simulation and by real world experiment with the same controller settings. The *lower plot* shows the actuating signal u during the real world experiment. Note that no anti-chattering method is realized

simulation,[3] see Table 1. Figure 2 shows the reference signal r utilized for numerical simulation and real world experiments. Although control law (11) requires a smooth reference signal r the control-loop is excited with an nonsmooth reference signal (from $t = 15\,\text{s}$ until $t = 20\,\text{s}$) as well. Especially this excitation reveals the outstanding tracking performance of the control-loop achieved by the desired dynamics introduced by (5), see Figs. 3 and 4.

[3]For further details concerning structure, convergence properties or tuning of robust exact differentiators the interested reader may consult e.g. [3, 5, 6].

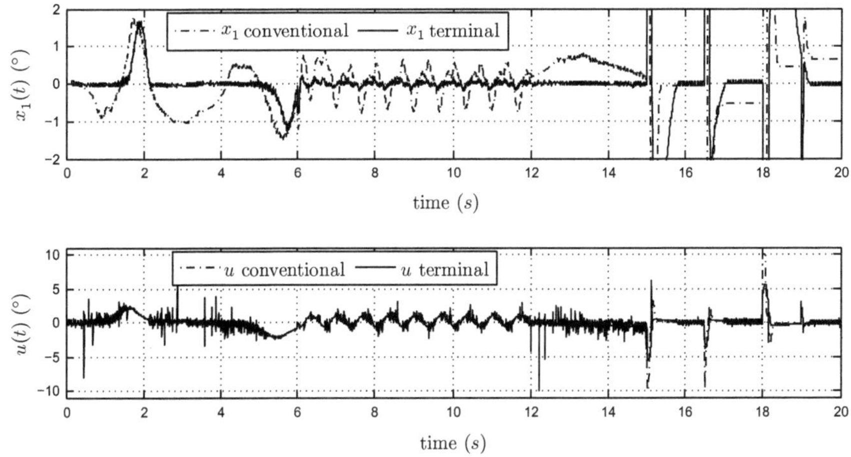

Fig. 4 The performance of the feedback loop is compared to a conventional "1-sliding" control concept, see [8]. The experiment demonstrates the superior tracking behavior of the discussed concept (labeled as terminal)

5 Conclusion

A control law based on an uncertain nonlinear plant model in regular form is designed. The approach is motivated by the well known backstepping procedure, hence a Lyapunov-function to proof closed-loop stability is provided. The main idea of the proposed concept is the introduction of a terminal attractor as desired motion of the controlled system. Numerical simulation and real world experiments demonstrate the effectiveness of the resulting feedback-loop.

Appendix

Computation of the Time-Derivative of $w(x_1)$

for $x_1 > 0$:

$$\frac{d}{dt}\left(\alpha x_1^{\beta}\right) = \frac{\alpha\beta}{x_1^{1-\beta}}\frac{dx_1}{dt} = \frac{\alpha\beta}{x_1^{1-\beta}}\varepsilon - \alpha^2\beta x_1^{2\beta-1} \tag{18}$$

for $x_1 < 0$:

$$\frac{d}{dt}\left\{\alpha\left[(-1)\,x_1\right]^{\beta}(-1)\right\} = (-1)^{1+\beta}\,\frac{\alpha\beta}{x_1^{1-\beta}}\frac{dx_1}{dt} =$$
$$= \frac{\alpha\beta\,(-1)^{1+\beta}}{x_1^{1-\beta}}\varepsilon - \alpha^2\beta\,(-1)^{2+2\beta}\,x_1^{2\beta-1} = \tag{19}$$
$$\frac{\alpha\beta}{\left[(-1)\,x_1\right]^{1-\beta}}\varepsilon - \alpha^2\beta\left[(-1)\,x_1\right]^{2\beta-1}(-1)$$

Combining (18) and (19) yields the function $d(x_1, \varepsilon)$ given in (9).

References

1. Filippov, A.F.: Differential Equations with Discontinuous Righthand Sides. Kluwer Academic, Dordrecht (1988)
2. Krstić, M., Kanellakopoulos, I., Kokotović, P.: Nonlinear and Adaptive Control Design. Wiley, New York (1995)
3. Levant, A.: Universal single-input-single-output (siso) sliding-mode controllers with finite-time convergence. IEEE Trans. Autom. Control **46**(9), 1447–1451 (2001)
4. Levant, A.: Principles of 2-sliding mode design. Automatica **43**, 576–586 (2007)
5. Moreno, J.A.: Lyapunov approach for analysis and design of second order sliding mode algorithms. In: 11th IEEE Workshop on Variable Structure Systems, Plenaries and Semiplenaries, Mexico City (2010)
6. Reichhartinger, M., Horn, M.: Finite-time stabilization by robust backstepping for a class of mechanical systems. In: Proceedings of 2011 IEEE International Conference on Control Applications (CCA), Denver, pp. 1403–1409 (2011)
7. Reichhartinger, M., Horn, M.: Robust position control of an electromechanical actuator for automotive applications. In: International Conference on Automation and Control Engineering (https://www.waset.org/journals/waset/v59/v59-320.pdf, 2011), vol. 59, World Academy of Science, Engineering and Technology, pp. 1688–1692
8. Reichhartinger, M., Horn, M.: Cascaded sliding-mode control of permanent magnet synchronous motors. In: Proceedings of 2012 12th International Workshop on Variable Structure Systems (VSS), Mumbai, pp. 173–177 (2012)
9. Utkin, V.: Sliding mode control design principles and applications to electric drives. IEEE Trans. Ind. Electron. **40**(1), 23–36 (1993)
10. Utkin, V., Guldner, J., Shi, J.: Sliding Mode Control in Electromechanical Systems. Taylor and Francis Ltd, Boca Raton, Florida, (2009)
11. Wai, R.-J., Wang, W.-H., Lin, C.-Y.: High-performance stand-alone photovoltaic generation system. IEEE Trans. Ind. Electron. **55**(1), 240–250 (2008)
12. Yu, S., Yu, X., Zhihong, M.: Robust global terminal sliding mode control of siso nonlinear uncertain systems. In: Proceedings of the 39th IEEE Conference on Decision and Control, Sydney, vol. 3, pp. 2198–2203 (2000)
13. Zak, M.: Terminal attractors in neural networks. Neural Netw. **2**(4), 259–274 (1989)

Convex Design for Lateral Control of a Blended Wing Body Aircraft

Alexander Schirrer, Martin Kozek, and Stefan Jakubek

Abstract This work demonstrates the application of convex control design methods for robust aeroelastic flight control design for a large, flexible blended-wing-body (BWB) passenger aircraft. High demands on control performance (robust stabilization, rigid-body dynamics shaping, vibration reduction, loads alleviation) have to be met over large physical parameter variations (airspeed, altitude, and mass). Pole assignment is achieved by an initial low-order controller. The remaining heterogenous set of time- and frequency-domain performance requirements is then directly addressed by convex control design for feedback control (based on the Youla parametrization) and feed-forward control. Algorithms improving design efficiency and allowing approximation of non-convex strong-stabilization constraints are proposed. High performance control laws are obtained which fulfill all posed goals. A critical discussion of limitations and strengths of the design method is given.

1 Introduction

Aerospace industries face steadily increasing demands with respect to effectiveness and efficiency. One onset to further improve fuel efficiency considers the shift to alternative aircraft configurations, such as blended-wing-body (BWB) aircraft as studied in the EU FP 7 research project ACFA 2020 [1]. While these configurations provide certain advantages, they also give rise to new challenges for control engineering. Active control concepts ("fly-by-wire") are employed to optimally trade off a multitude of conflicting control goals:

A. Schirrer (✉) · M. Kozek · S. Jakubek
Division of Control and Process Automation, Institute of Mechanics and Mechatronics, Vienna University of Technology, Vienna, Austria
e-mail: alexander.schirrer@tuwien.ac.at

Alexander K. Belyaev et al. (eds.), *Mechanics and Model-Based Control of Advanced Engineering Systems*, DOI 10.1007/978-3-7091-1571-8_28,
© Springer-Verlag Wien 2014

- Robustly ensure stabilization and fulfill handling qualities requirements
- Minimize vibrations due to gusts or turbulence excitation
- Maximize structural loads alleviation induced by gusts or maneuvers

These tasks are complicated further by the strong parameter dependency of the plant dynamics (e.g. on airspeed, altitude, and mass).

One fundamental question arising in the control design process is how to efficiently and successfully map these high-level objectives and constraints into the chosen control design algorithms. Most of the listed control objectives can readily be cast into or approximated by constraints (objectives) in the time and frequency domains: handling qualities can be defined by characteristic time constants or reference time responses; vibration damping objectives can be given in terms of damping ratio or weighted signal norm bounds; and structural loads can be quantified by peaks of time-domain load responses as a result of defined input signals. However, while the well-established $\mathscr{H}_2$ and $\mathscr{H}_\infty$-optimal control design methods allow an efficient, closed, optimal or robust controller design for complex multivariable control loops, the listed goals typically have to be approximated or transformed to frequency weighting functions which requires significant design and tuning effort.

Alternatively, "convex control design" or "convex controller synthesis" methods can be utilized. Despite their specific limitations, they allow more flexibility with respect to constraint and objective formulations and will be demonstrated in this work in the context of flight control design. In particular, convex synthesis allows to freely mix different time- and frequency-domain objectives and constraints within the same optimization problem formulation. Also, locally convex approximations of fundamentally non-convex, complicated constraints (such as strong stabilization or multi-model feedback constraints) can be introduced directly.

Convex control design and the Youla parametrization are detailed in Refs. [2, 3], and [10]. BWB aircraft control designs related to the application herein are studied in Refs. [5] (LQ-based), [8] (robust design by DGK-iteration), and [9] (scheduled LPV design via an LMI formulation). This work integrates and presents results from [7] (convex feedback design) and [6] (convex feed-forward design) and extends the application examples by a specific discussion of the convex control design method as a whole. A thorough treatment of existing and newly developed robust and optimal control design methods, including convex control design, is given in Ref. [4].

2 Methodology

When closed-loop transfer functions (or, equivalently, time-domain responses) can be represented in affine form in N free parameters θ_i, $i = 1, \ldots, N$, such as

$$\mathbf{T}(s) = \mathbf{T}_0(s) + \sum_{i=1}^{N} \mathbf{T}_i(s)\theta_i, \qquad \mathbf{z}(t) = \mathbf{z}_0(t) + \sum_{i=1}^{N} \mathbf{z}_i(t)\theta_i, \tag{1}$$

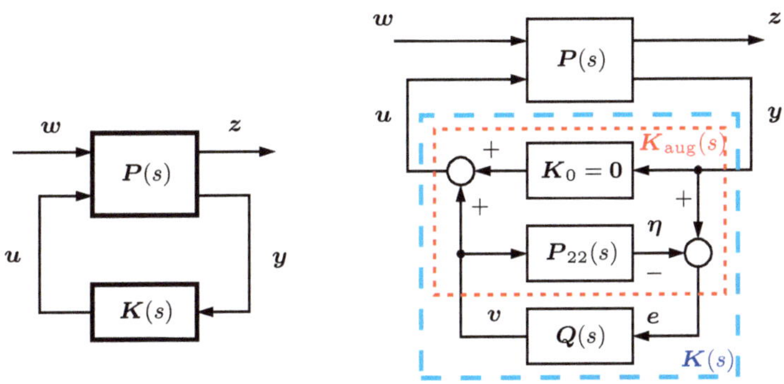

Fig. 1 Youla parametrization for a stable plant **P** [4]

this form can be utilized to formulate convex optimization objectives or constraints directly in the frequency- or time-domains (see [2] for a thorough treatment). Some important formulations are summarized in the following which lead to LP, QP, or LMI problems which can efficiently be solved.

For feed-forward control design, the form (1) is directly found for a controller onset $\mathbf{K}_{\mathrm{ff}}(s) = \sum_{i=1}^{N} \mathbf{K}_{\mathrm{ff},i}(s)\theta_i$ (with fixed, predefined basis functions $\mathbf{K}_{\mathrm{ff},i}(s)$), stable plant $\mathbf{P}(s)$ and objectives/constraints for the controlled system $\mathbf{T}(s) = \mathbf{P}(s)\mathbf{K}_{\mathrm{ff}}(s)$.

In the feedback control design case, the so-called Youla parametrization is utilized to obtain an affine closed-loop representation, see Sect. 2.1.

2.1 Youla Parametrization for a Stable Plant $\mathbf{P}(s)$

For the feedback interconnection of a stable plant $\mathbf{P}(s)$ and controller $\mathbf{K}(s)$ depicted in Fig. 1 (*left*), the closed-loop transfer from exogenous input $\mathbf{w}$ to exogenous output $\mathbf{z}$ is given by the lower Linear Fractional Transform (LFT) $\mathscr{L}_l$:

$$\mathbf{T}_{zw}(s) = \mathscr{L}_l\left(\mathbf{P}(s), \mathbf{K}(s)\right) = \mathbf{P}_{11}(s) + \mathbf{P}_{12}(s)\underbrace{\mathbf{K}(s)\left(\mathrm{I} - \mathbf{P}_{22}(s)\mathbf{K}(s)\right)^{-1}}_{\mathbf{Q}(s)}\mathbf{P}_{21}(s) \quad (2)$$

$$= \mathbf{T}_1(s) + \mathbf{T}_2(s)\mathbf{Q}(s)\mathbf{T}_3(s) \quad (3)$$

Thereby, $\mathbf{T}_{zw}(s)$ is nonlinearly dependent on $\mathbf{K}(s)$. However, the Youla- or $\mathbf{Q}$-parametrization of the system hides this nonlinearity in the sense that $\mathbf{T}_{zw}(s)$ is affine in the Youla parameter $\mathbf{Q}(s)$ of free dynamic order. Figure 1 (*right*) shows the block diagram equivalent to (2). It turns out (see Ref. [10]) that any stable MIMO transfer function $\mathbf{Q}(s)$ of compatible dimensions yields an internally stable closed loop, and that it parametrizes all internally stabilizing controllers for $\mathbf{P}(s)$.

Choice of Basis Functions For a finite-dimensional onset $\mathbf{Q}(s) = \sum_{i=1}^{N} \mathbf{Q}_i(s)\theta_i$, the optimal choice of basis functions $\mathbf{Q}_i(s)$ is not straightforward. Various strategies are reviewed in Ref. [3]. For control design it is important to limit the controller dynamic order, so ad-hoc chosen low-order transfer functions with their poles and zeros within control bandwidth will be chosen here.

2.2 Optimization Problem Formulation

Time-Domain l_∞-bounds Any closed-loop time-domain response constraint defined on a time gridding $t_k \in \mathscr{T} = \{t_1, t_2, \dots, t_n\}$ is representable by n scalar, linear inequalities,

$$z(t_k) = z_0(t_k) + \sum_{i=1}^{N} z_i(t_k)\theta_i < \bar{z}(t_k), \qquad k = 1, \dots, n. \tag{4}$$

Frequency-Domain $\mathscr{H}_\infty$-bounds The $\mathscr{H}_\infty$ constraint $\|\mathbf{G}(s)\|_\infty \leq \gamma$ for the stable MIMO transfer function $\mathbf{G}(s)$ is approximated on the frequency grid $\omega_k \in \Omega = \{\omega_1, \dots, \omega_n\}$ by bounding the maximum singular value $\bar{\sigma}\left(\mathbf{G}\left(\mathrm{j}\omega_k\right)\right) \leq \gamma$. This is equivalent to the symmetric linear matrix inequality (LMI) constraint

$$\left.\begin{bmatrix} \gamma\mathrm{I} & \Re(\mathbf{G}) & \mathbf{0} & \Im(\mathbf{G}) \\ & \gamma\mathrm{I} & \Im(\mathbf{G}^{\mathrm{H}}) & \mathbf{0} \\ & & \gamma\mathrm{I} & \Re(\mathbf{G}) \\ \star & & & \gamma\mathrm{I} \end{bmatrix}\right|_{\mathrm{j}\omega_k} \succeq 0, \tag{5}$$

where $\succeq 0$ denotes positive-semidefiniteness of the left-hand side.

Frequency-Domain $\mathscr{H}_2$-bounds Let a MIMO transfer function $\mathbf{G}(s)$ be stable and strictly proper, so that it possesses a finite $\mathscr{H}_2$-norm $\|\mathbf{G}(s)\|_2 = h$. A Riemann sum approximates h over a sufficiently fine and broad frequency gridding $\omega_k \in \{\omega_0, \dots, \omega_n\}$:

$$h^2 = \frac{1}{2\pi}\int_{-\infty}^{\infty} \mathrm{trace}\left[\left.\left(\mathbf{G}^{\mathrm{H}}\mathbf{G}\right)\right|_{\mathrm{j}\omega}\right] \mathrm{d}\omega \cong \tilde{h}^2 = \frac{1}{\pi}\sum_{k=1}^{n} \mathrm{trace}\left[\left.\left(\mathbf{G}^{\mathrm{H}}\mathbf{G}\right)\right|_{\mathrm{j}\omega_k}\right](\omega_k - \omega_{k-1}), \tag{6}$$

where $\left.\left(\mathbf{G}^{\mathrm{H}}\right)\right|_{\mathrm{j}\omega}$ denotes the Hermitian transpose of the complex matrix $\mathbf{G}(\mathrm{j}\omega)$. With the affine representation $\mathbf{G}(s) = \mathbf{G}_0(s) + \sum_{i=1}^{N} \mathbf{G}_i(s)\theta_i$, eq. (6) yields

$$\tilde{h}^2 = \frac{1}{\pi}\left(\beta + \boldsymbol{\gamma}^{\mathrm{T}}\boldsymbol{\theta} + \boldsymbol{\theta}^{\mathrm{T}}\boldsymbol{\Gamma}\boldsymbol{\theta}\right) \tag{7}$$

with constant coefficients $\beta \in \mathbb{R}, \boldsymbol{\gamma} \in \mathbb{R}^N$, and $\boldsymbol{\Gamma} \in \mathbb{R}^{N\times N}$ independent of $\boldsymbol{\theta}$ [4]. For $\boldsymbol{\Gamma} \succ 0$ (i.e., all basis functions $\mathbf{G}_i(s)$ affect h), using the Cholesky factorization $\boldsymbol{\Gamma} = \mathbf{L}^{\mathrm{T}}\mathbf{L}$ and Schur's complement, the following LMI condition is obtained:

$$\|\mathbf{G}\|_2^2 \cong \tilde{h}^2 \leq \delta^2 \Leftrightarrow \begin{bmatrix} \mathbf{I} & \mathbf{L}\boldsymbol{\theta} \\ \boldsymbol{\theta}^{\mathrm{T}}\mathbf{L}^{\mathrm{T}} & \delta^2 - \beta - \boldsymbol{\gamma}^{\mathrm{T}}\boldsymbol{\theta} \end{bmatrix} \succeq 0 \tag{8}$$

Note that (8) is of size $(N+1) \times (N+1)$, so high-precision precomputation of $\beta, \boldsymbol{\gamma}$, or $\boldsymbol{\Gamma}$ (large frequency grid size n) does not enlarge the LMI constraint (8).

Efficient Adaptive Constraint Refinement Time and frequency griddings are needed for constraints (4) and (5) which can strongly inflate the optimization problem. An iterative, heuristic constraint refinement methodology has been developed to keep the problem small by only actually formulating those constraints which are likely to be members of the final active constraint set [4]:

1. Start with coarse design grids (yielding a small optimization problem)
2. Formulate and solve the optimization problem
3. Check the solution on fine validation grids
4. Valid? ⇒ DONE. Else: add worst violated points to the design grids, go to 2.

Strong Stabilization Constraint Strong stabilization is the act of designing a stabilizing feedback controller which itself is a stable transfer function – a property useful for example if actuator/sensor signal clipping (saturation) occurs which effectively opens the control loop. However, this constraint is non-convex in the Youla parameter. With $\mathbf{A}_K(s, \boldsymbol{\theta})$ as the system matrix of $\mathbf{K}$ from (2),

$$\mathbf{K}(s) \text{ strongly stab.} \quad \Leftrightarrow \quad \max_i \Re\left(\lambda_i(\mathbf{A}_K(\boldsymbol{\theta}))\right) < 0 \quad \Leftrightarrow \quad \boldsymbol{\theta} \in \boldsymbol{\Theta}_{\mathrm{sstab}}. \tag{9}$$

To also consider this constraint, conservative, local convex approximations $\boldsymbol{\Xi}$ of the admissible parameter set $\boldsymbol{\Theta}_{\mathrm{sstab}}$ can be utilized (see Ref. [4] for details).

3 Aircraft System Model & Control Problem Statement

The EU FP7 research project ACFA2020 [1] studied active control concepts for large, flexible, ultra-efficient blended-wing-body (BWB) aircraft configurations. The complex multi-disciplinary dynamic modeling process is depicted in Fig. 2.

Two lateral control designs are considered in the following. The underlying models represent the lateral dynamics of a considered BWB configuration (see Fig. 3) as a set of 18 linearized state-space LTI models corresponding to a fuel mass parameter gridding (empty to full) at fixed cruise flight conditions. Each LTI system has 58 states (4 rigid-body states, 24 elastic states (12 modes), 10 lag states, and 20 actuator and sensor dynamic states).

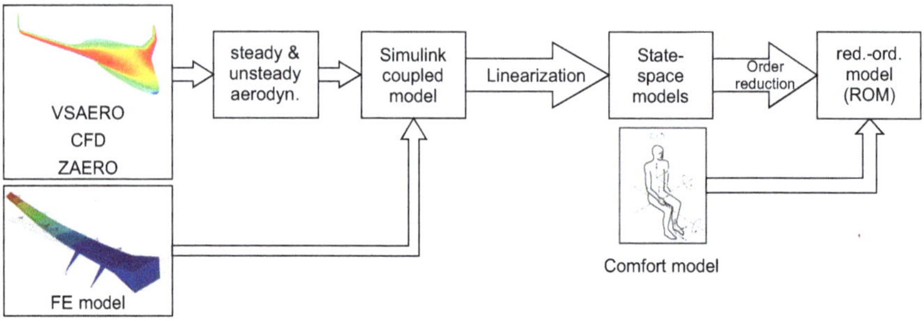

Fig. 2 Overview of the ACFA 2020 [1] modeling process of a BWB aircraft for aeroelastic control design studies. Illustrations are courtesy of Dept. of Mech., Biomech. & Mechatronics, CTU Prague (comfort, FE) and Inst. of lightweight structures, TU Munich (CFD)

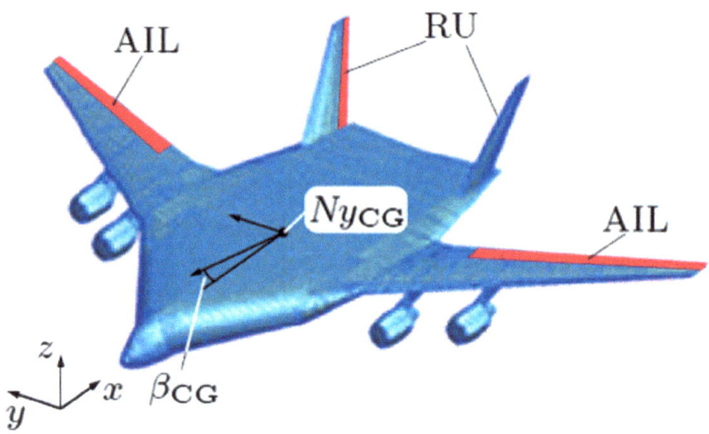

Fig. 3 BWB aircraft model for lateral aeroelastic control design [1]. Inputs: Lateral wind disturbance d, commanded aileron (AIL) and rudder (RU) deflections; Outputs: Structural loads My_{fin}, measurements of roll angle ϕ and rate p, yaw rate r, side-slip angle β, and lateral acceleration Ny_{CG}

Control design goals are robust stabilization (feedback), rigid-body dynamics shaping (feedback and feed-forward, according to pole placement and decoupling requirements), disturbance rejection (feedback), and maneuver loads alleviation (feed-forward via pilot command shaping).

Feedback Control Design Results An initial controller is obtained by partial eigenstructure assignment (see [7]) to fulfil robust stabilization and basic rigid-body mode shaping. The Youla parametrization (3) is constructed with an ad-hoc basis of $N = 40$ SISO basis functions $Q_i(s)$ of first and second order each, and an LMI problem is formulated and solved. Constraints include step response decoupling, lateral acceleration bounds, input deflection and rate limits, as well as a convex approximation of the strong stabilization constraint (see Fig. 4 and Ref. [7]).

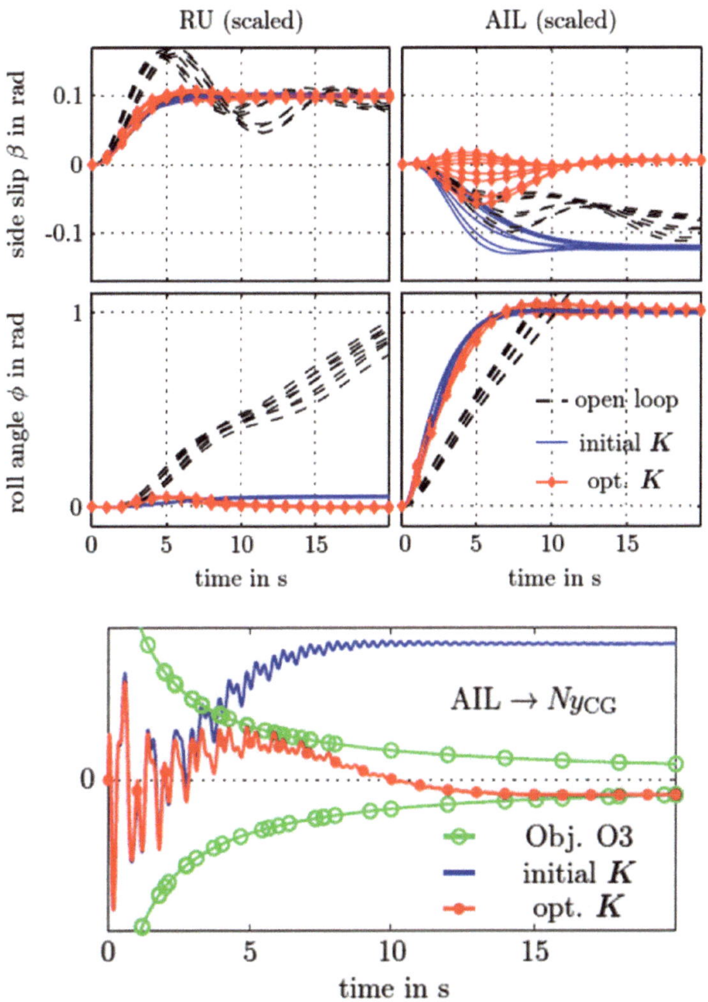

Fig. 4 Feedback control design results [7]

The circles (lower plot) indicate the small number of constraints ($10^{2\ldots3}$) to achieve successful final validation (corresponding to $10^{5\ldots6}$ constraints). The final controller order is high (94 states), so order reduction steps are necessary before implementation [4].

4 Results

Feed-Forward Control Design Results Loads alleviation in maneuvers is achieved by a pilot command shaping feed-forward control law (see Ref. [6]). Its design, also in the multi-model case, can directly be formulated as convex

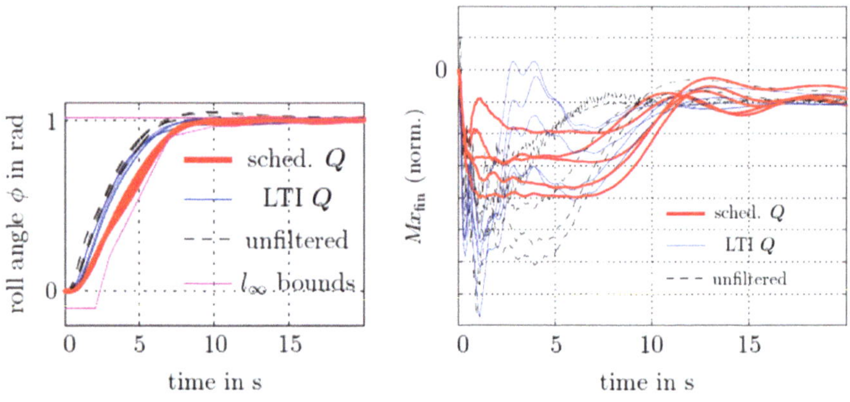

Fig. 5 Feed-forward control design results [6]

optimization problem. The constructed control law takes the reference angle as single input and produces 8 outputs which are fed into the aircraft feedback control loop. Here, only 8 ad-hoc basis functions have been utilized, and an LP problem results with the objective to minimize load peaks in a specified reference roll maneuver. Figure 5 shows that the optimized control law obeys the roll angle trajectory constraints. Assuming that the fuel mass is known online and available for controller scheduling, significant peak loads reductions are achieved, while it becomes clear that the critical peaks cannot be reduced when this information is not available during operation.

5 Discussion & Conclusions

To sum up, the convex control design method has been outlined and applied to an industrial application – the design of lateral aeroelastic flight control laws for a novel large, flexible blended-wing-body (BWB) aircraft. Signal constraints in time- and frequency-domain have been formulated, and advanced methods to incorporate non-convex constraints as well as to reduce optimization problem size have been presented. The method is successfully applied to the flight control problem and yields highly performing control laws addressing a diverse set of goals.

Robustness considerations Convex synthesis is by nature an optimal control design method, and the Youla-parametrized feedback formulation does not allow the (direct) consideration of perturbed plant dynamics. However, several attempts exist to introduce ideas to robustify the optimization: $\mathscr{H}_\infty$-bounds on the closed-loop sensitivity function can be defined (compare the mixed sensitivity design [10]), strong stabilization can be interpreted as simultaneous stabilization problem (see Ref. [4] and references therein), and finally the structured singular value μ [10]

can be considered in various relaxations. Similar to the well-known DK-iteration (or DGK-iteration) algorithm, an alternating series of convex synthesis designs and μ-analyses can be employed, known as "Q-μ-synthesis" [3]. Thereby, the obtained μ-bounds can be translated into singular value bounds on the closed loop and considered as constraint or via suitable objectives in the subsequent convex synthesis.

Strengths and weaknesses of convex controller synthesis Indeed convex control design has a series of limitations one should be aware of: It typically produces controllers of very high dynamic order (thus controller order reduction is inevitable). Moreover, pole placement cannot be done in the traditional sense, which renders the method unsuitable or inefficient for widely common pole placement tasks. The optimal choice of basis functions remains unclear and "global" optimality is only achieved within this finite-dimensional domain. Finally, the resulting optimization problems may grow to very large sizes, and pure $\mathscr{H}_2$ and $\mathscr{H}_\infty$ optimization is performed more efficiently by the existing standard methods.

However, remedies for these limitations have been proposed, and, in turn, important and distinct advantages are offered by the convex control design method: Convex optimization problems are globally tractable [2] and can be solved with extreme efficiency. The onset shows great design flexibility in terms of supported constraints and objectives, and mixing of heterogenous constraints is possible. This is especially true for time-domain constraints/objectives which are hardly supported (directly) by any other design method (except for Model Predictive Control).

Acknowledgements This work was financially supported by EU-FP7 grant 213321. The authors are grateful for the models provided by the ACFA2020 consortium.

References

1. Active Control of a Flexible 2020 Aircraft, EU FP 7 proj.no. 213321, www.acfa2020.eu
2. Boyd, S., Barratt, C.: Linear Controller Design: Limits of Performance. Prentice Hall, Englewood Cliffs (1991)
3. Dardenne, I.: Développement de méthodologies pour la synthèse de lois de commande d'un avion de transport souple. PhD thesis, Ecole Nationale Supérieure de l'Aéronautique et de l'Espace (SUPAERO) (1999). (Title (engl.): Development of methodologies for control law synthesis for a flexible transport aircraft)
4. Schirrer, A.: Efficient robust control design and optimization methods for flight control. PhD thesis, Inst. Mech. & Mechatronics, Vienna University of Technology, Vienna (2011)
5. Schirrer, A., Westermayer, C., Hemedi, M., Kozek, M.: LQ-based design of the inner loop lateral control for a large flexible BWB-type aircraft. In: IEEE International Conference on Control Applications (CCA), Yokohama, pp. 1850–1855 (2010)
6. Schirrer, A., Westermayer, C., Hemedi, M., Kozek, M.: Multi-model convex design of a scheduled lateral feedforward control law for a large flexible BWB aircraft. In: 18th IFAC World Congress, Milan, vol. 18-1, pp. 2126–2131 (2011)

7. Schirrer, A., Westermayer, C., Hemedi, M., Kozek, M.: Robust convex lateral feedback control synthesis for a BWB aircraft. In: 18th IFAC World Congress, Milan, vol. 18-1, pp. 7262–7267 (2011)
8. Schirrer, A., Westermayer, C., Hemedi, M., Kozek, M.: Robust feedback lateral control using a parameterized LFR model and DGK-iteration. In: Proceedings of the 4th EUCASS (European Conference for Aerospace Sciences), St. Petersburg (2011)
9. Westermayer, C., Schirrer, A., Hemedi, M., Kozek, M.: Linear parameter-varying control of a large blended wing body flexible aircraft. In: Automatic Control in Aerospace, Nara, vol. 18-1, pp. 19–24 (2010)
10. Zhou, K., Doyle, J., Glover, K.: Robust and Optimal Control. Prentice Hall, Upper Saddle River (1996)

Observability and Reachability, a Geometric Point of View

Kurt Schlacher and Markus Schöberl

Abstract The terms observability and reachability stem from the very beginning of modern control theory for linear, time-invariant systems. These are properties of the system, they are independent of the choice of solutions. This picture changes, if we turn to the nonlinear, time-variant or even time-invariant case. We present a short review of observability and reachability investigations for the latter class of systems based on functional analysis and differential geometry.

1 Introduction

The terms observability and reachability stem from the very beginning of modern control theory for linear, time-invariant systems, e.g. see [6], as well as [3, 7] and all the citations therein. Roughly speaking, these properties are system properties, which are independent of the choice of solutions. This picture changes totally, if we turn to the nonlinear time-variant or even nonlinear time-invariant case, e.g. see [5,9] and all the citations therein for approaches to different classes of nonlinear systems.

We discuss an approach, where we check the existence of "inverse maps" in the neighborhood of given solutions such that now observability and reachability depend also on the choice of these solutions. Nevertheless, one derives the well known Kalman conditions for the linear time-invariant case. We use methods from functional analysis, which provides us with the existence of these inverses, and from differential geometry, which allows us to determine the conditions for the existence of these inverses in a straightforward manner. Therefore, we present the system class under investigation in Sect. 2, Sect. 3 gives a short review about

K. Schlacher (✉) · M. Schöberl
Institute of Automatic Control and Control Systems Technology, Johannes Kepler University Linz, Altenberger Straße 69, 4040 Linz, Austria
e-mail: kurt.schlacher@jku.at; markus.schoeberl@jku.at

Alexander K. Belyaev et al. (eds.), *Mechanics and Model-Based Control of Advanced Engineering Systems*, DOI 10.1007/978-3-7091-1571-8_29,
© Springer-Verlag Wien 2014

the functional analysis based methods, whereas Sect. 4 is devoted to a differential geometric approach. Some conclusion complete this contribution.

2 Nonlinear ODE Systems

Let us consider the time-variant nonlinear system

$$\begin{aligned} x_t &= f(t, x, u) \\ y &= h(t, x) \end{aligned} \tag{1}$$

with n-dimensional state x, m-dimensional input u, l-dimensional output y. To refine the description we introduce the bundles $\mathscr{I} \to \mathscr{X} \to \mathbb{R}$ with local coordinates $(t, x, u) \mapsto (t, x) \mapsto (t)$ as well as $\mathscr{O} \to \mathbb{R}$ with local coordinates $(t, y) \mapsto (t)$, see e.g. [2] for an introduction to bundles, jet bundles, etc. The first jet bundle of $\mathscr{X}$ is denoted by $J_0^1(\mathscr{X})$ with coordinates $(t, x, x_t) \mapsto (t, x) \mapsto (t)$. This structure is chosen such that its bundle isomorphisms, $(\hat{t}, \hat{x}, \hat{u}) = \phi(t, x, u)$ and extensions to the first jet bundle,

$$\begin{aligned} \hat{t} &= \phi_{\hat{t}}(t) \\ \hat{x} &= \phi_{\hat{x}}(t, x) \\ \hat{x}_{\hat{t}} &= \left(\partial_t \phi_{\hat{x}} x(t, x) + \partial_x \phi_{\hat{x}}(t, x)\, x_t\right) / \partial_t \phi_{\hat{t}}(t) \\ \hat{u} &= \phi_{\hat{u}}(t, x, u) \end{aligned}$$

preserve the structure of (1). Furthermore these morphisms include static feedback as well. Next, we choose an initial value $(t_0, \bar{x}_0) \in \mathscr{X}$ and an input $\bar{u} \in L_2^m([t_0, t_1])$ such that not only the solution

$$\begin{aligned} \bar{x}(t) &= F_{t,t_0}(\bar{x}_0, \bar{u}) \\ \bar{y}(t) &= h(t, \bar{x}(t)) \end{aligned} \tag{2}$$

but all solutions $x(t) = F_{t,t0}(x_0, u)$ are unique on $[t_0, t_1]$ and meet $y \in L_2^l([t_0, t_1])$ provided x_0, u are close enough to $\bar{x}_0$, $\bar{u}$. *Observability* deals with the problem, whether the map

$$\begin{aligned} (\mathrm{id}, y, u)(t) &= \left(t, h \circ F_{t,t_0}(x_0, u), u\right) \\ &= F_o(t_0, x_0) \end{aligned} \tag{3}$$

with $t \in [t_0, t_1]$ from x_0 to the data $(t, y(t), u(t))$ available by measurement admits a left inverse F_o^{-1} such that

$$\begin{aligned} F_o^{-1}(\mathrm{id}, y, u) &= (t_0, x_0) \\ F_o^{-1} \circ F_o(t_0, x_0) &= (t_0, x_0) \end{aligned} \tag{4}$$

is met. A left inverse exists F_o^{-1}, iff F_o is *injective*. *Reachability* investigates the problem, whether the map

$$\begin{aligned}(t_1, x_1) &= F_{t_1.t_0}(x_0, u) \\ &= F_r(t_0, x_0, u)\end{aligned} \tag{5}$$

admits a right inverse F_r^{-1} such that

$$\begin{aligned}F_r^{-1}(t_1, x_1) &= (t_0, x_0, u) \\ (t_1, x_1) &= F_r \circ F_r^{-1}(t_1, x_1)\end{aligned} \tag{6}$$

is met. A right inverse exists F_r^{-1}, iff F_r is *surjective*.

To check the existence of the inverses of F_o, F_r, we will investigate the linearized or tangent map

$$\begin{aligned}\dot{x}(t) &= \dot{F}_{t,t_0}(x_0, u, \dot{x}_0, \dot{u}) \\ \dot{y}(t) &= \dot{h}\left(t, x(t), \dot{x}(t)\right)\end{aligned} \tag{7}$$

for the special choice $\dot{t} = 0$. It can easily be derived, if we restrict the linearized system

$$\begin{aligned}\dot{x}_t &= \partial_x f(t, x, u)\, \dot{x} + \partial_u f(t, x, u)\, \dot{u} = A(t, x, u)\, \dot{x} + B(t, x, u)\, \dot{u} \\ \dot{y} &= \partial_x h(t, x)\, \dot{x} \qquad\qquad\qquad\qquad\quad\; = C(t, x)\, \dot{x}\end{aligned} \tag{8}$$

to the solution of (1). The solution of (8) follows as

$$\begin{aligned}\dot{x}(t) &= \Phi(t, t_0)\, \dot{x}_0 + \int_0^t \Phi(t, \tau)\, B(\tau)\, \dot{u}(\tau)\, d\tau \\ \dot{y}(t) &= C(t)\, \dot{x}(t)\end{aligned} \tag{9}$$

with the transition matrix $\Phi(t, t_0)$. Therefore we assume, the solutions of (1) are at least continuously differentiable with respect to x_0, u. Here we use the standard notation of differential geometry, where $\dot{t}$, $\dot{x}$, $\dot{u}$, $\dot{y}$, $\dot{x}_t$ are additional coordinates for the tangent bundles $\mathcal{T}(\mathcal{X})$, $\mathcal{T}(\mathcal{I})$, $\mathcal{T}(\mathcal{O})$, $\mathcal{T}\left(J_0^1(\mathcal{X})\right)$, see e.g. [2].

3 A Functional Analysis Based Approach

The approach here is, one constructs the linearized maps $\dot{F}_o$, $\dot{F}_r$ of F_o, F_r for fixed $x_0 = \bar{x}_0$, $u = \bar{u}$. If it is possible to show that these maps are invertible, or even better one derives the corresponding left or right inverse explicitly, then one gets the existence of the left or right inverse F_o^{-1}, F_r^{-1}, see (4) and (5) by help of the inverse function theorem, see e.g. [1, 8].

Let us start with the observability of (1) first. We consider the linearized map $\dot{F}_o : \mathbb{R}^n \to \mathscr{Y} = L_2^l\left([t_0, t_1]\right)$, see (9),

$$\dot{y}(t) = C(t)\,\Phi(t, t_0)\,\dot{x}_0 \tag{10}$$

of (3) together with its adjoint $\dot{F}_o^* : \mathscr{Y}^* = L_2^{l,*}\left([t_0, t_1]\right) \to \mathbb{R}^{n,*}$,

$$\dot{x}_0^* = \int_{t_0}^{t_1} \dot{y}^*(t)\,C(t)\,\Phi(t, t_0)\,\mathrm{d}t\ , \tag{11}$$

see e.g. [1,8] for the adjoint. Obviously, the linear subspace $\operatorname{im}\left(\dot{F}_o\right) \subset \mathscr{Y}$ is closed because it is finite dimensional. Let us exploit the well known property $\mathscr{Y}^* \simeq \mathscr{Y}$, see e.g. [8], and choose the relation $\dot{y}^T = \dot{y}^*$ as an isomorphism between $\mathscr{Y}$ and $\mathscr{Y}^*$. Now, we construct the map $\dot{F}_o^* \circ \dot{F}_o : \mathbb{R}^n \to \mathscr{Y} \to \mathscr{Y}^* \to \mathbb{R}^{n,*}$ given by

$$\dot{x}_0^* = \dot{x}_0^T \int_{t_0}^{t_1} \Phi^T(t, t_0)\,C^T(t)\,C(t)\,\Phi(t, t_0)\,\mathrm{d}t = \dot{x}_0^T G_o(t_0, t_1)$$

and derive the observability Gramian $G_o(t_0, t_1)$. It is straightforward to see, that the Gramian G_o is regular, iff $\dot{F}_o$ is injective, and we derive a linearized left inverse $\dot{F}_o^{-1} = \left(\dot{F}_o^* \circ \dot{F}_o\right)^{-1} \circ \dot{F}_o^*$ of (4) as

$$\dot{x}_0 = G_o^{-1}(t_0, t_1) \int_{t_0}^{t_1} \Phi^T(t, t_0)\,C^T(t)\,\dot{y}(t)\,\mathrm{d}t\ .$$

Now we are able to deduce the existence of the left inverse F_o^{-1} from the implicit function theorem.

To investigate the reachability of (1) we consider the linearized map $\dot{F}_r : \mathscr{U} = L_2^m\left([t_0, t_1]\right) \to \mathbb{R}^n$, see (9),

$$\dot{x}_1 = \int_{t_0}^{t_1} \Phi(t_1, t)\,B(t)\,\dot{u}(t)\,\mathrm{d}t \tag{12}$$

of (5) together with its adjoint $\dot{F}_r^* : \mathbb{R}^{n,*} \to \mathscr{U}^* = L_2^{m,*}\left([t_0, t_1]\right)$,

$$\dot{u}^*(t) = \dot{x}_1^* \Phi(t_1, t)\,B(t)\ . \tag{13}$$

The linear subspace $\operatorname{im}\left(\dot{F}_r^*\right) \subset \mathscr{U}^*$ is closed because it is finite dimensional. Again we exploit the property $\mathscr{U}^* \simeq \mathscr{U}$ and choose the relation $\dot{u}^T = \dot{u}^*$ as an isomorphism between $\mathscr{U}$ and $\mathscr{U}^*$. We construct the map $\dot{F}_r \circ \dot{F}_r^* : \mathbb{R}^{n,*} \to \mathscr{U}^* \to \mathscr{U} \to \mathbb{R}^n$ by

$$\dot{x}_1 = \int_{t_0}^{t_1} \Phi(t_1, t)\,B(t)\,B^T(t)\,\Phi^T(t_1, t)\,\mathrm{d}t\,\dot{x}_1^{*,T} = G_r(t_0, t_1)\,\dot{x}_1^{*,T}$$

with the reachability Gramian $G_r(t_0, t_1)$. The matrix G_r is regular, iff $\dot{F}_r^*$ of (13) is injective. We derive a linearized right inverse $\dot{F}_r^{-1} = \dot{F}_r^* \circ \left(\dot{F}_r \circ \dot{F}_r^*\right)^{-1}$ of (6) as

$$\dot{u}(t) = B^T(t)\, \Phi^T(t_1, t)\, G_r^{-1}(t_0, t_1)\, \dot{x}_1 \; .$$

The existence of the right inverse F_r follows from the implicit function theorem. It is worth mentioning that the following relations are met: $\dot{F}_r$ is surjective, iff $\dot{F}_r^*$ is injective and vice versa.

Applying the results from above to the class of linear time-variant or invariant systems we get the well known simplifications, since the Gramians depend on the interval $[t_0, t_1]$ only. Even better, their rank is constant for any choice $t_1 > t_0$ in the time-invariant case. Additionally, one can use the relations to determine the initial condition x_0 or to steer the system to a desired terminal position x_1. Also the well known Kalman conditions for observability and reachability are derived in a straightforward manner. But these pleasant properties disappear in the nonlinear case. Therefore we need further investigations and other tools to derive more applicable results apart from the fact, that a problem is solvable in principle.

4 A Geometric Approach

Given any fiber[1] $\mathscr{F}_t$ of $\mathscr{X}$ over $t \in [t_0, t_1]$, see Sect. 3, the map $F_{t.t_0}$, see (1) and (2), is an isomorphism between the fibers $\mathscr{F}_{t_0}$ and $\mathscr{F}_t$ for a fixed input according to the assumption about the solutions of (1). This fact allows us to determine the pull back $F_{t.t_0}^*$ from $\mathscr{F}_t$ to $\mathscr{F}_{t_0}$ or push forward $F_{t.t_0,*}$ from $\mathscr{F}_{t_0}$ to $\mathscr{F}_t$ in a straightforward manner.

Let us begin with the observability problem, where we consider the case of a fixed input $u = \bar{u}$ first. We choose a fiber $\mathscr{F}_t$ of $\mathscr{X}$ over $t \in [t_0, t_1]$; obviously the output map of (1) leads to a foliation[2] of $\mathscr{F}_t$, given by $y(t) = h(t, x)$ such that its leaves generate the same output. This foliation is determined by the codistribution[3]

$$\Delta^* = \operatorname{span}(\mathrm{d}t, \mathrm{d}h) \; , \tag{14}$$

which is involutive by construction. The map F_{t,t_0}^* allows us to pull back Δ^* from $\mathscr{F}_t$ to the fiber $\mathscr{F}_{t_0}$ over t_0. Therefore we introduce the observability codistribution Δ_o^* as

$$\Delta_o^*(t_0) = \bigcup_{\tau \in [t_0, t_1]} F_{\tau,t_0}^*(\Delta^*) = \operatorname{span}\left(\mathrm{d}t, F_{t,t_0}^*(\mathrm{d}h)\right) \; . \tag{15}$$

---

[1] A fiber of $\mathscr{X} \to \mathbb{R}$ over t is the set of points $(t, x) \in \mathscr{X}$.

[2] A foliation partitions a manifold into submanifolds called leaves.

[3] We use the abbreviation $\mathrm{d}h = \mathrm{d}h_1, \ldots, \mathrm{d}h_l$. etc. to keep the notation as simple as possible.

Some facts are worth to be mentioned. Although the distribution $\Delta_0^*(t_0)$ is defined over t_0 only, it depends on t because of F_{t,t_0}^*. The 1-forms $F_{t,t_0}^*(\mathrm{d}h)$, restricted to (2) with the fixed initial condition $x_0 = \bar{x}_0$ are nothing else then the geometric version of the map $\dot{F}_o$ of (10). As shown above the system (1) is observable, iff $\dot{F}_o$ is injective. The integral manifolds of $\Delta_o^*(t_0)$ are submanifolds of initial values (t_0, x_0), which generate the same output y for $u = \bar{u}$. But the derivation of this foliation is still a laborious task, since it requires the map F_{t,t_0}^*. From $F_{t+\tau,t}^*\left(\mathrm{d}h\left(F_{t+\tau,t}\right)\right), \mathrm{d}h \in \Delta_o^*(t)$ one gets

$$f_e(\mathrm{d}h) = \lim_{\tau\to 0}\frac{1}{\tau}\left(F_{t+\tau,t}^*\left(\mathrm{d}h\left(F_{t+\tau,t}\right)\right) - \mathrm{d}h\right) \in \Delta_0^*(t_0)$$

with the Lie derivative of the cotangent vectors $\mathrm{d}h$ along the tangent vector[4]

$$f_e = \partial_t + f(t, x, \bar{u})\,\partial_x \tag{16}$$

over $\mathscr{X}$. Since we can repeat this arbitrarily, we construct the following sequence of codistributions

$$\Delta_0^* = \Delta^*\,, \quad \Delta_{i+1}^* = \Delta_i^* + f_e\left(\Delta_i^*\right)\,, \quad \Delta_0^* = \Delta_{k+1}^* = \Delta_k^* \tag{17}$$

with Δ^* from (14). If these codistributions have constant rank in a neighborhood of (t_0, x_0), then we derive the remarkable property, if $\Delta_l^* = \Delta_{l+1}^*$ is met, then also $\Delta_l^* = \Delta_{l+i}^*$, $i = 1, 2, \ldots$ holds. Therefore, k remains finite, and we choose k as small as possible. If we evaluate Δ_o^* at $t = t_0$, then it is straightforward to see it coincides with the distribution of (15). Therefore, one can determine $\Delta_o^*(t_0)$ in a simple algorithmic manner. Obviously, the map (10) is injective, if $\operatorname{rank}\left(\Delta_o^*\right) = n + 1$ is met. But now the chosen input $\bar{u}$ must be sufficiently often differentiable with respect to t.

The proposed machinery allows us to investigate further problems like, does there exist a trajectory through (t_0, x_0) with x_0 close enough to $\bar{x}_0$ such that Δ_o^* of (17) is of rank $n + 1$, or even that this is met for any trajectory through (t_0, x_0). Therefore, we replace the distribution of (14) by

$$\Delta^* = \operatorname{span}(\mathrm{d}t, \mathrm{d}h, \mathrm{d}u)\,,$$

where the output is no fixed function of t any more. In addition we replace the tangent vector field f_e, see (16) by the tangent vector field

$$f_{e,N} = \partial_t + f(t, x, u)\,\partial_x + \sum_{k=0}^{N-1} u_{k+1}\partial_{u_k}$$

[4] $f\partial_x$ is a shortcut for $\sum_{i=1}^{n} f^i \partial_{x_i}$.

on the N-th order jet-space of $\mathscr{I}$ for $0 \leq N < \infty$ with the derivative coordinates $u = u_0$, $u_1 = u_t, \ldots$ for the input and construct the sequence

$$\Delta_0^* = \operatorname{span}(\mathrm{d}t, \mathrm{d}h, \mathrm{d}u) \ , \quad \Delta_{i+1}^* = \Delta_i^* + f_{e,N}\left(\Delta_i^*\right) \ , \quad \Delta_{o,k}^* = \Delta_{k+1}^* = \Delta_k^* \ .$$

The rank condition $\operatorname{rank}\left(\Delta_{o,k}^*\right) = n+1+(k+1)\,m$ may be checked for a particular choice or all admissible choices of $u_0, \ldots, u_i$, $i = 0, \ldots, k$. Let us assume there exist at least one choice of u such that the rank condition is met. In this case an input u exists, such that the system is observable, otherwise we get a foliation of the state space and its leaves are sets of initial values (t_0, x_0), which lead to the same output for any choice of u. If we can meet the rank condition for any choice of u, then the system is observable for any choice of u. This corresponds to the linear case, where observability is independent of u. Otherwise we get a foliation of the state space and its leaves are sets of initial values (t_0, x_0), which lead to the same output for some choices of u.

Let us turn to reachability, where our considerations are mainly based on Chow's theorem, see e.g. [4]. Again the following considerations are restricted to initial conditions x_0 and inputs u sufficiently small to the nominal values $\bar{x}_0$ and $\bar{u}$. Let us fix the input $u = \bar{u}$ first. A set of functions $g\,(t, x)\,; \mathbb{R} \times \mathscr{X} \to \mathbb{R}^k$ is said to be invariant, if it meets $g\,(t, x) = g\,(t_0, \bar{x}_0)$ or equivalently $f_e(g) = 0$ with f_e from (16). These functions are said to be u-invariant iff the previous relation is met for any admissible input u. They meet[5]

$$\partial_u\left(g\right) = 0\ , \quad f_{e,u}\left(g\right) = 0 \tag{18}$$

with the tangent vector field $f_{e,u}$,

$$f_{e,u} = \partial_t + f\,(t, x, u)\,\partial_x \tag{19}$$

on $\mathscr{I}$, see Sect. 3. If one finds u-invariant functions, then one derives a foliation of the fibers $\mathscr{F}_t$ over t. In particular the submanifold of $\mathscr{F}_t$ defined by $g\,(t_1, x_1) = g\,(t_0, x_0)$ contains all points, which are reachable from x_0 at t_1. Obviously, the system (1) cannot be reachable, if such functions exist. Now, Chow's theorem states, that (1) is reachable, iff these functions do not exist. According to the considerations from above, the Lie derivative along elements of the distribution Δ,

$$\Delta = \operatorname{span}\left(\partial_u, f_{e,u}\right) \tag{20}$$

must vanish. One way to proceed would be to determine the push forward of Δ from $\mathscr{F}_t$ to $\mathscr{F}_{t_1}$. But now the push forward $F_{t.t_0,*}$ has to be considered for all admissible choices of u, since the input u is not fixed any more. A simpler way follows

[5] ∂_u is a shortcut for $\partial_{u_1}, \ldots, \partial_{u_l}$, $\partial_u\left(g\right)$ is a shortcut for $\partial_u\left(g_1\right), \ldots, \partial_u\left(g_l\right)$.

from the observation that the relations (18) contain further "hidden" constraints, since they imply $f_{e,u}\left(\partial_u\left(g\right)\right) - \partial_u\left(f_{e,u}\left(g\right)\right) = \left[f_{e,u}, \partial_u\right]\left(g\right) = 0$ with the Lie bracket[6] $\left[f_{e,u}, \partial_u\right]$. Since we can repeat this arbitrarily, we construct the sequence of distributions

$$\Delta_0 = \Delta\ , \quad \Delta_{i+1} = \Delta_i + [\Delta_i, \Delta_i]\ , \quad \Delta_r = \Delta_{k+1} = \Delta_k\ , \tag{21}$$

with Δ from (19). These sequence is called also sequence of derived flags, where the last flag Δ_r is closed with respect to the Lie bracket. If these distributions are of constant rank in a neighborhood of (t_0, x_0), then we have again the remarkable property, if $\Delta_l = \Delta_{l+1}$ is met, then also $\Delta_l = \Delta_{l+i}, i = 1, 2, \ldots$ holds. Therefore, k remains finite, and we choose k as small as possible. With rank $(\Delta_r) = n_r + 1$ we may distinguish the cases $n_r = n$, where no non trivial u-invariant functions exist and $n_r < n$, where $n - n_r$ u-invariant functions according to the Theorem of Frobenius, see e.g. [1, 4], can be found. In the latter case all trajectories, which emanate from (t_0, x_0) are restricted to an $n_r + 1$ submanifold of $\mathscr{X}$ given by $g\left(t, x\right) = g\left(t_0, \bar{x}_0\right)$. Roughly speaking, we succeeded this way to replace the push forward of the distribution (20) along all possible trajectory by the last flag of (21). But now, the functions f of (19) must be sufficiently often differentiable with respect to u.

But there is a significant difference between the results of Sect. 4 and the geometric ones. A short look to the map (12) shows that only the matrix B of (8) evaluated along the chosen trajectory enters the relation. Since the columns of B are equivalent to the bracket $\left[f_{e,u}, \partial_u\right]$, only the elements Δ_0, Δ_1 of the sequence (19) are taken into account to determine the existence of u-invariant functions. Therefore, the results of Sect. 4 about reachability are sufficient only. This is a remarkable difference to observability.

5 Conclusions

This contribution gave a short overview about observability and reachability for nonlinear time variant systems described by ordinary differential equations, where classical methods from functional analysis and differential geometry have been applied. In contrast to the linear case these properties depend also on the special choice of the trajectory. This seems to be unavoidable, like it was demonstrated in the functional analysis and differential geometry based approach. Additionally, it is worth mentioning that the latter one can be reformulated by help of transformation groups, see [12]. Of course, this tool reproduces all results presented here, but its main advantage is, that one can apply it to other classes of systems in

[6] We use also the abbreviation $\left[f_{e,u}, \partial_u\right] = \left[f_{e,u}, \partial_{u_1}\right], \ldots, \left[f_{e,u}, \partial_{u_m}\right]$.

a straightforward manner. The case of systems, described by implicit ordinary differential equations is discussed in [12], one finds sampled data systems in [11], and [10] presents a rigorous approach for complex systems of partial differential equations.

Acknowledgements The first author want to thank ACCM (Austrian Center of Competence in Mechatronics) for the partial support of this contribution. Markus Schöberl is an APART fellowship holder of the Austrian Academy of Sciences.

References

1. Choquet-Bruhat, Y., DeWitt-Morette, C.: Analysis, Manifolds and Physics. North-Holland, New York (1982)
2. Giachetta, G., Sardanashvily, G., Mangiarotti, L.: New Lagrangian and Hamiltonian Methods in Field Theory. World Scientific, Singapore/River Edge (1997)
3. Gilbert, E.G.: Controllability and observability in multivariable control systems. SIAM J. Control Optim. **1**(2), 128–151 (1963)
4. Hermann, R.: Differential Geometry and Calculus of Variations. Academic, New York (1968)
5. Isidori, A.: Nonlinear Control Systems, 3rd edn. Springer, London (1995)
6. Kalman, R.E.: On the general theory of control systems. In: Proceedings of the First International Congress of Automatic Control, Moscow, USSR, 1960, pp. 481–493
7. Kalman, R.E.: Mathematical description of linear dynamical systems. J. Soc. Ind. Appl. Math. A Control **1**(2), 152–192 (1963)
8. Kreyszig, E.: Introductory Functional Analysis with Applications. Wiley, New York (1989)
9. Nijmeijer, H., van der Schaft, A.: Nonlinear Dynamical Control Systems. Springer, New York (1990)
10. Rieger, K.: Analysis and Control of Infinite-Dimensional Systems: A Geometric and Functional-Analytic Approach. Shaker, Aachen (2002)
11. Rieger, K., Schlacher, K., Holl, H.: On the observability of discrete-time dynamic systems – a geometric approach. Automatica **44**(8), 2057–2062 (2008). Elsevier
12. Schlacher, K., Kugi, A., Zehetleitner, K.: A lie-group approach for nonlinear dynamic systems described by implicit ordinary differential equations. In: Proceedings 15th International Symposium on Mathematical Theory of Networks and Systems, South Bend, 2002

The Model of a Deformable String with Discontinuities at Spatial Description in the Dynamics of a Belt Drive

Yury Vetyukov and Vladimir Eliseev

Abstract We study dynamics of a belt drive with a nonlinear model of an extensible string at contour motion, in which the trajectories of particles of the belt are pre-determined. Writing the equations of string dynamics in a spatial frame and assuming the absence of slip of the belt on the surface of the pulleys we arrive at a new model with a discontinuous velocity field and concentrated contact forces. It is demonstrated how the model may be applied for analytical study of stationary regimes, and differences in the behavior of synchronous and friction belt drives are pointed out. We also present a novel approach for modeling transient behavior of the system, e.g. for the analysis of start up of the belt drive or its frequency response. Both numerical and analytical solutions of the resulting problem are possible.

1 Introduction

Reynolds [16] has introduced the concept of the creep of the belt near points, where it leaves the pulleys: the velocity of the belt and its strain are varying, which leads to sliding friction between the belt and the pulleys at these zones. This scheme is adopted in engineering and technical literature, see e.g. [4, 18]. Common assumptions for such computations ignore dynamic effects due to the acceleration of the belt; other hypotheses concerning redistribution of tension forces between the tight and the slack sides of the drive are often involved [4, 14].

Y. Vetyukov (✉)
Institute of Technical Mechanics, Johannes Kepler University, Altenberger Str. 69, A-4040 Linz, Austria
e-mail: yury.vetyukov@jku.at

V. Eliseev
Faculty of Mechanics and Mechanical Engineering, St. Petersburg State Polytechnical University, Polytechnicheskaya ul. 29, 195251 St. Petersburg, Russia
e-mail: yeliseyev@inbox.ru

Alexander K. Belyaev et al. (eds.), *Mechanics and Model-Based Control of Advanced Engineering Systems*, DOI 10.1007/978-3-7091-1571-8_30,
© Springer-Verlag Wien 2014

In the framework of the creep theory, Rubin [17] and Bechtel et al. [3] used a model of extensible string for describing steady operation of a belt drive. Solving the equations of string mechanics [2] for the case of steady motion (see e.g. [8, 15]), Rubin [17] consistently analyzes the fields of velocity and strain of the belt at three qualitatively different domains: at the free spans, at the zones of perfect contact between the belt and the pulley, as well as at the zones of sliding friction contact. Resulting relations for the rotational moments at the pulleys, their angular velocities and coefficient of efficiency of the belt drive are shown to be different from the results in the technical literature. The model has been further extended to a more complicated case of viscoelastic behavior of the belt [13] as well as to a model of contact between the belt and the pulley, which is more complicated than the classical law of dry friction [10]. Kong and Parker [11] analytically study stationary regimes with a rod model; bending stiffness makes the free spans to be no longer straight, and positions of the contact points are a priori unknown, which requires special numerical procedure for finding the solution. Leamy [12] considered perturbations of stationary motion, which should allow studying transient regimes with sufficiently small and slow deviations from a given steady one.

While traditional creep theory treats belts as extensible strings, modern V-belts are often reinforced with steel threads at the outer contour. This makes the belt as a whole almost inextensible, and its shear deformation plays a dominant role in the mechanism of contact with the pulley. Firbank [7] was the first who consistently treated such structures; for a comparative analysis between creep and shear belt drive models see [1].

In the present work we continue developing the approach, which was suggested by the authors for the case of extensible belt in [5,6]. Assuming that the particles of the belt are moving along a given contour, we transform the equations of nonlinear string dynamics to spatial (Eulerian) description. This brings the equations of dynamics to a domain with fixed boundaries, and preserves the level of generality, which is sufficient for a consistent analysis of unsteady operation of the drive. We are considering belt dynamics both on the pulleys and on the straight paths simultaneously with the same system of equations for the case of a perfect contact between the belt and the pulleys. Such a perfect contact is usually achieved in synchronous drives by means of a timing (toothed) belt, but may also be considered as a limiting case of the above mentioned creep model when the zone of slip degenerates to a point. As a consequence, contact forces between the pulley and the belt are concentrated in points, and discontinuities in the velocity and strain of the belt need to be treated. The corresponding jump conditions are new for the analysis of belt drives dynamics; see [9] for general relations of jump of mass and momentum. In Sect. 3 of the present work we shortly recall the results of the analysis of steady operation of timing and friction belt drives, presented by the authors in [6].

An essential part of the present contribution is Sect. 4, in which we are using the benefits of the spatial description and of the idealized model of belt-pulley contact for developing a novel strategy for simple and consistent transient analysis. The latter strategy may be used for both, numerical and analytical studies of e.g. steady acceleration or periodic vibrations of the drive in the vicinity of stationary motion. Important characteristics like frequency response of the system may be determined analytically, and the numerical scheme will allow for more complicated simulations.

2 Equations of Belt Dynamics at Contour Motion

A two-pulley belt drive is schematically presented in Fig. 1. Time variation of the material coordinates of the points 1...4, which come to the boundaries of the contact regions is unknown in advance and is a part of solution of the problem. This makes spatial description of the dynamics of the belt as an extensible string advantageous in the present case; in this section we recall and revise the equations of string dynamics at spatial description, which were derived in [6].

The trajectory, which is plotted by a thick line in Fig. 1, can be parametrically specified by the dependence of the position vector $\boldsymbol{R}$ on the arc coordinate σ with the following geometrical relations:

$$\boldsymbol{R} = \boldsymbol{R}(\sigma), \quad \boldsymbol{R}'(\sigma) = \boldsymbol{\tau}, \quad \boldsymbol{\tau}'(\sigma) = k\boldsymbol{n}, \tag{1}$$

in which k is the curvature, $\boldsymbol{\tau}$ is the unit tangent vector and $\boldsymbol{n}$ is the unit vector of normal to the trajectory. In Fig. 1 the domains $0 \le \sigma \le \sigma_1$ and $\sigma_2 \le \sigma \le \sigma_3$ answer correspondingly to the pulleys A and B, while the straight paths $\sigma_1 \le \sigma \le \sigma_2$ and $\sigma_3 \le \sigma \le \sigma_4$ are the slack and tight spans respectively; the values $\sigma = 0$ and $\sigma = \sigma_4$ answer to the same point.

At spatial description, our aim is to find the variation of the material coordinate s of the string for a given point in space with the time t:

$$s = S(\sigma, t). \tag{2}$$

In the reference configuration $s = \sigma$. Stretch of the string equals to $\partial\sigma/\partial s$, and an expression for the strain measure follows from the rule for an inverse function:

$$\varepsilon = \left(S'\right)^{-1} - 1; \tag{3}$$

we denote the derivative with respect to the spatial coordinate σ with a prime. The tension force Q is related to the strain by a linear constitutive relation

$$Q = Q_0 + b\varepsilon, \tag{4}$$

in which Q_0 is the tension in the reference configuration (pre-tension) and b is the stiffness of the belt.

Fig. 1 Scheme of a two-pulley drive with the domains of motion of the belt: 12, 34 – straight paths (slack span and tight span respectively), 41, 23 – regions of contact with the driving pulley A and with the driven one B

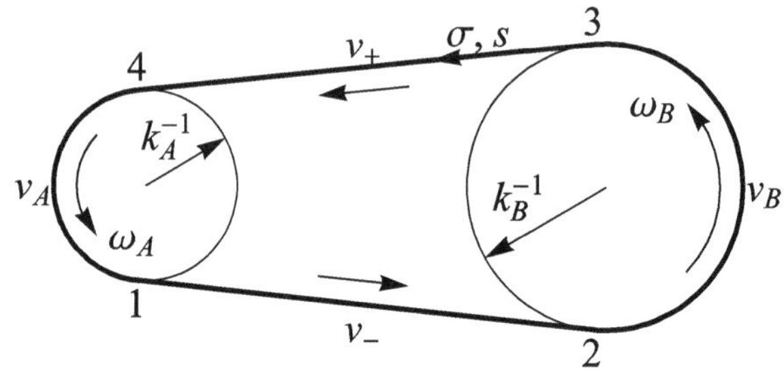

The equations of balance projected on the tangent and normal directions read

$$Q' + S'q_\tau = \rho S' w_\tau, \quad kQ + S'q_n = \rho S' w_n, \tag{5}$$

here q_τ and q_n are the components of the external force $\boldsymbol{q}$ per unit length in the reference configuration, ρ is the mass of the belt per unit length in the reference configuration, and the components of the vector of acceleration $\boldsymbol{w}$ are

$$w_\tau = vv' + \dot{v}, \quad w_n = kv^2. \tag{6}$$

The derivative with respect to time t at a given point in space $\sigma = \text{const}$ is denoted with a dot, and the velocity of a particle of the belt is

$$\boldsymbol{v} = v\boldsymbol{\tau}, \quad v = -\dot{S}\left(S'\right)^{-1}. \tag{7}$$

Introducing force factors $\tilde{\boldsymbol{q}}$ and belt density m per unit length of the contour in space according to

$$\tilde{\boldsymbol{q}} \equiv \boldsymbol{q}S', \quad m \equiv \rho S', \tag{8}$$

we may rewrite the equations of balance in a fully spatial form:

$$Q' + \tilde{q}_\tau = mw_\tau, \quad kQ + \tilde{q}_n = mw_n, \tag{9}$$

Equation of balance of mass at spatial description has the following form:

$$\int_{s_1}^{s_2} \rho\, ds = \int_{\sigma_1}^{\sigma_2} m\, d\sigma \quad \Rightarrow \quad \dot{m} + (mv)' = 0. \tag{10}$$

At the driven pulley the loading moment is specified:

$$M_B = -k_B^{-1} \int_{S(\sigma_2,t)}^{S(\sigma_3,t)} q_\tau\, ds = -k_B^{-1} \int_{\sigma_2}^{\sigma_3} \tilde{q}_\tau\, d\sigma. \tag{11}$$

3 Steady Operation

In a steady regime the velocity, force and strain at a given point in space do not change in time. We have $v = v(\sigma)$, $Q = Q(\sigma)$, $\varepsilon = \varepsilon(\sigma)$, but $S = S(\sigma, t)$. In [6] it is shown that

$$S' = \frac{c}{v}, \quad S = c\left(\int \frac{d\sigma}{v(\sigma)} - t\right), \quad c = \text{const.} \tag{12}$$

This relation between the material and spatial coordinates at steady motions of strings was established earlier in [8, 15], and was applied in the analysis of steady operation of belt drives by Rubin [17]. From (10) we find the "mass flow rate" $mv = \rho c = \text{const.}$

Variation of σ in the range from 0 to $L \equiv \sigma_4$ answers to the growth of s from 0 to L, in which L is the length of the string in the reference configuration. For a given field of velocities the last equality in (12) allows to relate L and the constant c (this is the "condition of compatibility" in the terminology of [17]).

The strain and the tension force are now functions of the velocity:

$$\varepsilon = \frac{v}{c} - 1, \quad Q = Q_0 + b\left(\frac{v}{c} - 1\right). \tag{13}$$

The tangential projection of the balance equations (9) leads to

$$\tilde{q}_\tau = mvv' - Q'. \tag{14}$$

From (14) it follows that the velocity $v = \text{const}$ at a straight path, where $\boldsymbol{q} = 0$ [6]. At the tight span we have $v = v_+$, $Q(v_+) = Q_+$, at the slack one $v = v_-$, $Q(v_-) = Q_-$ (Fig. 1).

At steady motion the moment at the driven pulley can be related to the velocities at the straight paths. Substituting (14) in (11) and integrating, we obtain

$$M_B = k_B^{-1}\left((Q_+ - \rho c v_+) - (Q_- - \rho c v_-)\right). \tag{15}$$

At the pulleys we assume perfect contact:

$$v = \text{const} \quad \Rightarrow \quad Q = \text{const} \quad \Rightarrow \quad q_\tau = 0. \tag{16}$$

The contradiction between (11), (15) and (16) is resolved by admitting discontinuities at the points $\sigma = \sigma_i$ and corresponding concentrated forces: only the "outermost" zones on the pulleys are loaded. For the loading moment we get

$$M_B = -k_B^{-1}(F_2 + F_3), \quad F_i = [\rho c v - Q]_i = [mv^2 - Q]_i,$$
$$[\ldots]_i \equiv \ldots|_{\sigma=\sigma_i+0} - \ldots|_{\sigma=\sigma_i-0}. \tag{17}$$

Table 1 Differences in the characteristics of steady operation of a belt drive depending on the belt-pulley contact model

Type of the drive	*Synchronous*	*Friction*
Points of discontinuity	4	2
$Q_+ + Q_- = 2Q_0$	Yes (if $\varepsilon \ll 1$)	No (if $k_A \neq k_B$)
Coefficient of efficiency	1	$v_-/v_+ < 1$

According to (14), the concentrated forces result from the jumps of the tension force and of the velocity in the corresponding points; see [9] for a further discussion.

Two particular cases of belt-pulley contact laws are considered in the work of the authors [6]. For a *synchronous*, or timing belt drive the teeth on the belt and on the pulleys impose a kinematic constraint: there is no stretch of the belt at the contact zone, i.e. $\varepsilon = 0$ and $v = c$; the velocity field $v(\sigma)$ jumps in all four points of contact σ_i. For a *friction* belt drive a continuity condition needs to be imposed in the points, where the belt comes into contact with the pulleys, and $v(\sigma)$ jumps only in σ_1 and σ_3. Particular properties of the two working regimes are determined in [6], and the essential differences are summarized in Table 1. A common assumption from the technical literature that the mean value of the tension forces at the straight spans equals to the pre-tension (see e.g. [4,14]) generally holds only for synchronous drives at small strains, and the coefficient of efficiency of a friction drive decreases with the strain of the belt in the working regime.

4 Transient Dynamics of a Friction Belt Drive

For the free spans with $\boldsymbol{q} = 0$ a single quasilinear wave equation follows from the equalities (3)–(7):

$$\ddot{S}S'^2 + \dot{S}^2 S'' = 2S'\dot{S}\dot{S}' + \frac{b}{\rho}S''. \tag{18}$$

For a friction drive the particles of the belt immediately adhere to the pulley surface at the points 2 and 4, and in these points we demand the velocity of the belt to be continuous:

$$\dot{S} + k_A^{-1}\omega_A \, S'\big|_{\sigma=\sigma_4} = 0; \tag{19}$$

for the slack span 12 the indices in the equations of the present section need to be adapted accordingly. Further, the solution is "transported" by the pulleys, and in the point 3 we have the solution from the point 2 with a certain delay:

$$S(\sigma_3, t) \equiv S_3(t) = S(\sigma_2, t - \tau_B(t)); \tag{20}$$

the particle with the material coordinate $S_3(t)$ came into contact with the pulley at the time instance $t - \tau_B(t)$. Since then the pulley has rotated to a certain angle:

$$\int_{t-\tau_B(t)}^{t} \omega_B(\tilde{t})d\tilde{t} = k_B^{-1}(\sigma_3 - \sigma_2) = \text{const.} \tag{21}$$

Differentiating this equality with respect to time, we find the relation of the delay time to the angular velocity of the driven pulley:

$$\dot{\tau}_B(t) = 1 - \frac{\omega_B(t)}{\omega_B(t - \tau_B(t))}. \tag{22}$$

Now we need to compute the moment, which acts on the driven pulley from the side of the belt. We substitute (9) in (11):

$$k_B M_B = -\int_{\sigma_2-0}^{\sigma_3+0} \left(m w_\tau - Q'\right) d\sigma = Q|_{\sigma_2-0}^{\sigma_3+0} - \int_{\sigma_2-0}^{\sigma_3+0} m\left(vv' + \dot{v}\right) d\sigma. \tag{23}$$

The fields m, v and $\dot{v}$ have at σ_3 a discontinuity of the first kind. "Concentrated" terms under the integral appear only due to v':

$$v = -\frac{\dot{S}}{S'}, \quad mvv' = \rho\left(\frac{\dot{S}\dot{S}'}{S'} - \frac{S''\dot{S}^2}{S'^2}\right);$$

$$\int_{\sigma_3-0}^{\sigma_3+0} mvv'd\sigma = -\rho\dot{S}_3^2\int_{\sigma_3-0}^{\sigma_3+0} \frac{S''}{S'^2}d\sigma = \rho\dot{S}_3^2\left[(S')^{-1}\right]_3. \tag{24}$$

Inside the domain $\dot{v} = k_B^{-1}\dot{\omega}_B$, $v' = 0$, and

$$k_B M_B = Q|_{\sigma_2-0}^{\sigma_3+0} - \rho\dot{\omega}_B k_B^{-1}(S_3 - S_2) - \rho\dot{S}_3^2\left[(S')^{-1}\right]_3. \tag{25}$$

Here S_3 is again determined by the solution in the point 2 with a delay. In the expression for the jump of $(S')^{-1}$ the value at $\sigma_3 + 0$ is taken from the solution on the span 34, and the value at $\sigma_3 - 0$ is "transported" by the pulley:

$$(S')^{-1}\Big|_{\sigma=\sigma_3-0,\ t=\tilde{t}} = (S')^{-1}\Big|_{\sigma=\sigma_2,\ t=\tilde{t}-\tau_B(\tilde{t})}. \tag{26}$$

A simple interpretation can be given to all three terms at the right hand side of (25). The first term is just the moment due to the tension forces and is taken into account by all technical approaches. The second term represents additional inertia of the pulley due to the attached part of the belt. And the third non-trivial term is the reactive force at the point, where the velocity undergoes a jump.

It should be noted, that the expression (25) could be derived using relations of jump of mass and momentum known from the literature. Following [9], we write the concentrated force, which acts on the belt in the point 3:

$$F_3 = [mv^2 - Q]_3 = \left[\rho S' \dot{S}^2 (S')^{-2}\right]_3 - [Q]_3. \tag{27}$$

The force F_3 contributes to the moment M_B as in (17), and the jump of mv^2 evidently leads to the third term (reactive force) at the right hand side in (25).

The problem is closed by the equations of dynamics for the pulleys

$$I_A \dot{\omega}_A = M_A + M_A^{ext}, \quad I_B \dot{\omega}_B = M_B + M_B^{ext}; \tag{28}$$

the external moments M_A^{ext}, M_B^{ext} may be prescribed functions of time, but other practically important cases including control loops may be considered.

5 Conclusion

Numerical analysis can be based on finite difference discretization of the wave equations (18) in space. Time integration of the resulting ordinary differential equations should be coupled with the solution of the equations for the delay times $\tau_{A,B}$ (22) and for the angular velocities of the pulleys $\omega_{A,B}$ (28). It is easier to start the simulation of a start up from a slowly moving regime to avoid infinite values of $\dot{\tau}_B$ in (22) when ω_B turns into zero at $t < 0$. The problems are coupled by the expressions for the moments (25) and by the boundary conditions (19) and (20). Particular implementation of this numerical scheme is left for the future work.

Time integration of delay differential equations with solution dependent delay times is a challenging computational task. But the system of equations for transient behavior allows for analytical studies for simple cases, like e.g. steady acceleration of the drive or periodic excitation in the vicinity of a known stationary regime: external moments in (28) have a harmonically varying component, and by searching harmonic solutions we determine the frequency response of the structure. One can also address the question of stability of stationary regimes. Together with numerical studies, these solutions will provide a valuable insight into dynamics of belt drives.

Acknowledgements Support of Yu. Vetyukov from the K2 Austrian Center of Competence in Mechatronics (ACCM) is gratefully acknowledged.

References

1. Alciatore, D., Traver, A.: Multipulley belt drive mechanics: creep theory vs shear theory. ASME J. Mech. Des. **117**, 506–511 (1995)
2. Antman, S.: Nonlinear Problems of Elasticity. Springer, New York (1995)
3. Bechtel, S., Vohra, S., Jacob, K., Carlson, C.: The stretching and slipping of belts and fibers on pulleys. ASME J. Appl. Mech. **67**, 197–206 (2000)
4. Brar, J., Bansal, R.: A Text Book of Theory of Machines. Laxmi Publications, New Delhi (2004)
5. Eliseev, V.: A model of elastic string for transmissions with flexible coupling (in Russian). Scientific and Technical Bulletin of St. Petersburg State Polytechnical University 84, 192–195 (2009)
6. Eliseev, V., Vetyukov, Y.: Effects of deformation in the dynamics of belt drive. Acta Mech. **223**, 1657–1667 (2012)
7. Firbank, T.: Mechanics of the belt drive. Int. J. Mech. Sci. **12**(12), 1053–1063 (1970)
8. Healey, T., Papadopoulos, J.: Steady axial motions of strings. ASME J. Appl. Mech. **57**, 785–787 (1990)
9. Irschik, H.: On rational treatments of the general laws of balance and jump, with emphasis on configurational formulations. Acta Mech. **194**, 11–32 (2007)
10. Kim, D., Leamy, M., Ferri, A.: Dynamic modeling and stability analysis of flat belt drives using an elastic/perfectly plastic friction law. ASME J. Dyn. Syst. Meas. Control **133**, 1–10 (2011)
11. Kong, L., Parker, R.: Steady mechanics of belt-pulley systems. ASME J. Appl. Mech. **72**, 25–34 (2005)
12. Leamy, M.: On a perturbation method for the analysis of unsteady belt-drive operation. ASME J. Appl. Mech. **72**(4), 570–580 (2005)
13. Morimoto, T., Iizuka, H.: Rolling contact between a rubber ring and rigid cylinders: mechanics of rubber belts. Int. J. Mech. Sci. **54**, 234–240 (2012)
14. Niemann, G., Winter, H.: Maschinenelemente (in German), vol. III, 2 edn. Springer, Berlin-Heidelberg-NewYork-Tokyo (1986)
15. Nordenholz, T., O'Reilly, O.: On kinematical conditions for steady motions of strings and rods. ASME J. Appl. Mech. **62**, 820–822 (1995)
16. Reynolds, O.: On the efficiency of belts or straps as communicators of work. Engineer **38**, 396 (1874)
17. Rubin, M.: An exact solution for steady motion of an extensible belt in multipulley belt drive systems. J. Mech. Des. **122**, 311–316 (2000)
18. Stolarski, T.A.: Tribology in Machine Design. Butterworth-Heinemann, Oxford (1990)

Doppler Effects for Dispersive Waves in Beam and Plate with a Moving Edge

Kazumi Watanabe

Abstract Doppler effects for dispersive deflection waves are discussed for 1D beam and 2D plate with a moving edge. The frequency shift, reflection angle and amplitude are derived exactly. The Doppler effects for the dispersive wave are much different from those for non-dispersive elastic and electromagnetic waves. But their approximate frequency shifts are identical to each other in the case of slow edge motion. However the reflection angle for the non-dispersive wave can exceed $\pi/2$, that for the dispersive wave cannot exceed and the reflected dispersive wave with the critical angle runs along the moving edge as if it were the evanescent wave.

1 Introduction

Doppler effects are produced by a moving source or reflector. The latter Doppler effects for EM waves [4] are well-known as the moving mirror problem and are dominant sensing principles for radar systems. In order to develop the deformation sensor, the Doppler effect for elastic waves in solids is a hopeful candidate for sensing the moving boundary [1] and thus much attention should be paid to "Elastodynamic Doppler effects [2]."

Deflection waves in the beam and plate are dispersive, but the EM wave and elastic SH/SV/P-waves are non-dispersive. The Doppler effects have been discussed for the non-dispersive wave, so far. The work on the effects for the dispersive waves is scarcely. The Doppler effects produced by a moving load on the beam and plate were discussed in [3] and it was found that the frequency shift was much different from that for the non-dispersive elastic waves. The present paper considers the Doppler effects for the dispersive waves in beam and plate with the moving edge.

K. Watanabe (✉)
Yamagata University, Yonezawa, Yamagata 992-8510, Japan
e-mail: kazy_watanabe_470921@hotmail.co.jp

Alexander K. Belyaev et al. (eds.), *Mechanics and Model-Based Control of Advanced Engineering Systems*, DOI 10.1007/978-3-7091-1571-8_31,
© Springer-Verlag Wien 2014

2 Beam

Let us consider an elastic beam and take the x-axis along the beam. The beam length is semi-infinite, but its edge which is simply-supported moves with uniform velocity V. The beam deflection is governed by the well-known beam equation,

$$\alpha^4 \frac{\partial^4 w(x,t)}{\partial x^4} + \frac{\partial^2 w(x,t)}{\partial t^2} = 0, \quad \alpha^4 = \frac{EI}{\rho A} \tag{1}$$

where EI is the bending rigidity and ρA is the mass per unit length. We assume that an incident wave with frequency ω and amplitude w_0 is running toward the moving edge, i.e.

$$w^{(i)}(x,t) = w_0 \exp\left(+\mathrm{i}\omega t - \mathrm{i}x\frac{\sqrt{\omega}}{\alpha}\right) = w_0 \exp\left\{\left(-\mathrm{i}\frac{\sqrt{\omega}}{\alpha}\right)\left(x - c_{fl}t\right)\right\} \tag{2}$$

where the velocity of the flexure wave depends on the frequency, $c_{fl} = \alpha\sqrt{\omega}$. The reflected wave runs toward the negative x-axis and is assumed as the sum of two wave terms,

$$w^{(r)}(x,t) = B \exp\left(+\mathrm{i}\omega_1 t + \mathrm{i}x\frac{\sqrt{\omega_1}}{\alpha}\right) + C \exp\left(+\mathrm{i}\omega_2 t + x\frac{\sqrt{\omega_2}}{\alpha}\right) \tag{3}$$

where ω_1 and ω_2 are unknown frequencies and B and C unknown amplitudes. These four unknowns are to be determined by the condition at the moving edge. Further, in order to guarantee the convergence at the negative infinity, it should be understood that $\mathrm{Re}(\sqrt{\omega_j}) \geq 0$; $j = 1, 2$. The total deflection is given by

$$\begin{aligned} w(x,t) &= w_0 \exp\left(+\mathrm{i}\omega t - \mathrm{i}x\frac{\sqrt{\omega}}{\alpha}\right) \\ &\quad + B \exp\left(+\mathrm{i}\omega_1 t + \mathrm{i}x\frac{\sqrt{\omega_1}}{\alpha}\right) + C \exp\left(+\mathrm{i}\omega_2 t + x\frac{\sqrt{\omega_2}}{\alpha}\right) \end{aligned} \tag{4}$$

and the bending moment by

$$M(x,t) = -EI\frac{\partial^2 w(x,t)}{\partial x^2} \tag{5}$$

The moving edge is denoted by $x = Vt$ and is simply-supported, i.e.

$$w(x = Vt, t) = 0, \quad M(x = Vt, t) = 0 \tag{6}$$

The supporting condition gives

$$w_0 \exp\left(+\mathrm{i}\omega t - \mathrm{i}Vt\frac{\sqrt{\omega}}{\alpha}\right)$$
$$+B\exp\left(+\mathrm{i}\omega_1 t + \mathrm{i}Vt\frac{\sqrt{\omega_1}}{\alpha}\right) + C\exp\left(+\mathrm{i}\omega_2 t + Vt\frac{\sqrt{\omega_2}}{\alpha}\right) = 0, \tag{7}$$

$$-\frac{\omega}{\alpha^2} w_0 \exp\left(+\mathrm{i}\omega t - \mathrm{i}Vt\frac{\sqrt{\omega}}{\alpha}\right)$$
$$-\frac{\omega_1}{\alpha^2} B\exp\left(+\mathrm{i}\omega_1 t + \mathrm{i}Vt\frac{\sqrt{\omega_1}}{\alpha}\right) + \frac{\omega_2}{\alpha^2} C\exp\left(+\mathrm{i}\omega_2 t + Vt\frac{\sqrt{\omega_2}}{\alpha}\right) = 0 \tag{8}$$

These two equations must be hold for all time. In order to equate both sides, all arguments of the exponential function must be same, i.e.

$$\omega - \frac{V}{\alpha\sqrt{\omega}}\omega = \omega_1 + \frac{V}{\alpha\sqrt{\omega}}\sqrt{\omega\omega_1} = \omega_2 - \mathrm{i}\frac{V}{\alpha\sqrt{\omega}}\sqrt{\omega\omega_2} \tag{9}$$

and the simple algebraic equations for the amplitude is obtained as

$$\left\{\begin{aligned} B + C &= -w_0 \\ \omega_1 B - \omega_2 C &= -\omega w_0 \end{aligned}\right. \tag{10}$$

Equation (9) determines two frequencies, ω_1 and ω_2,

$$\omega_1 = \omega(1-M)^2, \quad \omega_2 = \omega\left\{1 - M - 2\left(\frac{M}{2}\right)^2 + \mathrm{i}M\sqrt{1 - M - \left(\frac{M}{2}\right)^2}\right\} \tag{11}$$

where M is Mach number defined by $M = V/c_{fl}$. Solving the algebraic equations for B and C, we have the amplitudes

$$B = -\frac{2 - M - \frac{M^2}{2} + \mathrm{i}M\sqrt{1 - M - \left(\frac{M}{2}\right)^2}}{2 - 3M + \frac{M^2}{2} + \mathrm{i}M\sqrt{1 - M - \left(\frac{M}{2}\right)^2}} w_0 \tag{12}$$

$$C = \frac{M(2-M)}{2 - 3M + \frac{M^2}{2} + \mathrm{i}M\sqrt{1 - M - \left(\frac{M}{2}\right)^2}} w_0 \tag{13}$$

2.1 *Doppler Effects for 1D Beam*

The first term in the right hand side of Eq. (4) shows the incident wave with amplitude w_0 and frequency ω. The second and third terms with coefficients B and C respectively show two reflected waves. Discussing the argument of the exponential function in the third term, we easily learn that the term shows the exponential decay as the time goes on or as the space x approaches to the negative infinity. Due to the exponential decay, we exclude this term from the discussion of the Doppler effects.

The Doppler effects are discussed on the non-decaying reflected wave of the second term. The amplitude of the non-decaying reflected wave is complex, and thus the wave shows the phase shift. The normalized amplitude depends only on the Mach number,

$$\frac{|B|}{w_0} = \left| \frac{2 - M - \dfrac{M^2}{2} + \mathrm{i}M\sqrt{1 - M - \left(\dfrac{M}{2}\right)^2}}{2 - 3M + \dfrac{M^2}{2} + \mathrm{i}M\sqrt{1 - M - \left(\dfrac{M}{2}\right)^2}} \right| \tag{14}$$

The frequency and wave velocity of the reflected wave are derived from Eq. (11),

$$\text{frequency}: \quad (1 - M)^2\omega \tag{15}$$

$$\text{wave velocity}: \quad (1 - M)c_{fl}. \tag{16}$$

Then, the normalized frequency shift is given by

$$\frac{\Delta\omega}{\omega} = -(2 - M)M. \tag{17}$$

Due to the dispersive nature, the propagation velocity of the reflected wave is not same as that of the incident wave. Equation (16) states that the wave velocity is slower when the beam is stretching ($M > 0$), but is faster when the beam is contracting ($M < 0$).

3 Plate

Let us discuss the Doppler effects for 2D flexure wave in a plate. The plane coordinates (x, y) are placed on the neutral plane of the plate and its downward deflection is denoted by $w(x, y, t)$. The plate deflection is governed by

$$\alpha^4 \left(\frac{\partial^4 w}{\partial x^4} + 2\frac{\partial^4 w}{\partial x^2 \partial y^2} + \frac{\partial^4 w}{\partial y^4} \right) + \frac{\partial^2 w}{\partial t^4}, \quad \alpha = \left(\frac{D}{\rho h}\right)^{1/4} \tag{18}$$

where D is the bending rigidity and ρh is the mass per unit area of the plate. We assume that the plate edge is parallel to the y-axis and moves toward the positive x direction with velocity V. The non-decaying incident wave with incident angle ϕ is assumed as

$$w^{(i)}(x,y,t) = w_0 \exp\left\{+\mathrm{i}\omega\left(t - \frac{\cos\phi}{c_{fl}}x - \frac{\sin\phi}{c_{fl}}y\right)\right\} \tag{19}$$

where w_0, ω and $c_{fl}(=\alpha\sqrt{\omega})$ are the amplitude, frequency and velocity of the incident wave, respectively.

The reflected wave is the sum of two wave terms,

$$\begin{aligned} w^{(r)}(x,y,t) = &B\exp\left(+\mathrm{i}\omega_1 t + \mathrm{i}x\sqrt{\frac{\omega_1}{\alpha^2} - k_1^2} - \mathrm{i}k_1 y\right) \\ &+ C\exp\left(+\mathrm{i}\omega_2 t + x\sqrt{\frac{\omega_2}{\alpha^2} + k_2^2} - \mathrm{i}k_2 y\right) \end{aligned} \tag{20}$$

where frequencies ω_1 and ω_2, and wave numbers k_1 and k_2 are unknowns to be determined, and B and C are also unknown amplitudes. Further, in order to guarantee the convergence at the negative x infinity, we impose on the radicals,

$$\mathrm{Re}\left(\sqrt{\frac{\omega_j}{\alpha^2} - k_j^2}\right) > 0; \quad j = 1,2 \tag{21}$$

We apply the edge condition, which is the simple support,

$$w(x = Vt, y, t) = 0, \quad M_y(x = Vt, y, t) = 0 \tag{22}$$

where the bending moment M_y is derived by

$$M_y(x,y,t) = -D\frac{\partial^2 w(x,y,t)}{\partial x^2} \tag{23}$$

The total deflection is the sum of the incident and reflected waves, and then the edge condition yields to

$$\begin{aligned} w(x = Vt, y, t) \equiv\ &w_0 \exp\left\{+\mathrm{i}\omega\left(t - \frac{\cos\phi}{c_{fl}}Vt - \frac{\sin\phi}{c_{fl}}y\right)\right\} \\ &+ B\exp\left(+\mathrm{i}\omega_1 t + \mathrm{i}Vt\sqrt{\frac{\omega_1}{\alpha^2} - k_1^2} - \mathrm{i}k_1 y\right) \\ &+ C\exp\left(+\mathrm{i}\omega_2 t + Vt\sqrt{\frac{\omega_2}{\alpha^2} + k_2^2} - \mathrm{i}k_2 y\right) = 0 \end{aligned} \tag{24}$$

$$
\begin{aligned}
&-\frac{1}{D}M_y(x=Vt,y,t)\\
&\quad\equiv-\left(\omega\frac{\cos\phi}{c_{fl}}\right)^2 w_0\exp\left\{+\mathrm{i}\omega\left(t-\frac{\cos\phi}{c_{fl}}Vt-\frac{\sin\phi}{c_{fl}}y\right)\right\}\\
&\quad-\left(\frac{\omega_1}{\alpha^2}-k_1^2\right)B\exp\left(+\mathrm{i}\omega_1 t+\mathrm{i}Vt\sqrt{\frac{\omega_1}{\alpha^2}-k_1^2}-\mathrm{i}k_1 y\right)\\
&\quad+\left(\frac{\omega_2}{\alpha^2}+k_2^2\right)C\exp\left(+\mathrm{i}\omega_2 t+Vt\sqrt{\frac{\omega_2}{\alpha^2}+k_2^2}-\mathrm{i}k_2 y\right)=0
\end{aligned}
\tag{25}
$$

In order to equate both sides, equations for the frequencies and wave numbers are derived from the argument of the exponential function.

$$
\begin{aligned}
&\omega(1-M\cos\phi)=\omega_1+V\sqrt{\frac{\omega_1}{\alpha^2}-k_1^2}=\omega_2-\mathrm{i}V\sqrt{\frac{\omega_2}{\alpha^2}+k_2^2}\\
&\frac{\omega}{c_{fl}}\sin\phi=k_1=k_2
\end{aligned}
\tag{26}
$$

The simple algebraic equations for the amplitudes are also obtained as

$$
\begin{cases}
B+C=-w_0\\
\left(\frac{\omega_1}{\alpha^2}-k_1^2\right)B-\left(\frac{\omega_2}{\alpha^2}+k_2^2\right)C=-\left(\omega\frac{\cos\phi}{c_{fl}}\right)^2 w_0
\end{cases}
\tag{27}
$$

Solving Eq. (26) for ω_1 and ω_2, we have

$$
\omega_1=\omega(1-2M\cos\phi+M^2)
\tag{28}
$$

$$
\omega_2=\omega\left\{1-M\cos\phi-2\left(\frac{M}{2}\right)^2+\mathrm{i}M\sqrt{1-M\cos\phi+\sin^2\phi-\left(\frac{M}{2}\right)^2}\right\}
\tag{29}
$$

where M is Mach number defined by $M=V/c_{fl}$. The amplitudes are also obtained as

$$
B=-\frac{2-M\cos\phi-\dfrac{M^2}{2}+\mathrm{i}M\sqrt{1-M\cos\phi+\sin^2\phi-\left(\dfrac{M}{2}\right)^2}}{2-3M\cos\phi+\dfrac{M^2}{2}+\mathrm{i}M\sqrt{1-M\cos\phi+\sin^2\phi-\left(\dfrac{M}{2}\right)^2}}w_0
\tag{30}
$$

$$C = -\frac{(2\cos\phi - M)M}{2 - 3M\cos\phi + \dfrac{M^2}{2} + \mathrm{i}M\sqrt{1 - M\cos\phi + \sin^2\phi - \left(\dfrac{M}{2}\right)^2}}w_0 \tag{31}$$

3.1 Doppler Effects

Inspecting the argument of the exponential function in Eq. (20) with Eqs. (28) and (29), we learn that the third term with the coefficient C shows the exponential decay as the time goes or as the wave runs to the negative x direction. Due to this exponential decay, we exclude the third term from the further discussion. The second term with B shows the reflected wave with non-decaying nature, but phase change takes place due to the complex amplitude,

$$\frac{|B|}{w_0} = \left|\frac{2 - M\cos\phi - \dfrac{M^2}{2} + \mathrm{i}M\sqrt{1 - M\cos\phi + \sin^2\phi - \left(\dfrac{M}{2}\right)^2}}{2 - 3M\cos\phi + \dfrac{M^2}{2} + \mathrm{i}M\sqrt{1 - M\cos\phi + \sin^2\phi - \left(\dfrac{M}{2}\right)^2}}\right| \tag{32}$$

The conditional of Eq. (21) leads to the restriction on the incident angle,

$$\cos\phi > M. \tag{33}$$

This means that the horizontal velocity component of the incident wave must be larger than the edge velocity so that the incident wave can catch up to the moving edge. The reflection angle is also derived from the argument of its exponential function, i.e.

$$\psi = \tan^{-1}\left(\frac{\sin\phi}{\cos\phi - M}\right) \tag{34}$$

When the incident wave with the critical angle, $\cos\phi = M$, is coming, the reflection angle is $\pi/2$ and the reflected wave runs along the moving edge as if it were the evanescent wave. However, if Mach number is negative, the reflection angle is always smaller than the incident angle and all reflected waves come back to the incident wave side and no evanescent-like wave appears. The frequency of the reflected wave has been obtained by Eq. (28) and the normalized frequency shift is given by

$$\frac{\Delta\omega}{\omega} = -(2\cos\phi - M)M \tag{35}$$

4 Conclusions

Let us compare two Doppler effects for the dispersive and non-dispersive waves. The frequency shift for the non-dispersive wave such as electromagnetic wave [4] and elastic SH-wave [2] is given by

- Non-dispersive wave

$$\frac{\Delta\omega}{\omega} = -2M\frac{\cos\phi - M}{1 - M^2} \approx -2M\cos\phi + O(M^2) \tag{36}$$

  On the other hand, we have just obtained the Doppler frequency shift for the dispersive flexure wave,

- Dispersive wave

$$\frac{\Delta\omega}{\omega} = -2M\cos\phi + M^2 \approx -2M\cos\phi + O(M^2) \tag{37}$$

  In the above two equations, the far right equation shows an approximate expression for the small Mach number, $|M| \ll 1$. Then, the both frequency shifts are not same in the exact sense, but same when Mach number is sufficient small. Thus, when the edge motion is slow, the frequency shift is almost same for both waves, dispersive and non-dispersive.

As for the reflection angle, the two waves have different formulas

- Non-dispersive wave

$$\psi = \tan^{-1}\left\{\frac{(1 - M^2)\sin\phi}{(1 + M^2)\cos\phi - 2M}\right\} \approx \tan^{-1}\left(\frac{\sin\phi}{\cos\phi - 2M}\right); \quad |M| \ll 1 \tag{38}$$

- Dispersive wave

$$\psi = \tan^{-1}\left(\frac{\sin\phi}{\cos\phi - M}\right) \tag{39}$$

  Inspecting the reflection angle of Eq. (38), we learn that it is possible to exceed $\pi/2$ for the non-dispersive wave [2]. This excessed reflection angle states that the reflected wave behaves as if it were chasing the moving edge and never comes back to the incident wave side. However, for the dispersive wave, the reflection angle of Eq. (39) never exceeds $\pi/2$ and all reflected waves come back to the incident wave side.

References

1. Watanabe, K.: Measurability of beam vibrations by means of longitudinal ultrasonic waves. In: Proceedings of the 4th European Conference on Structural Control, St. Petersburg (Russia), 2008, vol. 2, pp. 834–850
2. Watanabe, K.: Elastodynamic Doppler effects by a moving stress-free edge. Int. J. Eng. Sci. **48**, 1995–2004 (2010)
3. Watanabe, K., Biwa, S.: Elastodynamic Doppler effects. Acta Mech. **195**, 27–59 (2008)
4. Yeh, C.: Reflection and transmission of electromagnetic waves by moving dielectric medium. J. Appl. Phys. **38**(11), 3513–3517 (1965)

Effect of Road Surface Roughness on Extraction of Bridge Frequencies by Moving Vehicle

Y.B. Yang, Y.C. Lee, and K.C. Chang

Abstract Measuring the bridge frequencies indirectly from a test vehicle passing the bridge is featured by its mobility, economy and promptness, compared with the conventional technique that requires vibration sensors to be mounted directly on the bridge. However, the road surface roughness may pollute the vehicle spectrum, rendering the bridge frequencies invisible. This paper is aimed at studying such an effect. First, a numerical analysis is conducted using the vehicle-bridge interaction element to demonstrate how the surface roughness affects the vehicle response. Then, an approximate theory in closed form is presented for physically interpreting the influence of surface roughness on the identification of bridge frequencies. To reduce the effect of surface roughness, two connected, identical vehicles are allowed to travel the bridge, with the response recorded and synchronized for one vehicle subtracted from the other. It was demonstrated that the visibility of the bridge frequencies can be greatly enhanced in the residue response.

1 Introduction

Conventionally, the identification of bridge frequencies requires the installation of vibration sensors directly on the bridge. Such an approach is called the direct approach and is the one frequently adopted by engineers. An alternative approach called the indirect approach was proposed by Yang et al. [6] for identifying the bridge frequencies, by using an instrumented test vehicle to travel over the bridge and then by extracting the bridge frequencies from the vehicle spectrum. With this

Y.B. Yang (✉)
Department of Civil Engineering, National Taiwan University, 10617 Taipei, Taiwan
e-mail: ybyang@ntu.edu.tw

Y.C. Lee · K.C. Chang
Department of Civil Engineering, National Taiwan University, 10617 Taipei, Taiwan

Alexander K. Belyaev et al. (eds.), *Mechanics and Model-Based Control of Advanced Engineering Systems*, DOI 10.1007/978-3-7091-1571-8_32,
© Springer-Verlag Wien 2014

approach, no vibration sensors need to be mounted on the bridge. The idea of indirect bridge frequency measurement was experimentally verified to be feasible by Lin and Yang [3], along with the potential applications of vehicle-bridge interaction (VBI) properties identified [4].

This paper is focused on the effect of road surface roughness critical to the extraction of bridge frequencies from the recorded acceleration response of the traveling test vehicle. First, numerical analyses are conducted using the VBI elements to demonstrate the pollution effect of road surface roughness. Then, an approximate theory in closed form is presented to physically assess the influence of road surface roughness. Finally, the responses obtained for two connected, identical vehicles are synchronized and then subtracted from each other to generate the *residue spectrum*, by which the visibility of bridge frequencies is greatly enhanced.

2 Roughness Profile Definition

The power spectral density (PSD) functions defined by ISO 8608 [2] for the road surface profiles is adopted, which are divided into eight classes, from Class A (best) to Class H (poorest). The PSD function $G_d(n)$ for the surface profile is

$$G_d(n) = G_d(n_0)\left(\frac{n}{n_0}\right)^{-w} \tag{1}$$

where n = the spatial frequency per meter, $w = 2$, $n_0 = 0.1$ cycle/m, and the functional value $G_d(n_0)$ is determined by the roughness class in ISO 8608 [2]. The amplitude d for each class of roughness selected is determined by

$$d = \sqrt{2G_d(n)\,\Delta n} \tag{2}$$

where Δn = the sampling interval of the spatial frequency. The road surface roughness can be superimposed as

$$r(x) = \sum_i d_i \cos\left(n_{s,i}x + \theta_i\right) \tag{3}$$

where $n_{s,i}$ = the ith spatial frequency, and d_i and θ_i denote the amplitude and the random phase angle, respectively, of the ith cosine function. In this study, the sampling interval Δn_s for the spatial frequency is taken as 0.04 cycle/m, and the range of spatial frequency n_s is taken as 1–100 cycle/m.

3 Finite Element Simulation

In the finite element simulation, only a simply supported bridge is considered, which is divided into 20 elements, each of 6 degrees of freedom (DOFs) for the two-dimensional case, and the vehicle is modeled as a single-DOF mass supported by a spring-dashpot unit. For an element of length l directly under the action of the vehicle, it is modeled as a VBI element with the effects of vehicle action and surface roughness included (see Fig. 1). The procedure for simulating the VBI system with surface roughness by the finite element method has been given in details elsewhere [1, 5, 7], which will not be recapitulated herein.

Three classes of surface roughness, A–C, are considered for the bridge traveled by the test vehicle. The properties adopted for the vehicle are: $m_v = 1{,}000\,\text{kg}$, $k_v = 3.947\,\text{MN/m}$, and $v = 2\,\text{m/s}$. The bridge has the following properties: elastic modulus $E = 27.5\,\text{GPa}$, moment of inertia $I_b = 0.175\,\text{m}^4$, mass per unit length $\overline{m} = 1{,}000\,\text{kg/m}$, and cross-sectional area $A = 2\,\text{m}^2$. In this study, a square root is taken of the geometric mean of the functional value provided by ISO 8608 [2] to reduce its magnitude. Besides, a very small value of 0.001×10^{-6} is assigned for the functional value (geometric mean) of Class A without taking the square root. Thus, the functional values adopted for the three classes of roughness are: A: $G_d^*(n_0) = 0.001 \times 10^{-6}\,\text{m}^3$; B: $G_d^*(n_0) = 8 \times 10^{-6}\,\text{m}^3$; and C: $G_d^*(n_0) = 16 \times 10^{-6}\,\text{m}^3$. With the above data, the first three frequencies computed of the bridge are 3.867, 15.27 and 34.3 Hz, and the vehicle frequency is 10 Hz, which is *larger* than the first frequency of the bridge.

The acceleration response and frequency spectrum solved of the test vehicle during its passage over the bridge for the three classes of roughness have been plotted in Fig. 2a–c. As can be seen, the amplitudes of the vehicle frequency and roughness frequencies are drastically amplified as the roughness level increases. Moreover, the third bridge frequency is hidden for Class B and C, while the first and second bridge frequency remain visible for all three classes.

We have actually conducted another case with the vehicle frequency adjusted as 2.067 Hz, i.e., less than the first bridge frequency. The observation for this case is similar, except that only the first frequency is visible for Class B roughness and all bridge frequencies are hidden for Class C roughness. The lesson here is that a test vehicle should be designed to have a frequency "greater" than the first frequency of the bridge, so as to enhance the visibility of bridge frequencies. In general, the above analyses indicate that bridge frequencies of higher modes may be polluted or hidden by the presence of road surface roughness.

To assess how the road surface roughness pollutes the bridge frequencies in the vehicle spectrum, one may consider *two extreme cases*: (i) The test vehicle moves over a bridge with smooth surface, meaning that the vehicle is excited exclusively by the vibration of the bridge. (ii) The vehicle moves over a bridge of an infinitely large flexural rigidity, but with rough surface. In this case, the vehicle is excited by the surface roughness only.

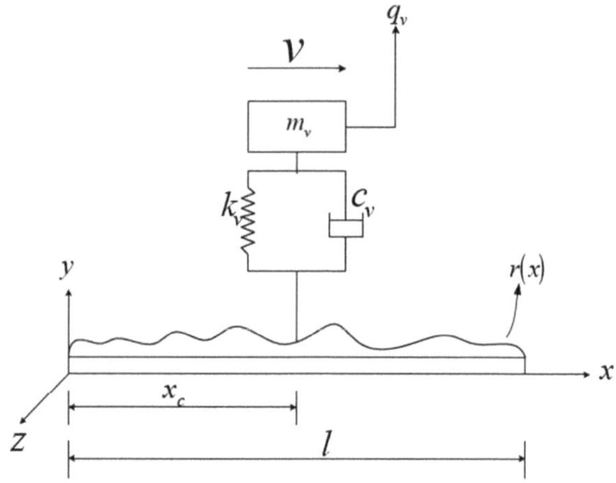

Fig. 1 Vehicle-bridge interaction element

Figure 3a, b show the vehicle acceleration response and Fourier spectrum, respectively, for the above two cases. As can be seen, the vehicle is excited much more dramatically by the surface roughness than by the bridge in vibration. This explains why substantial difficulty exists in extracting bridge frequencies from the vehicle spectrum, once the road surface roughness is taken into account. Two approaches can be adopted to resolve such a problem. One is to increase the vibration energy of the bridge by exposing the bridge to existing traffic or accompanying vehicles [1, 3]. The other is to suppress the effect of road surface roughness, which will be presented later on.

4 Vehicle Response in Closed Form with Roughness Included

Figure 4 shows the VBI model of concern, where the vehicle is modeled as a sprung mass m_v supported by a spring of stiffness k_v, and the simple bridge is modeled as a Bernoulli-Euler beam of length L, mass density $\overline{m}$ per unit length, and bending rigidity EI. The road surface roughness is denoted by $r(x)$. The damping effect of the system is neglected due to the transient nature of the problem considered.

Let the vehicle move over the bridge with speed v. The equations of motion for both the vehicle and bridge at time t can be expressed as follows:

$$m_v \ddot{q}_v(t) + k_v \left[q_v(t) - u_b(x,t)|_{x=vt} - r(x)|_{x=vt} \right] = 0 \tag{4}$$

$$\overline{m}\ddot{u}_b(x,t) + EIu''''_b(x,t) = f_c(t)\,\delta(x - vt) \tag{5}$$

where q_v and u_b denote the vertical displacement of the vehicle and bridge, respectively, and the contact force f_c is

$$f_c(t) = -m_v g + k_v \left[q_v(t) - u_b(x,t)|_{x=vt} \right] - k_v \left[r(x)|_{x=vt} \right] \tag{6}$$

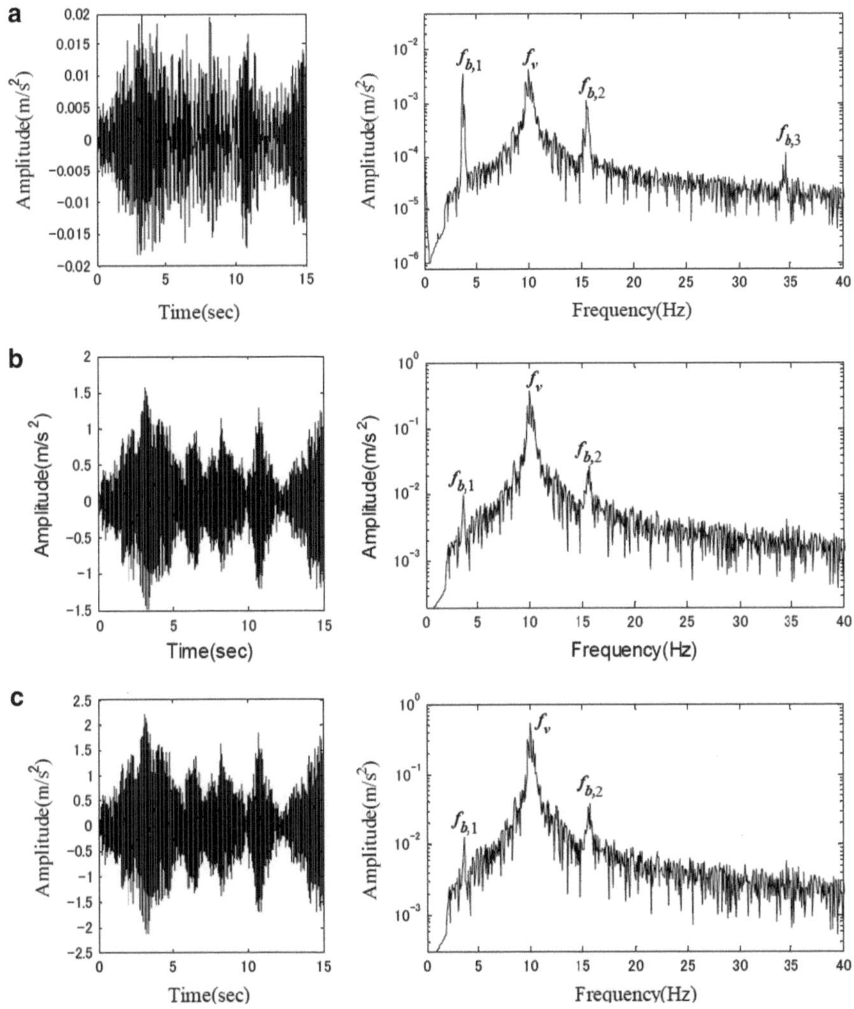

Fig. 2 Acceleration and frequency responses of test vehicle for Class: (**a**) A, (**b**) B, (**c**) C

The bridge response can be expressed in terms of the modal shapes $\sin(n\pi x/L)$ and generalized coordinates $q_{b,n}(t)$ as follows:

$$u(x,t) = \sum_{n=1}^{\infty} \sin\frac{n\pi x}{L} q_{b,n}(t) \tag{7}$$

Substituting Eq. (7) into Eq. (5), multiplying both sides of the equation by $\sin(n\pi x/L)$, and integrating from 0 to L, one obtains

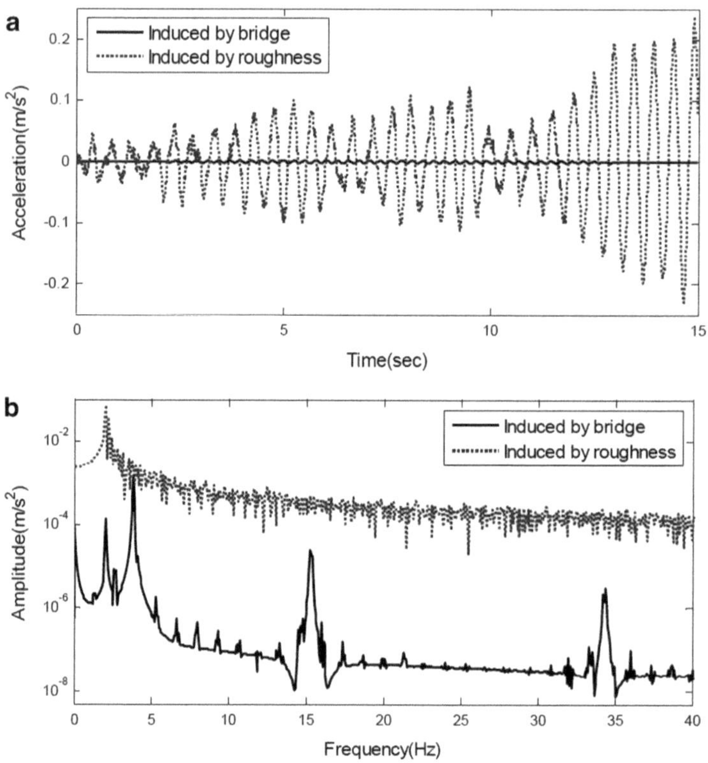

Fig. 3 Acceleration responses of the vehicle: (**a**) time history, (**b**) amplitude spectrum

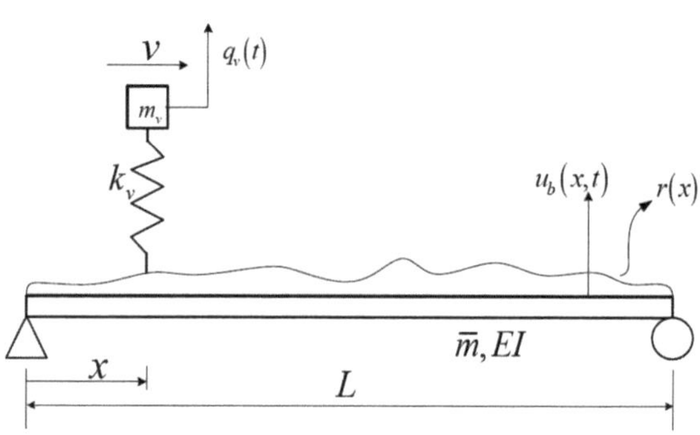

Fig. 4 Mathematical model

$$\ddot{q}_{b,n} + \omega_{b,n}^2 q_b = 2\sin\left(\frac{n\pi vt}{L}\right)\left\{\frac{-m_v g}{\overline{m}L} + \frac{m_v\omega_v^2}{\overline{m}L}\left[q_v - u_b(x,t)|_{x=vt}\right] - \frac{m_v\omega_v^2}{\overline{m}L} r(x)|_{x=vt}\right\} \tag{8}$$

where $\omega_{b,n}$ is the bridge frequency of the nth mode, $\omega_{b,n} = \left(n^2\pi^2/L^2\right)\sqrt{EI/\overline{m}}$, and ω_v is the vehicle frequency, $\omega_v = \sqrt{k_v/m_v}$.

Note the variations in the elastic force of the suspension and the inertial force of the vehicle caused by surface roughness, as represented by the last two terms of Eq. (8), have negligible effect on the bridge response [8]. As such, the equation of motion for the beam in Eq. (8) reduces to

$$\ddot{q}_{b,n} + \omega_{b,n}^2 q_b = 2\sin\frac{n\pi vt}{L}\left\{\frac{-m_v g}{\overline{m}L}\right\} \tag{9}$$

For zero initial conditions, one can solve Eq. (9) to obtain the following:

$$q_{b,n}(t) = \frac{\Delta_{st,n}}{1-S_n^2}\left[\sin\frac{n\pi vt}{L} - S_n\sin\omega_{b,n}t\right] \tag{10}$$

where $\Delta_{st,n} = -2m_v gL^3\big/(n^4\pi^4 EI)$ is the static deflection caused by the vehicle weight, and $S_n = n\pi v\big/(L\omega_{b,n})$ is a non-dimensional speed parameter. Substituting Eq. (10) into Eq. (7) yields the vertical displacement of the bridge as

$$u(x,t) = \sum_n \frac{\Delta_{st,n}}{1-S_n^2}\left\{\sin\frac{n\pi x}{L}\left[\sin\frac{n\pi vt}{L} - S_n\sin\omega_{b,n}t\right]\right\} \tag{11}$$

As for the vehicle, the equation of motion in Eq. (7) may be re-written as

$$\ddot{q}_v(t) + \omega_v^2 q_v(t) = \omega_v^2 r(x)|_{x=vt} + \omega_v^2 u_b(x,t)\Big|_{x=vt} \tag{12}$$

which indicates that the vehicle is excited by two sources, the bridge vibration and road surface roughness. Substituting Eq. (11) for $u(x,t)$ and Eq. (3) for $r(x)$ into Eq. (12), one obtains by Duhamel's integral the vehicle displacement as

$$q_v(t) = \sum_{n=1}^{\infty}\left\{A_{1,n}\cos\left(\frac{(n-1)\pi v}{L}\right)t + A_{2,n}\cos\left(\frac{(n+1)\pi v}{L}\right)tA_{3,n}\cos(\omega_v t) + A_{4,n}\cos\left(\omega_{b,n} - \frac{n\pi v}{L}\right)t + A_{5,n}\cos\left(\omega_{b,n} + \frac{n\pi v}{L}\right)t\right\}$$

$$+\sum_{i=1}\frac{\omega_v^2 d_i}{\omega_v^2-\left(n_{s,i}v\right)^2}\left\{\cos\left(n_{s,i}vt+\theta_i\right)-\cos\left(\theta_i\right)\cos\left(\omega_v t\right)+\frac{n_{s,i}v}{\omega_v}\sin\left(\theta_i\right)\sin\left(\omega_v t\right)\right\} \tag{13}$$

where the coefficients are

$$A_{1,n}=\frac{\Delta_{st,n}\omega_v^2}{2\left(1-S_n^2\right)\left(\omega_v+\frac{(n-1)\pi v}{L}\right)\left(\omega_v-\frac{(n-1)\pi v}{L}\right)},$$

$$A_{2,n}=\frac{-\Delta_{st,n}\omega_v^2}{2\left(1-S_n^2\right)\left(\omega_v+\frac{(n+1)\pi v}{L}\right)\left(\omega_v-\frac{(n+1)\pi v}{L}\right)}$$

$$A_{3,n}=\frac{2\Delta_{st,n}\omega_v^2\left(\frac{\pi v}{L}\right)^2 n}{2\left(1-S_n^2\right)\left(\omega_v+\frac{(n-1)\pi v}{L}\right)\left(\omega_v-\frac{(n-1)\pi v}{L}\right)\left(\omega_v+\frac{(n+1)\pi v}{L}\right)\left(\omega_v-\frac{(n+1)\pi v}{L}\right)}$$
$$-\frac{2\Delta_{st,n}S_n\omega_v^2\left(\frac{n\pi v}{L}\right)\omega_{b,n}}{\left(\omega_v-\omega_{b,n}+\frac{n\pi v}{L}\right)\left(\omega_v+\omega_{b,n}-\frac{n\pi v}{L}\right)\left(\omega_v+\omega_{b,n}+\frac{n\pi v}{L}\right)\left(\omega_v-\omega_{b,n}-\frac{n\pi v}{L}\right)}$$

$$A_{4,n}=\frac{-S_n\Delta_{st,n}\omega_v^2}{2\left(1-S_n^2\right)\left(\omega_v-\omega_{b,n}+\frac{n\pi v}{L}\right)\left(\omega_v+\omega_{b,n}-\frac{n\pi v}{L}\right)},$$

$$A_{5,n}=\frac{S_n\Delta_{st,n}\omega_v^2}{2\left(1-S_n^2\right)\left(\omega_v+\omega_{b,n}+\frac{n\pi v}{L}\right)\left(\omega_v-\omega_{b,n}-\frac{n\pi v}{L}\right)} \tag{14}$$

Differentiating Eq. (13) twice, the vehicle acceleration response is obtained as

$$\ddot{q}_v(t)=\sum_{n=1}^{\infty}\{\bar{\bar{A}}_{1,n}\cos\left(\frac{(n-1)\pi v}{L}\right)t+\bar{\bar{A}}_{2,n}\cos\left(\frac{(n+1)\pi v}{L}\right)t$$
$$+\bar{\bar{A}}_{3,n}\cos\left(\omega_v t\right)+\bar{\bar{A}}_{4,n}\cos\left(\omega_{b,n}-\frac{n\pi v}{L}\right)t+\bar{\bar{A}}_{5,n}\cos\left(\omega_{b,n}+\frac{n\pi v}{L}\right)t\}$$
$$+\sum_{i=1}\frac{\omega_v^2 d_i}{\omega_v^2-\left(n_{s,i}v\right)^2}\left[-\left(n_{s,i}v\right)^2\cos\left(n_{s,i}vt+\theta_i\right)\right.$$
$$\left.+\,\omega_v^2\cos\left(\theta_i\right)\cos\left(\omega_v t\right)-\left(n_{s,i}v\omega_v\right)\sin\left(\theta_i\right)\sin\left(\omega_v t\right)\right] \tag{15}$$

where the coefficients are

$$\bar{\bar{A}}_{1,n}=-\left(\frac{(n-1)\pi v}{L}\right)^2\times A_{1,n},\quad \bar{\bar{A}}_{2,n}=-\left(\frac{(n+1)\pi v}{L}\right)^2\times A_{2,n},$$

$$\bar{\bar{A}}_{3,n} = -\omega_v^2 \times A_{3,n}, \quad \bar{\bar{A}}_{4,n} = -\left(\omega_{b,n} - \frac{n\pi v}{L}\right)^2 \times A_{4,n},$$

$$\bar{\bar{A}}_{5,n} = -\left(\omega_{b,n} + \frac{n\pi v}{L}\right)^2 \times A_{5,n} \tag{16}$$

As can be seen from Eq. (15), the vehicle response is affected by the roughness-related frequencies $n_{s,i}v$, with $n_{s,i}$ indicating the spatial frequency $n_{s,i}$ of surface roughness and v the vehicle speed, in addition to the three groups of frequencies: driving frequencies $(n \pm 1)\pi v/L$, bridge frequencies $\omega_{b,n} \pm n\pi v/L$, and vehicle frequency ω_v. The solution for the vehicle response in Eq. (15) differs from the one by Yang and Lin [4] in the appearance of the terms due to road surface roughness, which are denoted as $\ddot{q}_{v,r}$,

$$\ddot{q}_{v,r} = \sum_{i=1} \frac{\omega_v^2 d_i}{\omega_v^2 - (n_{s,i}v)^2} \Big[-(n_{s,i}v)^2 \cos(n_{s,i}vt + \theta_i) + \omega_v^2 \cos(\theta_i)\cos(\omega_v t) - (n_{s,i}v\omega_v)\sin(\theta_i)\sin(\omega_v t) \Big] \tag{17}$$

The roughness term $\ddot{q}_{v,r}$ plays the role of introducing the roughness frequencies to the vehicle response, while amplifying the amplitudes of the vehicle frequency, especially when any of the roughness frequencies $n_{s,i}v$ is close to the vehicle frequency ω_v, as implied by the denominator. Both effects are unfavorable to identification of bridge frequencies from the vehicle response. Evidently, the present closed-form solution offers a theoretical basis for clearly interpreting the influence of road surface roughness on the vehicle response.

5 Reducing the Roughness Effect Using Two Connected Vehicles

One approach for reducing the effect of road surface roughness is to record the responses of two connected, identical vehicles during their passage over the bridge, and then to deduct the response recorded for one vehicle from that of the other [9]. The preceding closed-form solution can be extended to dealing with the case of two connected vehicles. By letting two connected, identical vehicles travel over the same bridge, and by deducting the (synchronized) response obtained for one vehicle from the other, one can obtain the *residue response*, which is basically free of the surface roughness effect. However, due to the restriction of paper length, all the theoretical derivations in this regard are just omitted herein. Only numerical simulations will be given below.

The properties adopted of the simple bridge are the same as those previously used: $L = 30\,\text{m}$, $E = 27.5\,\text{GPa}$, $I_b = 0.175\,\text{m}^4$, $\overline{m} = 1{,}000\,\text{kg/m}$, and $A = 2\,\text{m}^2$.

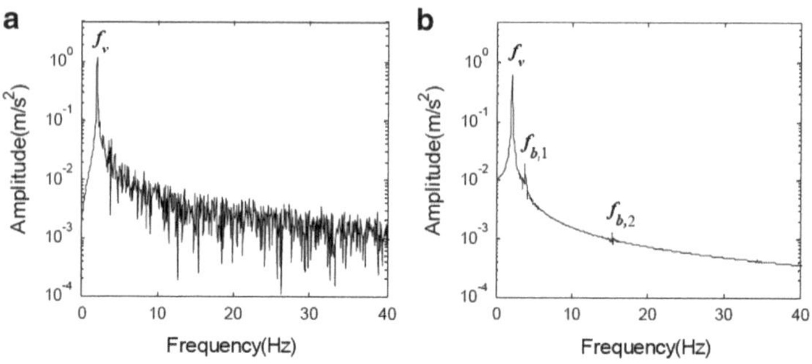

Fig. 5 Acceleration response spectrum of rear vehicle: (**a**) original; (**b**) subtracted

Two identical vehicles with the following properties are adopted: $m_{v,1} = m_{v,2} = 1{,}000\,\text{kg}$, $k_{v,1} = k_{v,2} = 170\,\text{kN/m}$. They move with speed $v = 2\,\text{m/s}$ and spacing $s = 3\,\text{m}$ over the bridge of road surface class A. Figure 5a shows the amplitude spectrum of the acceleration response calculated of the rear vehicle, in which only the vehicle frequency f_v, but no bridge-related frequencies, can be identified. As was stated previously, the difficulty of identifying bridge frequency from the vehicle spectrum is due to the fact that the roughness-induced responses are so large that the bridge-related responses are overshadowed.

To reduce the effect of surface roughness in time domain, one first synchronizes the responses of the two vehicles with respect to identical contact points, and then subtracts the synchronized response for one vehicle from the other. Figure 5b shows the subtracted or residue spectrum. As can be seen, the effect of roughness has been largely reduced, making it possible to identify the first and second bridge frequencies, $f_{b,1}$ and $f_{b,2}$.

Effort has also been carried out to study the various parameters of the two connected vehicles, such as the use of different frequencies for the two vehicles, the spacing between the two vehicles, and the effect of different roughness classes. It was concluded that the use of two identical vehicles can yield the best performance, while the vehicle spacing is not a crucial parameter.

6 Conclusions

In this study, the effect of road surface roughness was demonstrated in the finite element simulation with the vehicle frequency either greater or less than the first frequency of the bridge. From engineering practice, it is suggested that the vehicle frequency be made greater than the first bridge frequency for better resolution of the bridge frequencies. An approximate theory in closed form was then presented for the vehicle-bridge interaction system with the effect of surface roughness included. This theory allows us to examine physically the effect of road surface roughness on the vehicle response, and particularly on the identification of bridge frequencies from the vehicle response. The approximate theory was then extended for the case of two connected vehicles, from which the residue response is obtained as the difference between two synchronized responses. Due to the restriction of paper length, the theoretical formulation in this part was omitted. However, it was demonstrated theoretically and numerically that the effect of surface roughness can be basically suppressed using the residue spectrum.

Acknowledgements The study reported herein is sponsored partially by the National Science Council (ROC) via NSC 100-2221-E-224-045-MY2. The assistance from Tongji University (973 program, Ministry of Science and Technology, PRC) is also acknowledged.

References

1. Chang, K.C., Wu, F.B., Yang, Y.B.: Effect of road surface roughness on indirect approach for measuring bridge frequencies from a passing vehicle. Interact. Multiscale Mech. Int. J. **3**(4), 299–308 (2010)
2. ISO 8608: Mechanical vibration-road surface profiles – reporting of measured data. Geneva, 1995
3. Lin, C.W., Yang, Y.B.: Use of a passing vehicle to scan the bridge frequencies – an experimental verification. Eng. Struct. **27**(13), 1865–1878 (2005)
4. Yang, Y.B., Lin, C.W.: Vehicle-bridge interaction dynamics and potential applications. J. Sound Vib. **284**(1–2), 205–226 (2005)
5. Yang, Y.B., Yau, J.D.: Vehicle-bridge interaction element for dynamic analysis. J. Struct. Eng. ASCE **123**(11), 1512–1518 (1997). (Errata: 124(4), p. 479)
6. Yang, Y.B., Lin, C.W., Yau, J.D.: Extracting bridge frequencies from the dynamic response of a passing vehicle. J. Sound Vib. **272**(3–5), 471–493 (2004)
7. Yang, Y.B., Yau, J.D., Wu, Y.S.: Vehicle-Bridge Interaction Dynamics: With Applications to High-Speed Railways. World Scientific, Singapore (2004)
8. Yau, J.D., Yang, Y.B., Kuo, S.R.: Impact response of high speed rail bridges and riding comfort of rail cars. Eng. Struct. **21**(9), 836–844 (1999)
9. Yang, Y.B., Li, Y.C., Chang, K.C.: Using two connected vehicles to measure the frequencies of bridges with rough surface: a theoretical study. Acta Mech. **223**(8), 1851–1861 (2012)

Index

Alexander K. Belyaev et al. (eds.), *Mechanics and Model-Based Control of Advanced Engineering Systems*, DOI 10.1007/978-3-7091-1571-8,

© Springer-Verlag Wien 2014

C

D

Q

R

T

MIX
Papier aus verantwortungsvollen Quellen
Paper from responsible sources
FSC® C105338

If you have any concerns about our products,
you can contact us on
ProductSafety@springernature.com

In case Publisher is established outside the EU,
the EU authorized representative is:
Springer Nature Customer Service Center GmbH
Europaplatz 3, 69115 Heidelberg, Germany

Printed by Libri Plureos GmbH
in Hamburg, Germany